Advanced Satellite Remote Sensing Techniques for Meteorological, Climate and Hydroscience Studies

Advanced Satellite Remote Sensing Techniques for Meteorological, Climate and Hydroscience Studies

Guest Editors

Haobo Li
Suelynn Choy
Yuriy Kuleshov
Mayra Ivelisse Oyola-Merced
Xiaoming Wang

MDPI

Basel • Beijing • Wuhan • Barcelona • Belgrade • Novi Sad • Cluj • Manchester

Guest Editors

Haobo Li
School of Science
RMIT University
Melbourne
Australia

Suelynn Choy
School of Science
RMIT University
Melbourne
Australia

Yuriy Kuleshov
Australian Bureau of Meteorology
Melbourne
Australia

Mayra Ivelisse Oyola-Merced
Department of Atmospheric and Oceanic Sciences
University of Wisconsin-Madison
Madison
USA

Xiaoming Wang
Aerospace Information Research Institute
Chinese Academy of Sciences
Beijing
China

Editorial Office

MDPI AG
Grosspeteranlage 5
4052 Basel, Switzerland

This is a reprint of the Special Issue, published open access by the journal *Remote Sensing* (ISSN 2072-4292), freely accessible at: https://www.mdpi.com/journal/remotesensing/special_issues/7038CAMQ16.

For citation purposes, cite each article independently as indicated on the article page online and as indicated below:

Lastname, A.A.; Lastname, B.B. Article Title. *Journal Name* **Year**, *Volume Number*, Page Range.

ISBN 978-3-7258-3471-6 (Hbk)
ISBN 978-3-7258-3472-3 (PDF)
https://doi.org/10.3390/books978-3-7258-3472-3

Contents

Preface

The increasing impact of climate change has led to more frequent and intense severe weather events, posing significant challenges for disaster preparedness, climate resilience, and environmental sustainability. Understanding and monitoring these atmospheric processes require advanced remote sensing technologies which are capable of capturing high-resolution, multi-source data. This Special Issue brings together 11 high-quality studies that explore state-of-the-art remote sensing methodologies for severe weather detection, hydrometeorological analysis, and climate monitoring.

The research presented in this reprint covers a diverse array of topics, including the evaluation of near-real-time satellite-based precipitation data, the use of Global Navigation Satellite System (GNSS) for atmospheric water vapor measurement, and the role of mesospheric temperature and water vapor in polar cloud formation. Several studies also focus on integrating multi-source remote sensing datasets to enhance precipitation retrieval accuracy and improve extreme weather event detection. These contributions demonstrate the potential of cutting-edge remote sensing techniques in strengthening forecasting capabilities and supporting the advancement of climate research.

The motivation behind compiling this Special Issue is to bridge the gap between atmospheric remote sensing innovation and its practical applications in meteorology, hydrology, and climate science. By presenting a collection of studies that employ satellite observations, e.g., GNSS atmospheric parameters, we aim to provide researchers, policymakers, and other stakeholders with valuable insights into leveraging advanced technologies for environmental monitoring and disaster mitigation.

We would like to extend our sincere appreciation to all contributing authors for their dedication to high-quality research, as well as to the reviewers whose constructive feedback has been instrumental in enhancing the scientific rigor of these studies. Special thanks go to the Editorial Team for their support in bringing this Special Issue to publication. It is our hope that the findings presented here will inspire further research and contribute to the ongoing development of satellite remote sensing applications in weather and climate sciences.

Haobo Li, Suelynn Choy, Yuriy Kuleshov, Mayra Ivelisse Oyola-Merced, and Xiaoming Wang

Guest Editors

remote sensing

MDPI

Article

Comparative Assessment of Spire and COSMIC-2 Radio Occultation Data Quality

Cong Qiu [1,2], Xiaoming Wang [1,3,*], Kai Zhou [1], Jinglei Zhang [1], Yufei Chen [1,2], Haobo Li [4], Dingyi Liu [1,2] and Hong Yuan [1]

1 Aerospace Information Research Institute, Chinese Academy of Sciences, No.9 Dengzhuang South Road, Haidian District, Beijing 100094, China; qiucong@aircas.ac.cn (C.Q.); zhoukai@aircas.ac.cn (K.Z.); zhangjinglei@aircas.ac.cn (J.Z.); chenyufei20@mails.ucas.ac.cn (Y.C.); liudingyi21@mails.ucas.ac.cn (D.L.); yuanhong@aircas.ac.cn (H.Y.)
2 School of Electronic, Electrical and Communicating Engineering, University of Chinese Academy of Sciences, No. 19A Yuquan Road, Shijingshan District, Beijing 100049, China
3 College of Resources and Environment, University of Chinese Academy of Sciences, No. 19A Yuquan Road, Shijingshan District, Beijing 100049, China
4 School of Science (Geospatial), RMIT University, Melbourne, VIC 3001, Australia; haobo.li@rmit.edu.au
* Correspondence: wxm@aoe.ac.cn; Tel.: +86-10-82178896

Abstract: In this study, we investigate the performances of a commercial Global Navigation Satellite System (GNSS) Radio Occultation (RO) mission and a new-generation RO constellation, i.e., Spire and Constellation Observing System for Meteorology, Ionosphere, and Climate 2 (COSMIC-2), respectively. In the statistical comparison between Spire and COSMIC-2, the results indicate that although the average signal-to-noise ratio (SNR) of Spire is far weaker than that of COSMIC-2, the penetration of Spire is comparable to, and occasionally even better than, that of COSMIC-2. In our analysis, we find that the penetration depth is contingent upon various factors including SNR, GNSS, RO modes, topography, and latitude. With the reanalysis of the European Centre for Medium-Range Weather Forecasts and Radiosonde as the reference data, the identical error characteristics of Spire and COSMIC-2 reveal that overall, the accuracy of Spire's neutral-atmosphere data products was found to be comparable to that of COSMIC-2.

Keywords: GNSS-RO; Spire; COSMIC-2

Citation: Qiu, C.; Wang, X.; Zhou, K.; Zhang, J.; Chen, Y.; Li, H.; Liu, D.; Yuan, H. Comparative Assessment of Spire and COSMIC-2 Radio Occultation Data Quality. *Remote Sens.* **2023**, *15*, 5082. https://doi.org/10.3390/rs15215082

Academic Editor: Michael E. Gorbunov

Received: 11 September 2023
Revised: 14 October 2023
Accepted: 22 October 2023
Published: 24 October 2023

1. Introduction

For nearly two decades, Global Navigation Satellite System (GNSS) Radio Occultation (RO) has served as a remote sensing technique providing vast amounts of data for numerical weather prediction [1,2], space weather analyses [3], and climate change research [4,5]. During GNSS-RO measurements, spaceborne receivers on low-Earth-orbiting (LEO) satellites collect GNSS signals affected by refraction from the Earth's atmosphere when GNSS satellites disappear or emerge past the Earth's horizon, yielding atmospheric profiles with the advantages of high accuracy, high vertical resolution, all-weather capability, and global coverage [6,7].

With the successful GPS/MET [8] experiment, the GNSS-RO technique has been identified as a promising technique for the retrieval of atmospheric profiles. The GPS/MET experiment, launched in April of 1995, played a pivotal role in demonstrating the feasibility and effectiveness of GNSS-RO for atmospheric research. Thus, atmospheric monitoring research of Earth involving GNSS-RO observations has been in demand, contributing to a series of RO missions, including the Ørsted [9], Sunsat [10], Challenging Minisatellite Payload [11], Satellite for Scientific Applications C/D [12], Gravity Recovery and Climate Experiment [13], Constellation Observing System for Meteorology, Ionosphere, and Climate 1/2 (COSMIC-1/2) [14,15], Meteorological Operational satellite program-A/B/C [16], and

Feng Yun-3C/D [17] missions. Notably, COSMIC-2 is a joint mission launched by Chinese Taiwan and the U.S. involving multiple satellites designed to collect RO data using signals from GNSS. The mission aims to enhance global weather prediction, ionospheric research, and climate monitoring. Although the number of RO profiles produced by operational GNSS-RO missions is currently far below the minimum profile number required for global observing systems, notably, a frequency of 16,000–20,000 globally distributed occultations per day can achieve the requirement of assimilation in numerical weather prediction [18]. Recently, due to the advantages of small-satellite technologies, including their low economic costs and short research and development periods, RO techniques have been rapidly developed; thus, some commercial GNSS-RO missions have been performed with small satellites to complement the shortage of scientific GNSS-RO data, such as Spire [19]. Spire is a commercial space-based company that operates a constellation of nanosatellites for various purposes, including RO measurements. In the Spire program, a constellation of nanosatellites, known as the LEO Multi-Use Receiver Satellite Bus, has been built to generate a tremendous amount of RO data.

Many evaluations of GNSS-RO data have been implemented to verify the high quality of GNSS-RO retrievals, including GPS/MET [8], Challenging Minisatellite Payload [20], Gravity Recovery and Climate Experiment [13], Meteorological Operational satellite program [21], COSMIC-1 [14], etc. However, comprehensive investigations into the parallels and distinctions between scientific occultation missions and commercial nanosatellite occultation missions, such as COSMIC-2 and Spire, have been limited. Ho et al. (2020) [15] preliminarily investigated the COSMIC-2 neutral atmospheric profile quality using radiosonde data and other RO profiles in terms of profile precision, stability, accuracy, and uncertainty. They found that COSMIC-2 data had a relatively consistent quality with that of COSMIC-1, and the higher signal-to-noise ratio (SNR) of the COSMIC-2 measurements allowed RO signals to penetrate deeper into the lower troposphere, slightly improving the retrieval accuracy. Chen et al. (2021) [22] made a statistical comparison of COSMIC-2 with data from radiosonde, RO data from other missions, global analyses from ECMWF and the National Centers for Environmental Prediction final, and other satellite products, and all the datasets had consistent vertical variations. The temperature profiles showed a mean difference of <0.5 °C and a standard deviation (STD) of 1.5 °C, and the water vapor pressure showed deviations within 2 hPa in the lower troposphere. Spire has operated a constellation of more than 110 LEO nanosatellites since 2019, and Spire's RO refractivity profiles have comparable quality with that of COSMIC-2 [23]. Johnston et al. (2021) conducted a comparison between specific humidity profiles derived from COSMIC-2 RO data and those from ERA5 and MERRA-2 reanalysis datasets. The findings reveal a strong concordance between COSMIC-2 specific humidity and ERA5 while highlighting more pronounced discrepancies with MERRA-2, especially within the boundary layer [24]. Forsythe (2020) et al. validated the ionospheric electron density through Spire's CubeSats RO measurements, and the RO ionospheric inversion results showed significant consistency with the digisonde measurements and Arecibo incoherent scatter radar data [25].

Although the studies above already estimated the RO profiles of Spire and COSMIC-2 and obtained some preliminary results, the properties of the Spire and COSMIC-2 RO retrievals, such as their global coverage and SNR influence, discrepancies between their retrieval qualities, have not been studied. The GNSS-RO constellation observation distribution exhibits global coverage [26], and the SNR is the critical factor limiting the deeper penetration of the GNSS-RO observations [27]. Additionally, the limited payloads of the small satellites result in low power consumption and low gain antennas, thereby reducing the RO retrieval quality. In this study, we aimed to systematically analyze Spire and COSMIC-2 RO profiles from UCAR with other datasets, including ECMWF Reanalysis and radiosonde datasets.

This paper is organized as follows. An introduction to the data and methodology is given in Section 2. The systematic comparison of the Spire and COSMIC-2 retrievals with the ECMWF reanalysis and radiosonde data is discussed in Section 3. Finally, conclusions are provided in Section 4.

2. Data and Methodology

2.1. GNSS-RO Data

The Spire and COSMIC-2 RO data involved in this study, including the neutral atmospheric excess phase and "wet" profile products, are available to freely download from the COSMIC Data Analysis and Archive Center (CDAAC) (https://www.cosmic.ucar.edu/ accessed on 21 October 2023) [28]. The SNRs of RO events are recorded in the excess phase files and meteorological parameters, such as the refractivity, pressure, temperature, relative humidity, specific humidity, and water vapor pressure, are provided in the "wet" profile files [15]. In the current study, Spire and COSMIC-2 RO data from the day of year (DOY) 60 in 2022 to 059 in 2023 are used. It is important to emphasize that the Spire data comprises three navigation satellite systems: GPS, GLONASS, and GALILEO, whereas the COSMIC-2 data only includes GPS and GLONASS.

The equation representing refractivity (N), which is a function of pressure (P in hPa or mbar), temperature (T in K), and water vapor pressure (e in hPa or mbar) in the neutral atmosphere, is given as follows by Smith et al. (1953) [29].

$$N = 77.6\frac{P}{T} + 3.73 \times 10^5 \frac{e}{T^2}. \tag{1}$$

Based on the assumption that water vapor could be negligible, RO "dry" profiles, including "dry pressure" and "dry temperature", are obtained by Equation (1). However, this assumption is unreasonable because more moisture exists below the upper troposphere. Hence, RO "wet" profiles (level-2 products, named "wetPrf" or "wetPf2") including moisture information are extracted using the one-dimensional variational (1DAR) method from the RO bending angle profiles [30]. The vertical resolution of the "wet" profiles is 0.05 km from the surface to below 20 km altitude and 0.1 km from above 20 km to 60 km altitude.

2.2. ERA5 Datasets

ERA5 is the fifth-generation global atmospheric reanalysis product [31], and hourly ERA5 data representing pressure levels during 2019, used in this study, are among the most advanced three-dimensional global analyses available for estimating the quality of Spire and COSMIC-2 RO profiles as benchmark values. As shown in Table 1, the required variables in the ERA5 dataset related to this study, including the specific humidity (kg/kg), temperature (K), and geopotential (m^2/s^2), are available at a horizontal resolution of 0.25° × 0.25° on 37 pressure levels from 1000 hPa to 1 hPa. It can be downloaded publicly from the provided URL https://cds.climate.copernicus.eu/cdsapp#!/dataset/reanalysis-era5-pressure-levels?tab=form (accessed on 21 October 2023).

Table 1. ERA5 hourly data on pressure levels used in this study.

Projection	Regular Latitude-Longitude Grid
Horizontal coverage	Global
Horizontal resolution	0.25° × 0.25°
Vertical coverage	1000 hPa to 1 hPa
Vertical resolution	37 pressure levels
Temporal resolution	Hourly
Required variables	Specific humidity, temperature, and geopotential

2.3. Radiosonde Data

The Integrated Global Radiosonde Archive version 2 (IGRA2) provided by the National Centers for Environment Information is a radiosonde dataset [32] containing variables such as pressure, geopotential height, temperature, and relative humidity from high-quality sounding performed at more than 2800 globally distributed stations; these data are accessible at the website https://www.ncei.noaa.gov/pub/data/igra/ (accessed on 21 October 2023). Here, IGRA2 data are used as the other benchmark values to eliminate the effect of the assimilation of COSMIC-2 and Spire data in the ECMWF Integrated Forecasting System since March and May of 2020, respectively [33]. The IGRA2 observation data are very limited above the 30 km altitude due to the flight limits of radiosondes. Hence, the limit height for comparison between the GNSS-RO and radiosonde data is set to 30 km in this work.

2.4. Methodology

In the comparison of radiosonde and GNSS-RO data, data pairs are collocated within the spatiotemporal windows of 1 h and 100 km. Furthermore, the vertical resolutions of both the ERA5 and radiosonde data are not comparable to that of the GNSS-RO data Therefore, the ERA5 and radiosonde data are interpolated into the vertical resolutions of the GNSS-RO data.

In this study, the mean difference and STD used in the statistical calculations are defined using the following equations to evaluate the GNSS-RO product properties:

$$\overline{\Delta x_a} = \frac{1}{n}\sum\nolimits_{i=1}^{n}\left(x_i^{ro} - x_i^{t}\right), \tag{2}$$

$$STD_{\Delta x_a} = \sqrt{\frac{1}{n}\sum\nolimits_{i=1}^{n}\left(\left(x_i^{ro} - x_i^{t}\right) - \overline{\Delta x_a}\right)^2}, \tag{3}$$

$$\overline{\Delta x_r} = \frac{\frac{1}{n}\sum_{i=1}^{n}\left(x_i^{ro} - x_i^{t}\right)}{x_i^{t}}, \tag{4}$$

$$STD_{\Delta x_r} = \sqrt{\frac{1}{n}\sum\nolimits_{i=1}^{n}\left(\frac{\left(x_i^{ro} - x_i^{t}\right)}{x_i^{t}} - \overline{\Delta x_r}\right)^2}, \tag{5}$$

where x_i^{ro} and x_i^{t} represent the GNSS-RO and benchmark temperature, relative humidity, pressure, and refractivity, respectively, the subscript i stands for the ith GNSS-RO-benchmark collocation, and n is the number of collocations.

Emphasis should be placed on the fact that data quality control is conducted as part of the data quality assessment process. The reference values were derived from ERA5 or radiosonde data. RO refractivity profiles with relative errors surpassing 5%, as well as wet pressure profiles exceeding 900% or dropping below −90%, were eliminated.

3. Comparison Results

3.1. Initial Analysis

The GNSS-RO constellation pattern impacts the distribution of RO event observations over the globe. Without the specific configuration for Spire, consisting of a diverse set of orbits compounding Sun-Synchronous Orbits (SSO), 83–85° Orbits, Equatorial Orbit, 51.6° Orbits, and 37° Orbits and the continuous changing of satellites owing to their short operational lifetime of 2+ years [34]. As for COSMIC-2, six satellites orbit around the Earth at an altitude of 550 km with a 24° inclination [35]. Furthermore, given that GNSS constellations encompass diverse signal frequencies, constellation configurations, and modulation-demodulation techniques, potentially influencing RO events. Hence, separate investigations will be conducted for GPS, GLONASS, and GALILEO RO events.

The spatial distribution of the Spire RO events obtained from satellites with a different orbit type has an obvious difference. As shown in Figure 1, the Spire RO events observed on DOY 60 in 2022 are scattered globally. The red dots signify GPS RO events, the green dots indicate GLONASS RO events, whereas the blue dots represent GALILEO RO events. For COSMIC-2, Chen et al. (2021) [22] mention that RO events only cover the low-latitude area (±45°). Notably, few RO events recorded by COSMIC-2 occurred slightly beyond the edges of the ±45° region; these events are regarded as occurring on the scale within the latitude area of ±45° in this study. The coverage areas of the RO events by Spire are wider than those by COSMIC-2; thus, Spire can provide global RO data due to the constellation characteristics of LEMUR-2 consisting of a series of orbits.

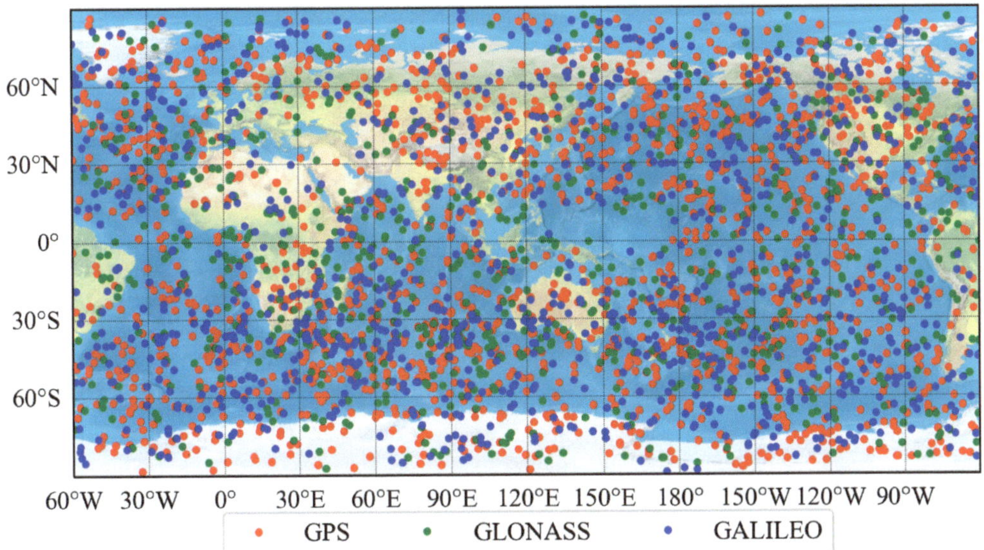

Figure 1. Spire RO event distribution on DOY 060 in 2022 (red: GPS, green: GLONASS, and blue: GALILEO).

Furthermore, the RO events can be classified into two modes, setting occultation or rising occultation, based on the relative movement trend on the occultation observation. Considering the different navigation satellite systems, here, the RO events were divided into six groups: GPS/Set, GPS/Rise, GLONASS/Set, GLONASS/Rise, GALILEO/Set, and GALILEO/Rise.

Table 2 shows the percentage for six groups of profiles (wetPf2) from the Spire and COSMIC-2 satellite data obtained from CDAAC from DOY 060 in 2022 to 059 in 2023 under quality control. Generally, there were more setting occultation events than rising occultation events (except GLONASS-derived RO events for Spire). The number of atmospheric profiles for Spire amounts to 1,663,197, surpassing COSMIC-2 with the number of 1,440,424, which is a great supplement to the high-latitude regions that COSMIC-2 data cannot cover. However, the Spire satellite constellation comprises dozens of satellites, whereas COSMIC-2 consists of only six. Daily data produced by individual Spire satellites from UCAR contain no more than 300 RO soundings, which is much smaller than those recorded by individual COSMIC-2 satellites, each providing over 700 RO soundings.

Table 2. The percentage for six groups of profiles (wetPf2) from the Spire and COSMIC-2 data obtained from CDAAC from DOY 060 in 2022 to 059 in 2023 under quality control.

GNSS-RO Mission	Number of Profiles	Mode	GPS	GLONASS	GALILEO
Spire	1,663,197	Set	26.46%	14.80%	12.14%
		Rise	19.97%	15.29%	11.34%
COSMIC-2	1,440,424	Set	34.08%	18.71%	None
		Rise	29.43%	17.78%	None

With the RO events observed from DOY 060 in 2022 to 059 in 2023 by Spire and COSMIC-2, we carried out an investigation on the RO events obtained with different navigation satellite systems, i.e., GPS, GLONASS, and GALILEO. As shown in Table 2, the percentages of the GPS-derived RO events are much larger than GLONASS- and GALILEO-derived RO events for both Spire and COSMIC-2. This is reasonable considering the number of operational GPS satellites is larger than GLONASS and GALILEO satellites. For Spire, GLONASS-derived RO events are slightly higher than those derived from GALILEO. Although there is no significant difference in the number of satellites in orbit for GLONASS and GALILEO, it is important to note that some Spire satellites do not receive GALILEO signals, including S128, S115, S117, and others.

3.2. SNR

The SNR is the key factor impacting deeper occultation, especially for surface and tropical RO soundings. Currently, Spire and COSMIC-2 have developed new-generation GNSS-RO payloads, i.e., STRATOS and the Tri-GNSS Radio-occultation System (TGRS), respectively, to improve their retrieval quality in the low troposphere. It should be noted that, compared to other payloads with high-gain antennas (e.g., the TGRS and CION [27]), STRATOS is equipped with relatively low-gain antennas to track GNSS signals, thus directly leading to relatively low SNR values in their measurements.

Figure 2 features a dual x-axis. The top x-axis represents the altitude in km, being the straight-line height between the GNSS and LEO satellites, while the bottom x-axis represents the time sequence in seconds. The y-axis corresponds to the SNR for two GPS frequency bands. In this context, there are two GPS frequency bands, L1 at 1575.42 MHz and L2 at 1227.60 MHz. In Figure 2, the SNR time series of two examples of a typical rising occultation, observed in an adjacent area in the tropics nearly simultaneously by STRATOS and the TRGS, are depicted. The 1-s average SNR of the two rising occultation events shown with the black and green lines increases with increasing altitude. Furthermore, obvious fluctuations or oscillations in the SNR curve can be observed at altitudes between −100 km and −50 km, resulting from signals being temporarily captured and then abruptly disappearing due to atmospheric ducting and super refraction [12]. It appears that the SNR time series of the Spire is lower than that of the COSMIC-2. For example, at altitudes above 0 km, the L1 SNR of the Spire is ~320 volts/volt, approximately one quarter of the L1 SNR of the COSMIC-2, which is ~1400 volts/volt. Similarly, the L2 SNR of the Spire is ~200 volts/volt, nearly half of the L2 SNR of the COSMIC-2, which is ~500 volts/volt. Moreover, according to the study by Sokolovskiy et al. (2014) [36], the L1 and L2 SNRs of the COSMIC-1 during several rising occultation events were ~600–800 volts/volt and ~200–600 volts/volt, respectively, at altitudes above 0 km. Therefore, the results that STRATOS, which has relatively low-gain antennas, has a slightly weaker ability to capture and track signals compared to IGOR, the payload of the COSMIC-1. Also, TRGS has an even stronger ability to capture and track signals compared to both STRATOS and IGOR.

To generally compare the capability of STRATOS and the TRGS in capturing and tracking signals, we also investigated the SNR in the altitude range of 60–80 km. The 60–80 km altitude range is optimal for evaluating signal strength, unaffected by atmospheric interference. It's sufficiently high to render attenuation from typical atmospheric refraction negligible, yet it doesn't extend to the E-layer where disturbances are more pronounced [21,27]. As a result, the average L1-signal SNR within the 60–80 km altitude range (hereafter referred to as the SNR average) is related to the signal strength of the RO event. Figure 3 shows the average SNR histograms of the normalized probabilities for the Spire, while Figure 4 displays the average SNR histograms of the normalized probabilities for the COSMIC-2. For both GPS and GLONASS, the SNR averages of all the Spire data (blue) range from ~200–1600 volts/volt with only one peak, which is much weaker than the COSMIC-2 averages (brown), which vary from ~200–2250 with two peaks (see Figures 3a and 4a). To investigate the influence of the navigation satellite systems and occultation mode (setting or rising occultation) received by LEO on the SNR, all the data were divided into six groups: GPS/Set, GPS/Rise, GLONASS/Set, GLONASS/Rise, GALILEO/Set, and GALILEO/Rise.

In Figure 3b,c and Figure 4b,c, the tops and bottoms of the x-axes represent the SNR averages of the setting and rising occultation-normalized probability histograms, respectively. Both the GPS/Set and GPS/Rise SNR averages of the Spire range from ~200–600 volts/volt, while those of COSMIC-2 range from ~300–2000 volts/volt. In Figure 3c, the SNR averages for both the GLONASS/Set and GLONASS/Rise of the Spire range from ~300 to 1500 volts/volt, while those of COSMIC-2 range from ~250–2500 volts/volt in Figure 4c. In Figure 3d, the SNR averages for both the GALILEO/Set and GALILEO/Rise of the Spire range from ~300 to 1500 volts/volt. The SNR averages for both the GALILEO/Set and GALILEO/Rise of the Spire span from ~300 to 750 volts/volt. It is possible that the broader range of the SNR averages for GPS- or GALILEO-derived RO data is due to its utilization of CDMA wireless data transmission, while GLONASS utilizes FDMA. It's worth noting that the occultation mode does not affect the scale of the SNR averages, while the navigation satellite system (either GPS, GLONASS, or GALILEO) has an outstanding effect on the range of the SNR averages. The SNR averages of the GPS-derived RO exhibit a sharper peak compared to that of the GLONASS-derived RO, as shown in the comparison between Figure 3b,c. The GPS- and GLONASS-derived RO data of the Spire produce two distinct peaks that are widely separated from each other, as shown in Figure 3a. In contrast, Figure 4a displays only one peak due to the proximity of the COSMIC-2 peaks, while the SNR averages for GALILEO-derived RO data do not exhibit a clear peak in Figure 4c.

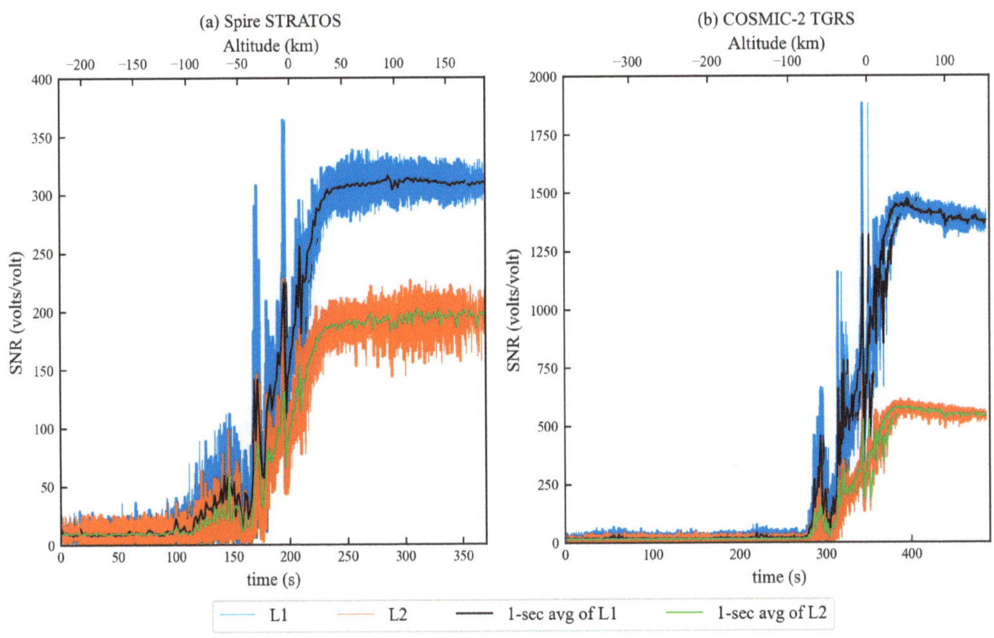

Figure 2. SNRs of two examples of a typical rising occultation in a tropical region with open-loop tracking (blue: L1, orange: L2, black: 1-s average of L1, and green: 1-s average of L2). (**a**) STRATOS, RO starting time 1847 UTC, 1 March 2022, located at 29.79°N, 141.45°W; (**b**) the TGRS, RO starting time 1845 UTC, 1 March 2022, located at 30.52°N, 142.44°W. (The top x-axis represents the altitude in km, being the straight-line height between the GNSS and LEO satellites, while the bottom x-axis represents the time sequence in second. The y-axis corresponds to the SNR for two GPS frequency bands).

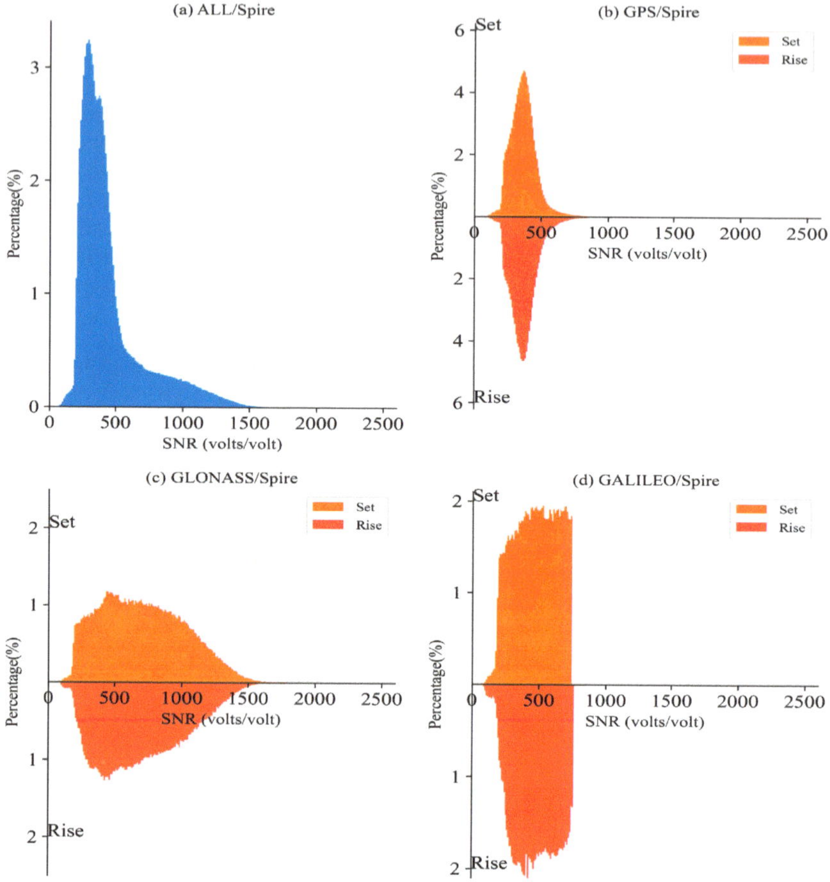

Figure 3. Histograms of the normalized probability of the average L1 SNR (volts/volt) values for the Spire between 60 and 80 km.

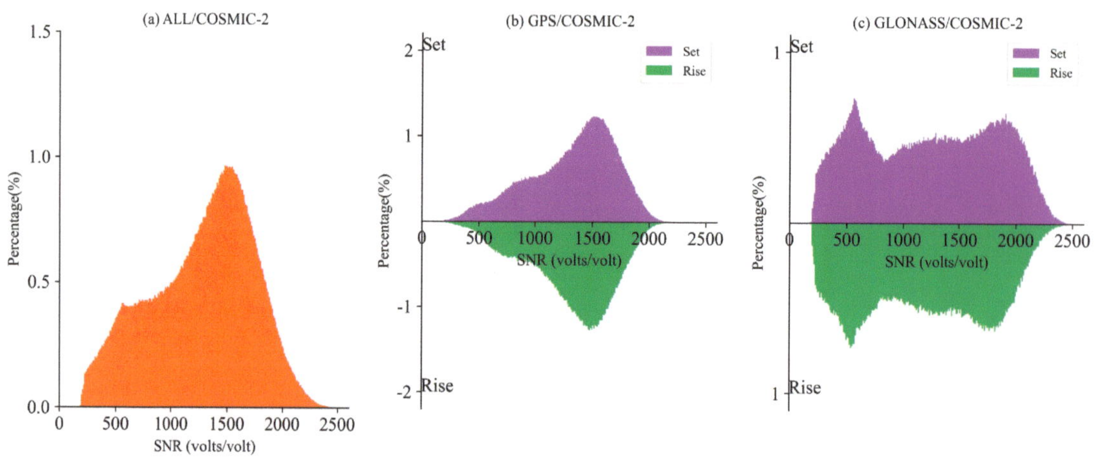

Figure 4. Histograms of the normalized probability of the average L1 SNR (volts/volt) values for the COSMIC-2 between 60 and 80 km.

Ho et al. (2020) mentions that enhancing SNR can improve penetration and data quality [15], and Jing et al. (2023) points out the correlation between the penetration of COSMIC-2 and latitude [37]. Therefore, based on these insights, to analyze the correlations between the SNR and penetration and between the SNR and data quality, it was necessary to compare the mean SNRs at different latitudes. Thus, a statistical comparison was performed on all the data and for the four data groups at latitudinal intervals of 15°. As shown in Figure 5, the influence of the RO mode on the mean SNR with latitudinal variations was not significant. Figure 5a shows that the mean SNR values of the GPS- and GALILEO-derived RO data of the Spire do not show a clear dependence on latitude, while the mean SNR values of the GLONASS-derived RO data of the Spire show some fluctuation with latitude. For COSMIC-2, Figure 5b shows that generally, the mean SNR values are much higher in the low latitudes than in the high latitudes.

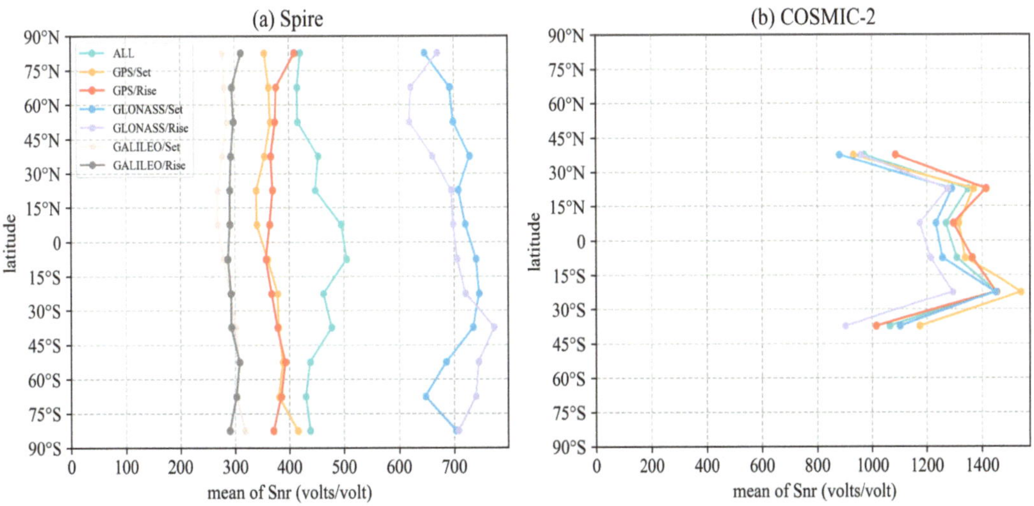

Figure 5. Mean SNR values of the Spire and COSMIC-2 at different latitudes: (**a**) Spire and (**b**) COSMIC-2. Light green: all data, orange: GPS/Set, Red: GPS/Rise, blue: GLONASS/Set, and purple: GLONASS/Rise, khaki: GALILEO/Set, and gray: GALILEO/Rise.

In summary, the mean SNR values of the Spire were significantly smaller than that of the COSMIC-2. In addition, the mean SNR values of the GPS-derived RO showed a sharper peak for both the Spire and COSMIC-2 than mean SNR values of the GLONASS-derived RO, with no effect of the occultation mode observed. Notably, the SNR averages of the GALILEO-derived RO data show no obvious peak. We then examined the penetrations of different missions in relation to the varying SNR strengths in Section 3.3.

Table 3 presents the mean SNR values for the Spire, COSMIC-2, and COSMIC-1. The mean SNR values of the Spire for the GPS-, GLONASS, and GALILEO-derived RO were 371, 708, and 480 volts/volts, respectively, and the total mean SNR for Spire was 503 volts/volts. As for the COSMIC-2, the mean SNR values for the GPS- and GLONASS-derived RO were 1315 and 1210 volts/volt, respectively, and the total mean SNR for COSMIC-2 was 1276 volts/volt. The mean SNR value for the COSMIC-1, obtained from [38], was 704 volts/volt. These results indicate that the ability of the Spire to track only GPS signals is significantly weaker than that of the COSMIC-1, and with the joint consideration of GLONASS and GALILEO, Spire can achieve a slightly weaker capability than COSMIC-1. The COSMIC-2 has a superior ability to track signals compared to both the COSMIC-1 and Spire.

Table 3. Mean L1 SNR values of Spire, COSMIC-2 and COSMIC-1 (unit: volts/volt).

GNSS-RO Mission	GPS	GLONASS	GALILEO	Total
Spire	371	708	480	503
COSMIC-2	1315	1210	None	1276
COSMIC-1	704	None	None	704 [38]

3.3. Penetration

The lower atmosphere is important for numerical weather prediction and atmospheric science research. Because of thick water vapor near the surface, RO observations are limited in their ability to provide atmospheric information in the lower troposphere. As a result, the penetration GNSS-RO detected is also an important indicator of the quality of RO sound. In this section, statistical analysis was conducted on the penetration of COSMIC-2 and Spire, and the effect of SNR on the penetration was analyzed.

As shown in Table 4, 76.60% of the Spire data can achieve penetration depths below 1 km in the lower troposphere. The COSMIC-2 was able to detect the surface atmosphere at a 1 km height in about 78.12% of RO events, which is better than the Spire. According to Ho et al. (2020), increasing the SNR can improve the RO penetration depth, which improves the ability of COSMIC-2 in penetrating deep into the lowest 100 m of the troposphere. Therefore, the conclusion that the COSMIC-2 can penetrate deeper than the Spire is consistent with previous experiments [15] that attribute this difference to the higher SNR of the COSMIC-2. Moreover, in comparison to the capability to detect the surface atmosphere at a 1 km height, it is noteworthy that setting occultation events can achieve greater penetration depths than rising occultation events for both Spire and COSMIC-2, independent of the navigation satellite system's influence.

Table 4. Penetration depth percentages of Spire and COSMIC-2 from DOY 060 in 2022 to 059 in 2023. (altitude of penetration H_p).

GNSS-RO Mission	Group	$H_p \leq 1$ km	1 km < $H_p \leq 5$ km	5 km < $H_p \leq 10$ km	H_p > 10 km
Spire	GPS/Set	78.62%	19.75%	1.55%	0.08%
	GPS/Rise	77.38%	21.08%	1.46%	0.09%
	GLONASS/Set	78.29%	20.16%	1.49%	0.06%
	GLONASS/Rise	74.31%	23.93%	1.64%	0.12%
	GALILEO/Set	79.05%	19.29%	1.56%	0.10%
	GALILEO/Rise	73.29%	24.57%	2.02%	0.12%
	Total	76.60%	22.20%	1.07%	0.13%
COSMIC-2	GPS/Set	79.20%	19.87%	0.90%	0.03%
	GPS/Rise	75.34%	23.61%	1.01%	0.04%
	GLONASS/Set	78.21%	20.57%	1.02%	0.21%
	GLONASS/Rise	75.09%	23.62%	1.21%	0.08%
	Total	78.12%	20.79%	1.01%	0.08%

However, it is important to note that the statistical method used above is not perfect due to the influence of topography (e.g., mountains) on the penetration depth. Taking terrain into account (using data from ETOPO2 v2), we plotted the Spire RO events in a global topographic map where areas at $H_t \leq 1$ km, 1 km $< H_t \leq 5$ km, and $H_t > 5$ km (terrain high H_t) are denoted with white, brown, and deep brown colors, as shown in Figure 6. Figure 6 shows that RO events at $H_p \leq 1$ km (altitude of penetration H_p) are scattered in the region at $H_t \leq 1$ km (Figure 6a), those at 1 km $< H_p \leq 5$ km (Figure 6b) are found mainly in the region at 1 km $< H_t \leq 5$ km (e.g., the South Pole and mountainous areas), those at 5 km $< H_p \leq 10$ km (Figure 6c) are concentrated mainly in the region at $H_t > 5$ km (e.g., the

Himalayan Mountains and Andes Mountains), and those at $H_p > 10$ km (Figure 6d) are few and dotted around the world.

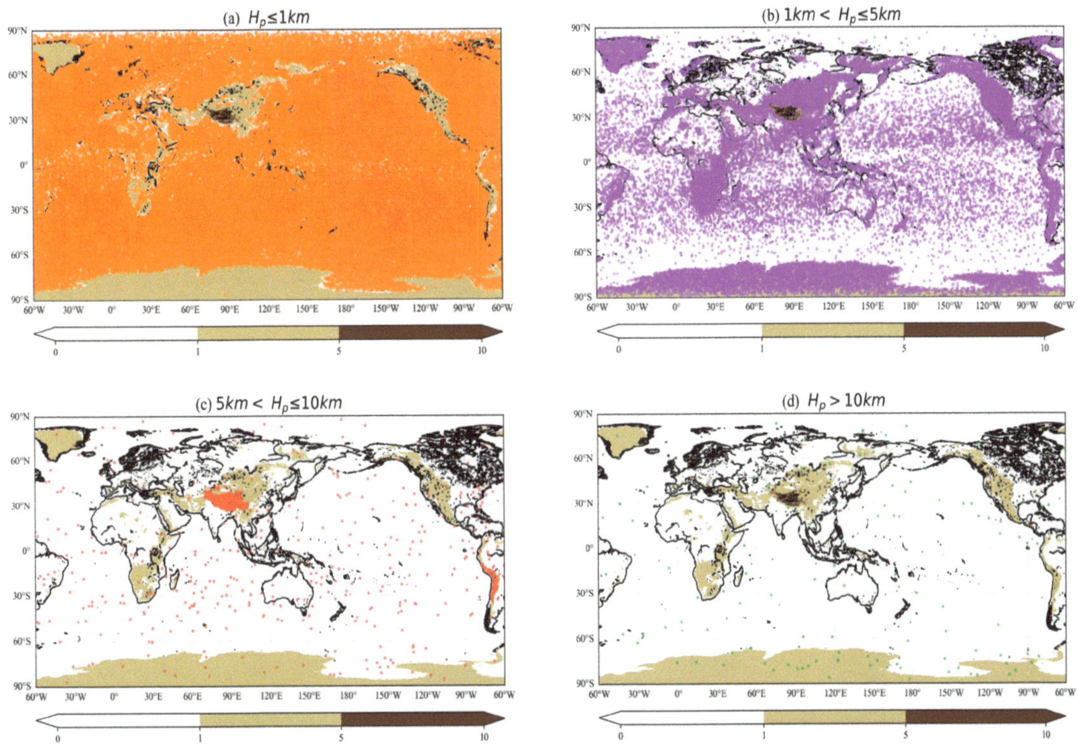

Figure 6. Penetration map of global-topography RO events recorded by Spire from DOY 060 in 2022 to 091 in 2022: (**a**) $H_p \leq 1$ km (altitude of penetration H_p), (**b**) 1 km $< H_p \leq 5$ km, (**c**) 5 km $< H_p \leq 10$ km, and (**d**) $H_p > 10$ km; white: $H_t \leq 1$ km, brown: 1 km $< H_t \leq 5$ km, and deep brown: $H_t > 5$ km (terrain high H_t).

Although the RO events recorded by the COSMIC-2 are limited to the lower-latitude area (±45°) (as mentioned in Section 3.1), similar phenomena between the COSMIC-2 and Spire records are shown in Figure 7. Due to the great number of RO events recorded by COSMIC-2, the influences of topography are clearly visible, particularly for RO events at 5 km $< H_p \leq 10$ km, which are concentrated in the Himalayan Mountains and Andes Mountains, as indicated by the red points in Figure 7c.

Table 4 summarizes the percentages of RO data recorded below 1 km for the Spire and COSMIC-2, using the division scheme described in Section 3.1. The table shows that, in general, the percentages of the setting occultation events (except for GPS-derived RO events for the Spire, highlighted in gray) are slightly higher than those of the rising occultation events. Additionally, the ability to penetrate the lower troposphere during rising occultation events is slightly weaker than during setting occultation events. When compared to the COSMIC-2 data, the percentages of the corresponding items in the Spire data are higher, indicating that the Spire has a superior ability to penetrate the lower troposphere globally compared to the COSMIC-2 in lower-latitude regions. By considering the information in Table 4, the penetration depth also depends on the navigation satellite system and occultation mode.

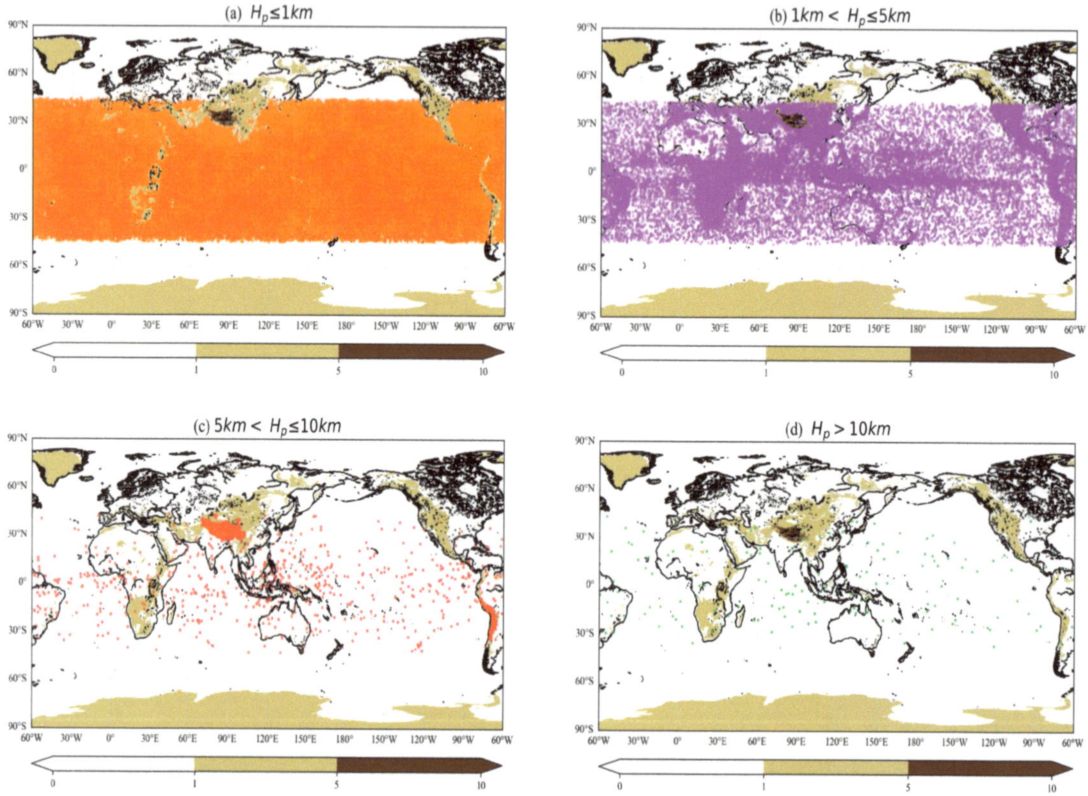

Figure 7. Penetration map of global-topography RO events recorded by COSMIC-2 from DOY 060 in 2022 to 091 in 2022: (**a**) $H_p \leq 1$ km (altitude of penetration H_p), (**b**) $1\ km < H_p \leq 5$ km, (**c**) 5 km $< H_p \leq 10$ km, and (**d**) $H_p > 10$ km; white: $H_t \leq 1$ km brown: 1 km $< H_t \leq 5$ km, and deep brown: $H_t > 5$ km (terrain high H_t).

To quantify the ability of the studied systems to detect the lower troposphere, the influences of topography were minimized by subtracting the terrain height from the penetration depth. Figure 8a shows the cumulative percentages of data below 1 km for the Spire and COSMIC-2 at different latitudes, indicating that the global Spire data are susceptible to topographic effects, particularly at the South Pole. With topographic effects accounted for, the COSMIC-2 surpasses the Spire within the latitude range of ±45°. In general, the ability of the Spire global data, with topographic effects fixed, to penetrate the lower troposphere becomes stronger with increasing latitude. Combining Figures 5–7, the penetration depth is also determined by topography and latitude. Figure 8b exhibits the cumulative percentages of RO events at various penetration depths. The Spire RO events recorded with fixed topographic data show better performance in detecting the near-surface atmosphere, and similar results are seen for the COSMIC-2 data, which perform slightly worse than the fixed Spire data (see Figure 8).

After fixation, the Spire and COSMIC-2 data below an altitude of 1 km make up 88.7% and 85.3% of all the data, respectively. Due to topographic changes and water vapor variations with increasing latitude, the penetration depth is affected, thus leading to the retrieval statistics. Through comparing the fixed Spire data within the lower-latitude range of ±45° to the fixed COSMIC-2 data, it is found that the fixed Spire (±45°) data below the 1 km altitude accounted for 84.2% of all the data. In ascending sequence, the ability of the systems to perform deeper soundings could be ranked as follows: Spire fixed (±45°), COSMIC-2 fixed, and Spire fixed.

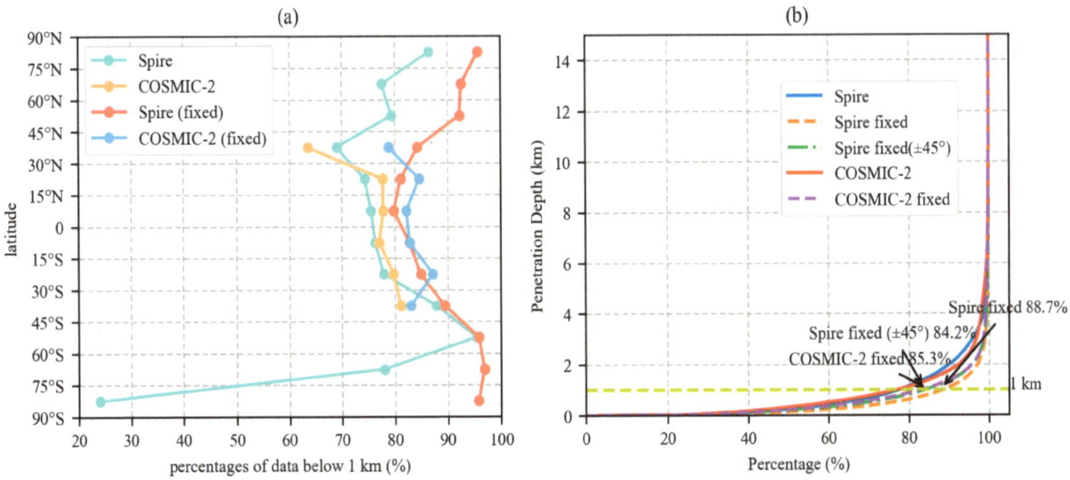

Figure 8. Cumulative percentages (**a**) of data below 1 km for Spire and COSMIC-2 at different latitudes (light green: Spire, orange: COSMIC-2, red: Spire with the topographic data fixed, and blue: COSMIC-2 fixed) and (**b**) cumulative percentages at each penetration for Spire and COSMIC-2 (blue solid line: Spire, orange dashed line: Spire fixed, green with dashed-dotted line: Spire fixed within the latitude area of ±45°, red solid line: COSMIC-2, and purple: COSMIC-2 fixed).

Therefore, COSMIC-2 has a better performance in sounding the deeper troposphere than Spire. Owing to topographic effects, especially the effects of mountains, the penetration depth is obviously affected. After removing the effects of terrain, the penetration of Spire is improved significantly, but under the same circumstances (e.g., at latitudes within ±45°), COSMIC-2 still has a better performance than Spire w.r.t deep penetration.

3.4. RO Retrieval Quality Assessment

The GNSS-RO product "wetPf2" includes atmospheric refractivity, temperature, specific humidity, and pressure. The other parameters are retrieved from the refractivity using the 1DAR method, so we preliminarily analyzed the refractivity characteristics. The COSMIC-2 RO data evaluated herein [15] have sufficient accuracy to assess the Spire data using stringent collocation criteria of a 1 h temporal window and a spatial distance of 100 km. Over 10 km, the differences were uniformly distributed on both sides of the y-axis at the zero point, as shown in Figure 9. Below 10 km, the differences between Spire and COSMIC-2 are largest, with more positive differences compared to the other heights, although these findings may have resulted from the increased water vapor or few considered collocation pairs.

Figure 10a shows that the root mean square (RMS) for the absolute difference of the refractivity below 30 km decreased with increasing latitude in comparison with the ERA5 dataset. In addition, the refractivity profiles of COSMIC-2 are chiefly in the tropics and diminish sharply as the latitude increases, while those of Spire are primarily situated in middle latitudes and few near the poles. Owing to the lower moisture at high latitudes, the Spire data was limited to the latitude range of ±45° to facilitate a comparison with the COSMIC-2 data.

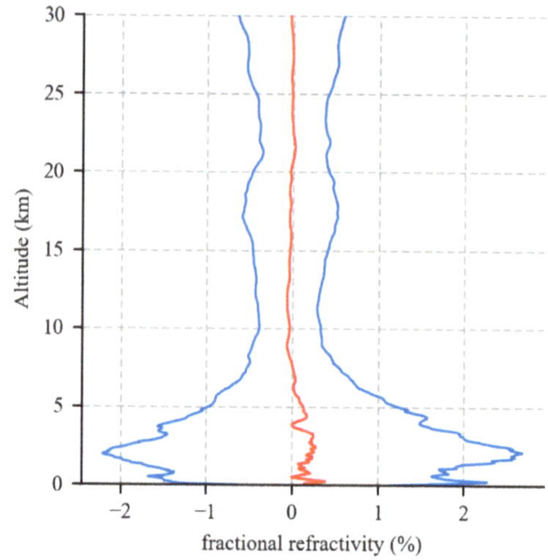

Figure 9. The fractional refractivity comparison between the Spire and COSMIC-2 wetPf2 data below 30 km (Red: mean fractional difference, and blue: STD).

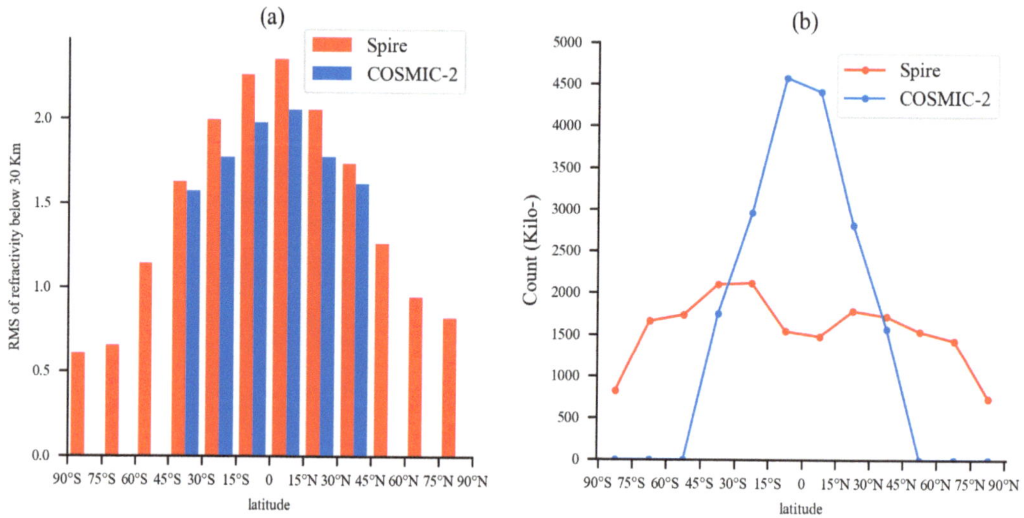

Figure 10. RMS of refractivity below 30 km and profile counts at different latitudes (red: Spire, blue: COSMIC-2). (**a**) RMS of refractivity below 30 km and (**b**) counts (Kilo-).

The evaluated GNSS RO data covers the entire year, enabling the analysis of the neutral atmospheric refractivity error seasonal characteristics below 30 km as shown in Figure 11. The solid lines represent the mean fractional differences in refractivity, while the dotted lines indicate the STD of fractional differences in refractivity. The black, red, green, and blue lines correspond to spring, summer, autumn, and winter, respectively. Figure 11a–c represent refractivity error seasonal characteristics for Spire and COSMIC-2 for Spire, Spire (±45°), and COSMIC-2, respectively. Based on the mean and STD of fractional differences in refractivity observed across different seasons, there does not appear to be any significant seasonal variation in refractivity errors. This may be due to the fact that GNSS uses L-Band navigation signals, which have the ability to penetrate clouds and rain, resulting in minimal weather-related interference.

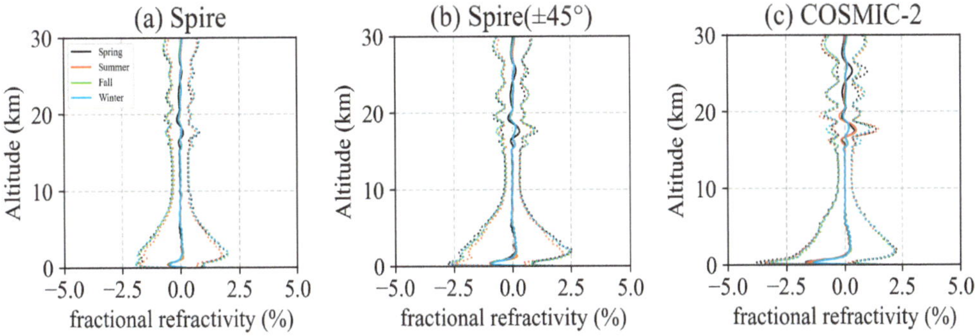

Figure 11. The neutral atmospheric refractivity error seasonal characteristics for Spire and COSMIC-2 from DOY 060 in 2022 to 059 in 2023 below 30 km. (black: Spring, red: Summer, green: Fall, and blue: Winter).

Due to the coarse coordinate scale in Figure 11 displayed, in Figure 12, the mean and STD of fractional differences in refractivity are presented separately to analyze the refractivity error characteristics. Also, it conducts an analysis of how different satellite navigation systems and RO event modes affect these refractivity error characteristics.

The neutral-atmosphere refractivity data acquired from Spire and COSMIC-2 were compared with the ERA5 dataset under the division scheme described in Section 3.1. The results obtained for all the data and for the six data groups are plotted in Figure 12. The statistics of those four groups (GPS/Set, GPS/Rise, GLONASS/Set, and GLONASS/Rise) have identical results, while GALILEO/Set and GALILEO/Rise show better statistics than the others, especially below 10 km shown in Figure 12d,e. Thus, the RO event modes did not impact the retrieval quality. The better statistics for GALILEO may be related to the fact that the precision of GALILEO code and phase observations outperforms those of the GPS and GLONASS ones [39,40]. Figure 12a–c show the statistical results obtained for Spire, revealing that below 10 km, the mean fractional differences for GPS/Set, GPS/Rise, GLONASS/Set, and GLONASS/Rise were negative within the magnitude of ~0.8% and maximal STD of ~2.0%, while the mean fractional differences for GALILEO/Set and GALILEO/Rise were negative within the magnitude of ~0.6% and maximal STD of ~1.8%; at heights between 10 and 25 km, the mean fractional differences were negative within the magnitude of ~0.20% and maximal STD of ~0.4%; and above 25 km, the mean fractional differences were positive within the magnitude of ~0.25% and maximal STD of ~0.75%. Within the latitude range of ±45°, a smaller proportion of Spire GNSS RO data were analyzed compared to the global data as shown in Figure 12d,e. Below 10 km, the mean fractional differences for GPS/Set, GPS/Rise, GLONASS/Set, and GLONASS/Rise were negative within the maximum value of ~1.2% and maximal STD of ~2.2%, while the mean fractional differences for GALILEO/Set and GALILEO/Rise were negative within the maximum value of ~0.5% and maximal STD of ~1.8%; at heights between 10 and 25 km, the mean fractional differences were within the magnitude of 0.2%, with a maximal STD of ~0.5%; and above 25 km, the mean fractional differences were positive within the maximum value of ~0.3% and maximal STD of ~0.8%. For the COSMIC-2 data shown in Figure 12g–i, below 10 km, the mean fractional differences were negative within the maximum value of ~1.8% and maximal STD of ~2.0%; at heights between 10 and 25 km, the mean fractional differences were within the magnitude of ~0.2% and maximal STD of ~0.7%; and above 25 km, the mean fractional differences were positive within the maximum value of ~0.2% and maximal STD of ~1.0%.

It is evident that the Spire data collected within latitudes of ±45° has comparable quality to the COSMIC-2 data. Beyond the ±45° latitude range, the retrieval quality of the Spire data was higher than that of the COSMIC-2 data, perhaps because less moisture is present at higher latitudes.

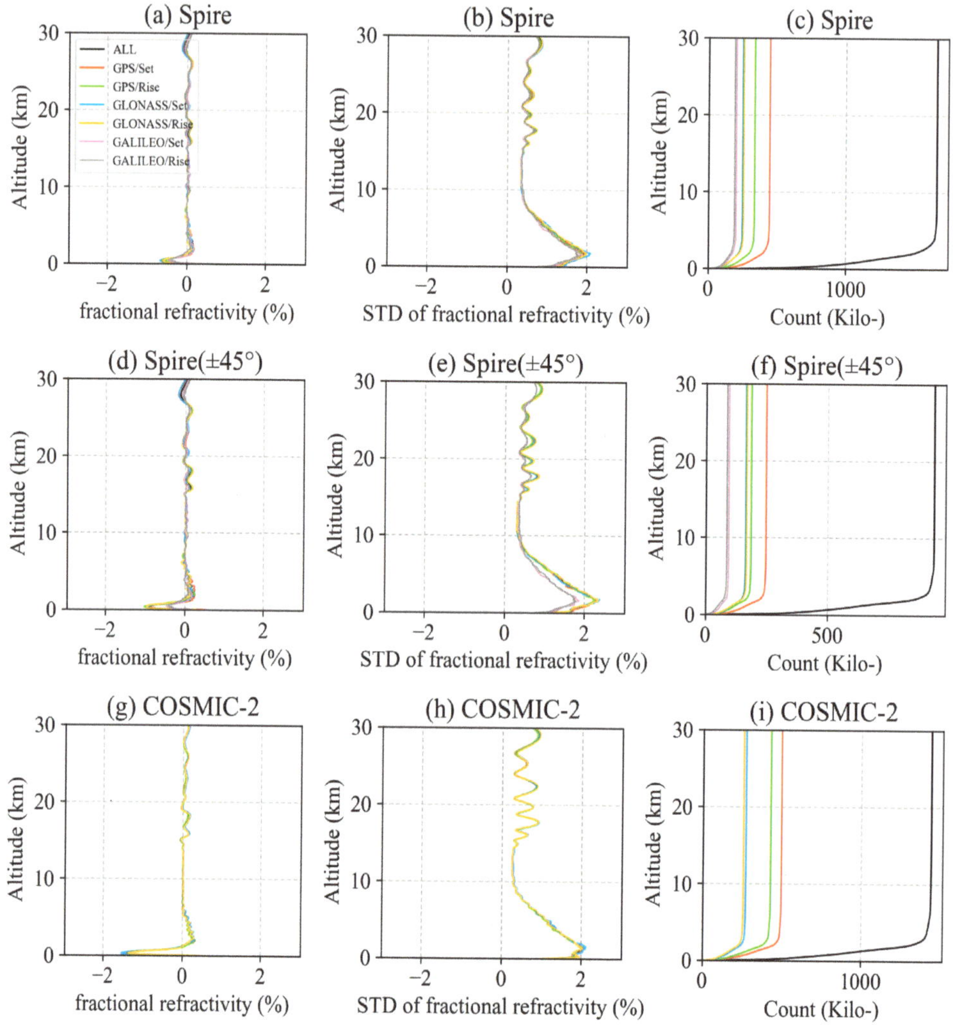

Figure 12. Neutral atmospheric refractivity comparison of the Spire and COSMIC-2 wetPf2 products with ERA5 data from DOY 351 in 2020 to 015 in 2021 below 30 km (black: all data, red: GPS/Set, green: GPS/Rise, blue: GLONASS/Set, yellow: GLONASS/Rise, purple: GALILEO/Set, and gray: GALILEO/Rise). (**a–c**) Mean fractional differences, STDs, and counts in the Spire record, (**d–f**) mean fractional differences, STDs, and counts in the Spire record (±45°), and (**g–i**) mean fractional differences, STDs, and counts in the COSMIC-2 record.

Furthermore, we performed a statistical comparison of the specific humidity, absolute temperature, and pressure obtained from RO events through the 1DAR method with those of the ERA5 dataset under the same circumstances. Figure 13 shows the mean differences and STDs in the meteorological parameters retrieved from the Spire (±45°) and COSMIC-2 data. The mean differences and STDs of both the Spire and COSMIC-2 data exhibit identical statistical results, especially above the 10 km altitude. Below 10 km, the mean differences in specific humidity for both the Spire and COSMIC-2 data were negative, with maximal values of ~0.45 g/kg and ~0.70 g/kg and maximal STDs of ~0.9 g/kg and ~0.8 g/kg, respectively; the mean differences in temperature in the Spire and COSMIC-2 data were positive within ~0.25 K and ~0.4 K with maximal STDs of ~0.7 K and ~0.7 K, respectively, and the mean differences in pressure in the Spire and COSMIC-2 data were negative within ~0.2 hPa and ~0.1 hPa, with maximal STDs of ~1.1 hPa and ~1.3 hPa, respectively. Above 10 km, according to the statistics obtained for the Spire and COSMIC-2 data, the specific humidity exhibited mean differences and STDs of nearly equal zero; the temperature exhibited mean differences and STDs that fluctuated near zero within a maximum value of ~0.3 K and a maximal STD of ~1.8 K, and the pressure exhibited mean differences and positive STDs within the maximum value of ~0.1 hPa and maximal STD of ~0.5 hPa, indicating very similar results between the two data sources. Compared to the ERA5 product, the specific humidity, temperature, and pressure of Spire (±45°) and COSMIC-2 indicated identical retrieval qualities.

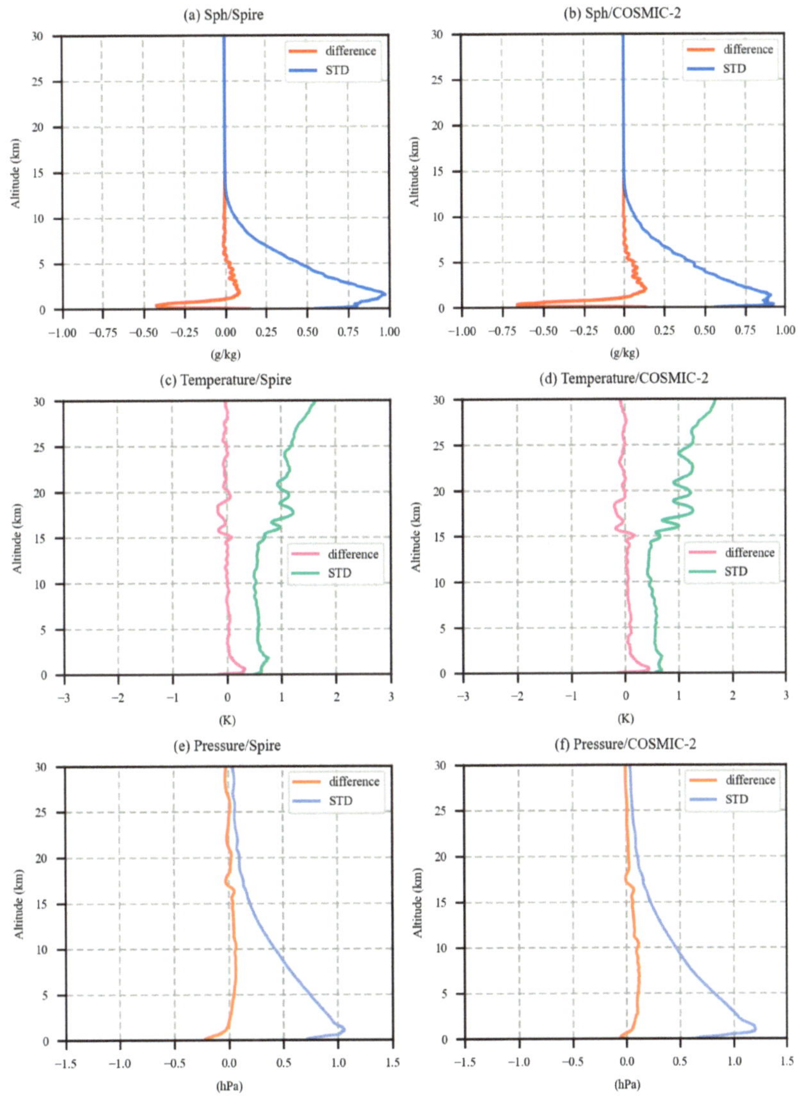

Figure 13. Meteorological parameter comparison of the Spire (±45°) and COSMIC-2 wetPf2 products compared with ERA5 data from DOY 351 in 2020 to 015 in 2021 below 30 km ((**a**,**b**) specific humidity for Spire and COSMIC-2, respectively; (**c**,**d**) absolute temperature for Spire and COSMIC-2, respectively; (**e**,**f**) pressure for Spire and COSMIC-2, respectively).

To remove the coupling effect that arises when assimilating data into the ERA5 dataset, Spire (±45°) and COSMIC-2 data were compared with radiosonde data. Above an altitude of approximately 25 km, insufficient radiosonde data were used to perform the statistical comparison; thus, at this height, the results were not credible. The mean differences in the refractivity, specific humidity, temperature, and pressure of the Spire data were negative within ~1.8%, ~0.80 g/kg, ~0.45 K, and ~1.0 hPa below 10 km, respectively, with maximal STDs of ~3.0%, ~1.8 g/kg, ~1.7 K, and ~2.2 hPa, respectively, as shown in Figure 14. Above the altitude of 10 km, the mean differences in refractivity, specific humidity, temperature, and pressure in the Spire data fluctuated near zero within 0.2%, 0.1 g/kg, ~0.2 K, and ~0.3 hPa, respectively, with maximal STDs of ~1.0%, ~0.1 g/kg, ~1.3 K, and ~0.7 hPa, respectively.

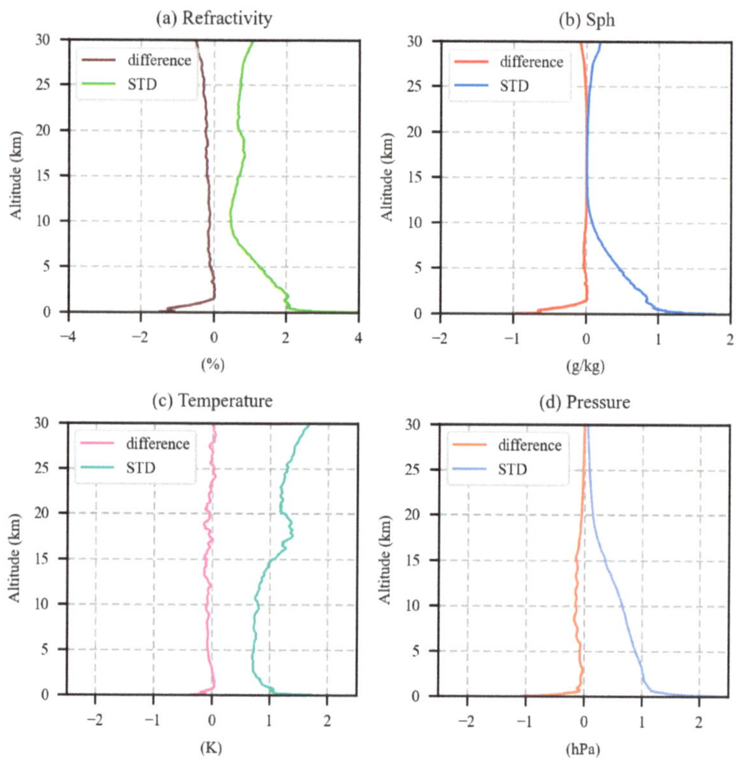

Figure 14. Mean differences and STDs of the Spire (±45°) wetPf2 product compared to radiosonde data from DOY 060 in 2022 to 059 in 2023 below 30 km: (**a**) refractivity, (**b**) specific humidity, (**c**) absolute temperature, and (**d**) pressure.

Figure 15 shows that the mean differences in refractivity, specific humidity, temperature, and pressure in the COSMIC-2 data were negative within ~2.2%, ~1.2 g/kg, ~0.2 K, and ~0.2 hPa below 10 km, with maximal STDs of ~2.8%, ~1.5 g/kg, ~1.7 K, and ~1.2 hPa, respectively. Above 10 km, the mean differences in refractivity, specific humidity, and pressure in the COSMIC-2 data fluctuated near zero within values of 0.1%, 0.1 g/kg, and ~0.1 hPa, with maximal STDs of ~0.3%, ~0.2 g/kg, and ~0.8 hPa below 10 km, respectively. The difference in temperature was negative above 10 km within ~0.3 K and with a maximal STD of ~1.3 K. Apparently, compared to the radiosonde data, the retrieval quality of the Spire and COSMIC-2 data exhibited similar statistics, and in the refractivity comparison, the COSMIC-2 data had slightly smaller differences than the Spire data.

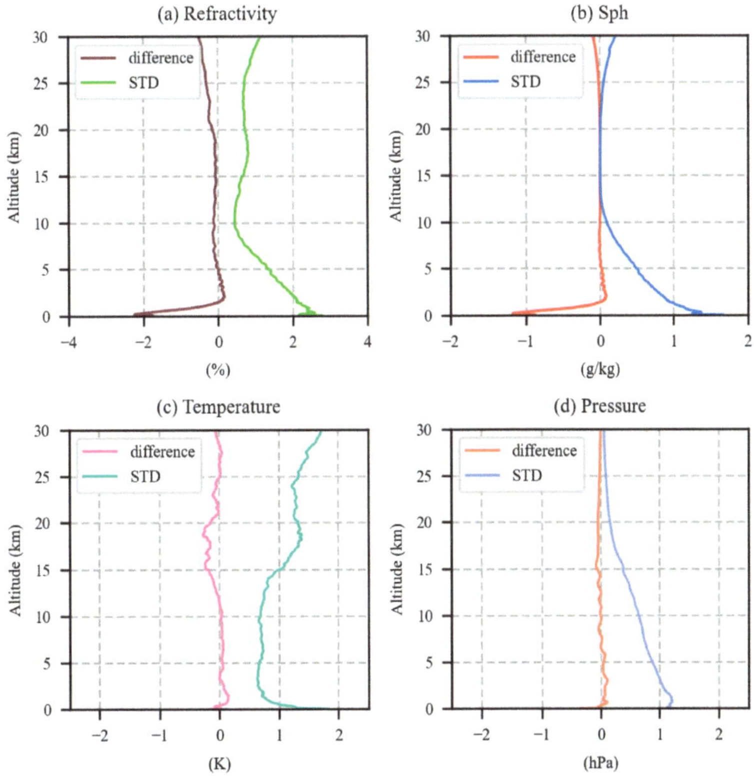

Figure 15. Mean differences and STDs of the COSMIC-2 wetPf2 product compared with radiosonde data from DOY 060 in 2022 to 059 in 2023 below 30 km: (**a**) refractivity, (**b**) specific humidity, (**c**) absolute temperature, and (**d**) pressure.

4. Discussion

The quality of Spire and COSMIC-2 RO data is assessed Radio Occultation (RO) data is assessed comprehensively.

Nowadays, Spire can receive three navigation systems: GPS, GLONASS, and GALILEO, while the COSMIC-2 is limited to GPS and GLONASS. The initial analysis of GNSS-RO data from Spire and COSMIC-2 shows that Spire can offer a broader global coverage of RO events because of consisting of a diverse set of orbits. GPS-derived RO events were more than GLONASS and GALILEO, due to the greater number of operational GPS satellites. Although there is no significant difference in the number of satellites in orbit for GLONASS and GALILEO, GLONASS-derived RO events for Spire slightly outnumber those derived from GALILEO due to some Spire satellites, such as S128, S115, S117, and others, not receiving GALILEO signals.

The results of SNR indicate that the ability of the Spire to track only GPS signals is significantly weaker than that of the COSMIC-1, and with the joint consideration of GLONASS and GALILEO, Spire can achieve a slightly weaker capability than COSMIC-1. The COSMIC-2 has a superior ability to track signals compared to both the COSMIC-1 and Spire.

COSMIC-2 outperforms Spire in achieving better penetration, primarily attributed to COSMIC-2's higher SNR. Moreover, setting occultation events consistently showed greater penetration depths than rising occultation events for both Spire and COSMIC-2, regardless of the satellite navigation system used. After fixation on the influence of topography, such as mountains, on penetration depth, the Spire and COSMIC-2 data below an altitude of

1 km make up 88.7% and 85.3% of all the data, respectively. Due to topographic changes and water vapor variations with increasing latitude, the penetration depth is affected, thus leading to the retrieval statistics. Through comparing the fixed Spire data within the lower latitude range of ±45° to the fixed COSMIC-2 data, it is found that the fixed Spire (±45° data below the 1 km altitude accounted for 84.2% of all the data. In ascending sequence, the ability of the systems to perform deeper soundings could be ranked as follows: Spire fixed (±45°), COSMIC-2 fixed, and Spire fixed. Therefore, COSMIC-2 has a better performance in sounding the deeper troposphere than Spire.

With the ERA5 and Radiosonde as the reference data, it is evident that the Spire data collected within latitudes of ±45° has comparable quality to the COSMIC-2 data Beyond the ±45° latitude range, the retrieval quality of the Spire data was higher than that of the COSMIC-2 data, perhaps because less moisture is present at higher latitudes Additionally, the analysis of mean and STD of fractional differences in refractivity across different seasons does not reveal significant seasonal variations in refractivity errors. In addition, Spire can produce a great number of atmospheric profiles with quality comparable to that of COSMIC-2 to complement the limitation of COSMIC-2 and cover the low-latitude area (±45°).

5. Conclusions

In this study, we mutually analyzed the coverage, SNR, and penetration characteristics of Spire and COSMIC-2 data and assessed the corresponding RO neutral-atmospheric products through comparisons with ERA5 and radiosonde datasets considering the division of GNSS and RO modes. Based on the above assessment and analysis, the conclusions are as follows:

- Spire's RO events demonstrated global coverage due to various orbiting geometries, while COSMIC-2 events were concentrated in the tropics and reduced at higher latitudes.
- GPS-derived RO events were generally more abundant than GLONASS-derived events in both Spire and COSMIC-2 datasets. And GLONASS-derived RO events slightly outnumbered those derived from GALILEO for Spire.
- STRATOS payload on Spire, equipped with lower-gain antennas, exhibited weaker signal capturing compared to IGOR (COSMIC-1) and significantly weaker than TRGS (COSMIC-2).
- The SNR averages of the GLONASS-derived RO events in the Spire data are much stronger than those of the GPS-derived events, while for COSMIC-2, the strengths of the SNR averages had the same magnitudes, with little difference observed between the GPS- and GLONASS-derived RO events.
- In the same coverage area (±45°), COSMIC-2 demonstrated better penetration ability than Spire.
- Based on the research by Gorbunov et al. (2022) [27], it has been revealed that the SNR serves as an indicator of signal strength and holds a crucial role in penetration. Penetration depth was found to be influenced by SNR, GNSS, RO modes, topography, and latitude, as revealed by combined results obtained in Sections 3.2 and 3.3.
- Compared to the ERA5 and radiosonde products, the Spire and COSMIC-2 datasets have identical retrieval qualities when considering the RO data of Spire and COSMIC-2. The accuracy of the neutral-atmosphere Spire data products acquired herein was comparable with those of COSMIC-2.

Corroborated the COSMIC-2 retrieval quality assessments made by Ho et al. (2020), our study contributes to the understanding of the capabilities and performance of Spire and COSMIC-2 RO retrievals. These findings emphasize the valuable role of nanosatellite GNSS-RO techniques like Spire in advancing atmospheric monitoring. Incorporating commercial initiatives such as Spire supplements scientific GNSS-RO data and addresses the need for global observing systems.

Author Contributions: Conceptualization, C.Q.; methodology, C.Q.; software, C.Q.; validation, C.Q., K.Z., and J.Z.; formal analysis, Y.C.; investigation, H.L.; resources, X.W.; data curation, D.L.; writing—original draft preparation, C.Q.; writing—review and editing, X.W.; supervision, H.Y.; project administration, X.W.; funding acquisition, X.W. All authors have read and agreed to the published version of the manuscript.

Funding: This research was supported by the funding program from the Aerospace Information Research Institute and CAS Pioneer Hundred Talents Program.

Data Availability Statement: The Spire and COSMIC-2 RO data involved in this study are available to freely download from the CDAAC at https://www.cosmic.ucar.edu/ accessed on 21 October 2023. ERA5 data can be downloaded publicly from the provided URL https://cds.climate.copernicus.eu/cdsapp#!/dataset/reanalysis-era5-pressure-levels?tab=form accessed on 21 October 2023. The IGRA2 data are accessible at the website https://www.ncei.noaa.gov/pub/data/igra/ accessed on 21 October 2023.

Conflicts of Interest: The authors declare no conflict of interest.

References

1. Bowler, N.E. Revised GNSS-RO observation uncertainties in the Met Office NWP system. *Q. J. R. Meteorol. Soc.* **2020**, *146*, 2274–2296. [CrossRef]
2. Bai, W.; Deng, N.; Sun, Y.; Du, Q.; Xia, J.; Wang, X.; Meng, X.; Zhao, D.; Liu, C.; Tan, G.; et al. Applications of GNSS-RO to Numerical Weather Prediction and Tropical Cyclone Forecast. *Atmosphere* **2020**, *11*, 1204. [CrossRef]
3. Yue, X.; Schreiner, W.S.; Pedatella, N.; Anthes, R.A.; Mannucci, A.J.; Straus, P.R.; Liu, J.-Y. Space Weather Observations by GNSS Radio Occultation: From FORMOSAT-3/COSMIC to FORMOSAT-7/COSMIC-2. *Space Weather* **2014**, *12*, 616–621. [CrossRef]
4. Gleisner, H.; Ringer, M.A.; Healy, S.B. Monitoring global climate change using GNSS radio occultation. *NPJ Clim. Atmos. Sci.* **2022**, *5*, 6. [CrossRef]
5. Ho, S.-P.; Anthes, R.A.; Ao, C.O.; Healy, S.; Horanyi, A.; Hunt, D.; Mannucci, A.J.; Pedatella, N.; Randel, W.J.; Simmons, A.; et al. The COSMIC/FORMOSAT-3 Radio Occultation Mission after 12 Years: Accomplishments, Remaining Challenges, and Potential Impacts of COSMIC-2. *Bull. Am. Meteorol. Soc.* **2020**, *101*, E1107–E1136. [CrossRef]
6. Elrahman Yassien, A.; El-Kutb Mousa, A.; Rabah, M.; Saber, A.; Zhran, M. Analysis of spatial and temporal variation of precipitable water vapor using COSMIC radio occultation observations over Egypt. *Egypt. J. Remote Sens. Space Sci.* **2022**, *25*, 751–764. [CrossRef]
7. Zhran, M.; Mousa, A. Planetary boundary layer height retrieval using GNSS Radio Occultation over Egypt. *Egypt. J. Remote Sens. Space Sci.* **2022**, *25*, 551–559. [CrossRef]
8. Rocken, C.; Anthes, R.; Exner, M.; Hunt, D.; Sokolovskiy, S.; Ware, R.; Gorbunov, M.; Schreiner, W.; Feng, D.; Herman, B.; et al. Analysis and validation of GPS/MET data in the neutral atmosphere. *J. Geophys. Res. Atmos.* **1997**, *102*, 29849–29866. [CrossRef]
9. Larsen, G.B.; Syndergaard, S.; Høeg, P.; Sørensen, M.B. Single frequency processing of Ørsted GPS radio occultation measurements. *GPS Solut.* **2005**, *9*, 144–155. [CrossRef]
10. Mostert, S.; Koekemoer, J.-A. The science and engineering payloads and experiments on sunsat. *Acta Astronaut.* **1997**, *41*, 401–411. [CrossRef]
11. Schmidt, T.; Wickert, J.; Marquardt, C.; Beyerle, G.; Reigber, C.; Galas, R.; König, R. GPS radio occultation with CHAMP: An innovative remote sensing method of the atmosphere. *Adv. Space Res.* **2004**, *33*, 1036–1040. [CrossRef]
12. Hajj, G.A.; Ao, C.O.; Iijima, B.A.; Kuang, D.; Kursinski, E.R.; Mannucci, A.J.; Meehan, T.K.; Romans, L.J.; de la Torre Juarez, M.; Yunck, T.P. CHAMP and SAC-C atmospheric occultation results and intercomparisons. *J. Geophys. Res. Atmos.* **2004**, *109*, D06109–D06132. [CrossRef]
13. Beyerle, G.; Schmidt, T.; Michalak, G.; Heise, S.; Wickert, J.; Reigber, C. GPS radio occultation with GRACE: Atmospheric profiling utilizing the zero difference technique. *Geophys. Res. Lett.* **2005**, *32*, L13806–L13810. [CrossRef]
14. Liou, Y.; Pavelyev, A.G.; Liu, S.; Pavelyev, A.A.; Yen, N.; Huang, C.; Fong, C. FORMOSAT-3/COSMIC GPS Radio Occultation Mission: Preliminary Results. *IEEE Trans. Geosci. Remote Sens.* **2007**, *45*, 3813–3826. [CrossRef]
15. Ho, S.-P.; Zhou, X.; Shao, X.; Zhang, B.; Adhikari, L.; Kireev, S.; He, Y.; Yoe, J.G.; Xia-Serafino, W.; Lynch, E. Initial Assessment of the COSMIC-2/FORMOSAT-7 Neutral Atmosphere Data Quality in NESDIS/STAR Using In Situ and Satellite Data. *Remote Sens.* **2020**, *12*, 4099. [CrossRef]
16. Gorbunov, M.E.; Lauritsen, K.B.; Benzon, H.H.; Larsen, G.B.; Syndergaard, S.; Sørensen, M.B. Processing of GRAS/METOP radio occultation data recorded in closed-loop and raw-sampling modes. *Atmos. Meas. Tech.* **2011**, *4*, 1021–1026. [CrossRef]
17. Li, X.; Zhang, K.; Meng, X.; Zhang, Q.; Zhang, W.; Li, X.; Yuan, Y. LEO–BDS–GPS integrated precise orbit modeling using FengYun-3D, FengYun-3C onboard and ground observations. *GPS Solut.* **2020**, *24*, 48. [CrossRef]
18. Bowler, N.E. An assessment of GNSS radio occultation data produced by Spire. *Q. J. R. Meteorol. Soc.* **2020**, *146*, 3772–3788. [CrossRef]

19. Ruston, B.; Swadley, S.; Baker, N. An evaluation of GNSS-RO data from NOAA Commercial Radio Occultation Purchase using the NRL global NWP system. In Proceedings of the 8th International Radio Occultation Working Group Meeting, Virtual Meeting, 7–13 April 2021.
20. Wickert, J.; Reigber, C.; Beyerle, G.; König, R.; Marquardt, C.; Schmidt, T.; Grunwaldt, L.; Galas, R.; Meehan, T.K.; Melbourne, W.G.; et al. Atmosphere sounding by GPS radio occultation: First results from CHAMP. *Geophys. Res. Lett.* **2001**, *28*, 3263–3266. [CrossRef]
21. Schreiner, W.; Sokolovskiy, S.; Hunt, D.; Rocken, C.; Kuo, Y.H. Analysis of GPS radio occultation data from the FORMOSAT-3/COSMIC and Metop/GRAS missions at CDAAC. *Atmos. Meas. Tech.* **2011**, *4*, 2255–2272. [CrossRef]
22. Chen, S.-Y.; Liu, C.-Y.; Huang, C.-Y.; Hsu, S.-C.; Li, H.-W.; Lin, P.-H.; Cheng, J.-P.; Huang, C.-Y. An Analysis Study of FORMOSAT-7/COSMIC-2 Radio Occultation Data in the Troposphere. *Remote Sens.* **2021**, *13*, 717. [CrossRef]
23. Masters, D.; Nguyen, V.; Irisov, V.; Nogues-Correig, O.; Tan, L.; Yuasa, T.; Ringer, J.; Rocken, C.; Gorbunov, M. Commercial RO Becomes Reality: Rapid Growth, External Validation, and Operational Assimilation of Spire RO. In Proceedings of the 8th International Radio Occultation Working Group Meeting, Virtual Meeting, 7–13 April 2021.
24. Johnston, B.R.; Randel, W.J.; Sjoberg, J.P. Evaluation of Tropospheric Moisture Characteristics Among COSMIC-2, ERA5 and MERRA-2 in the Tropics and Subtropics. *Remote Sens.* **2021**, *13*, 880. [CrossRef]
25. Forsythe, V.V.; Duly, T.; Hampton, D.; Nguyen, V. Validation of Ionospheric Electron Density Measurements Derived From Spire CubeSat Constellation. *Radio Sci.* **2020**, *55*, e2019RS006953. [CrossRef]
26. Juang, J.-C.; Tsai, Y.-F.; Chu, C.-H. On constellation design of multi-GNSS radio occultation mission. *Acta Astronaut.* **2013**, *82*, 88–94. [CrossRef]
27. Gorbunov, M.; Irisov, V.; Rocken, C. The Influence of the Signal-to-Noise Ratio upon Radio Occultation Retrievals. *Remote Sens.* **2022**, *14*, 2742. [CrossRef]
28. Irisov, V.; Nguyen, V.; Masters, D.; Sikarin, R.; Bloom, A.; Gorbunov, M.; Rocken, C. Comparison of Spire RO profiles Processed by UCAR and Spire. In Proceedings of the 8th International Radio Occultation Working Group Meeting, Virtual Meeting, 7–13 April 2021.
29. Smith, E.K.; Weintraub, S. The Constants in the Equation for Atmospheric Refractive Index at Radio Frequencies. *Proc. IRE* **1953**, *41*, 1035–1037. [CrossRef]
30. Palmer, P.I.; Barnett, J.J.; Eyre, J.R.; Healy, S.B. A nonlinear optimal estimation inverse method for radio occultation measurements of temperature, humidity, and surface pressure. *J. Geophys. Res. Atmos.* **2000**, *105*, 17513–17526. [CrossRef]
31. Hersbach, H.; Bell, B.; Berrisford, P.; Hirahara, S.; Horányi, A.; Muñoz-Sabater, J.; Nicolas, J.; Peubey, C.; Radu, R.; Schepers, D.; et al. The ERA5 global reanalysis. *Q. J. R. Meteorol. Soc.* **2020**, *146*, 1999–2049. [CrossRef]
32. Durre, I.; Yin, X.; Vose, R.S.; Applequist, S.; Arnfield, J. Enhancing the Data Coverage in the Integrated Global Radiosonde Archive. *J. Atmos. Ocean. Technol.* **2018**, *35*, 1753–1770. [CrossRef]
33. Katrin, L.; Neill, B.; Elias, H.; Sean, H. *Assimilating Spire and COSMIC-2 Data into the IFS*; ECMWF Newsletter: Reading, UK, 2021.
34. Spire Global, Inc. *Orbital Debris Assessment Report (ODAR) 2020*; Spire Global, Inc.: Vienna, VA, USA, 2020.
35. Schreiner, W.S.; Weiss, J.P.; Anthes, R.A.; Braun, J.; Chu, V.; Fong, J.; Hunt, D.; Kuo, Y.-H.; Meehan, T.; Serafino, W.; et al. COSMIC-2 Radio Occultation Constellation: First Results. *Geophys. Res. Lett.* **2020**, *47*, e2019GL086841. [CrossRef]
36. Sokolovskiy, S.V.; Schreiner, W.S.; Zeng, Z.; Hunt, D.C.; Kuo, Y.H.; Meehan, T.K.; Stecheson, T.W.; Mannucci, A.J.; Ao, C.O. Use of the L2C signal for inversions of GPS radio occultation data in the neutral atmosphere. *GPS Solut.* **2014**, *18*, 405–416. [CrossRef]
37. Jing, X.; Ho, S.-P.; Shao, X.; Liu, T.-C.; Chen, Y.; Zhou, X. Spire RO Thermal Profiles for Climate Studies: Initial Comparisons of the Measurements from Spire, NOAA-20 ATMS, Radiosonde, and COSMIC-2. *Remote Sens.* **2023**, *15*, 3710. [CrossRef]
38. Sokolovskiy, S.; Schreiner, W.; Weiss, J.; Zeng, Z.; Hunt, D.; Braun, J. Initial assessment of COSMIC-2 data in the lower troposphere (LT). In Proceedings of the Joint 6th ROM SAF User Workshop and 7th IROWG Workshop, Helsingør, Denmark, 19–25 September 2019.
39. Li, X.; Zhang, X.; Ren, X.; Fritsche, M.; Wickert, J.; Schuh, H. Precise positioning with current multi-constellation Global Navigation Satellite Systems: GPS, GLONASS, Galileo and BeiDou. *Sci. Rep.* **2015**, *5*, 8328. [CrossRef] [PubMed]
40. Paziewski, J.; Sieradzki, R.; Wielgosz, P. On the Applicability of Galileo FOC Satellites with Incorrect Highly Eccentric Orbits: An Evaluation of Instantaneous Medium-Range Positioning. *Remote Sens.* **2018**, *10*, 208. [CrossRef]

remote sensing MDPI

Review

Review Analysis of Irrigation and Application of Remote Sensing in the Lower Mekong River Basin

Guanghui Wang [1,2,*], Sadia Bibi [1], Tingju Zhu [1], Fuqiang Tian [3] and Marcelo A. Olivares [4]

1 ZJU-UIUC Institute, International Campus, Zhejiang University, Haining 314400, China
2 Henan Branch of China South-to-North Water Diversion Group Middle Route Corporation Limited, Zhengzhou 450000, China
3 Department of Hydraulic Engineering, State Key Laboratory of Hydroscience and Engineering, Tsinghua University, Beijing 100084, China
4 Department of Civil Engineering, University of Chile, Santiago 7830490, Chile
* Correspondence: guanghui.20@intl.zju.edu.cn

Abstract: Irrigated agriculture is indispensable to the Lower Mekong River Basin (LMB), which ensures food security and provides livelihoods for tens of millions of people. Irrigation, agricultural production, hydropower and aquatic ecosystem health are intertwined in LMB, so it is necessary to adopt a holistic approach to analyze irrigation problems. Here, we discuss the challenges and opportunities of LMB irrigation. Bibliometric analysis is carried out to determine the characteristics and patterns of watershed irrigation literature, such as the importance of authors, affiliated institutions, and their distribution in China. Based on bibliometric analysis, research topics are determined for thematic review. Firstly, we investigated the factors that directly affect the demand and supply of irrigation water and associated crop yield impacts. Secondly, we analyzed the influence of water availability, land use and climate change on agricultural irrigation. Thirdly, we analyzed the adverse effects of improper irrigation management on the environment, such as flow pattern change, ecosystem deterioration and land subsidence caused by groundwater overexploitation. Fourthly, the time–space mismatch between water supply and demand has brought serious challenges to the comprehensive water resources management in cross-border river basins. In each specific application area, we sorted out the technologies in which remote sensing technology is used. We hope that this review will contribute to in-depth research and decision analysis of remote sensing technology in agricultural irrigation.

Keywords: bibliometric analysis; irrigation; agriculture; Lower Mekong River Basin; water resource management; remote sensing

Citation: Wang, G.; Bibi, S.; Zhu, T.; Tian, F.; Olivares, M.A. Review Analysis of Irrigation and Application of Remote Sensing in the Lower Mekong River Basin. *Remote Sens.* **2023**, *15*, 3856. https://doi.org/10.3390/rs15153856

Academic Editors: Yuriy Kuleshov, Xiaoming Wang, Suelynn Choy, Haobo Li and Mayra I. Oyola-Merced

Received: 27 May 2023
Revised: 26 July 2023
Accepted: 2 August 2023
Published: 3 August 2023

1. Introduction

Stretching over 4900 km from the Tibetan Plateau to the South China Sea, the Mekong River drains an area of 795,000 km^2 and has an average annual discharge of 475 km^3. The Lower Mekong River Basin (LMB, Figure 1) includes parts of Cambodia, Lao PDR, Thailand and Vietnam, and it is home to more than 70 million people [1]. The distribution of precipitation in the LMB has a strong seasonality, which plays a decisive role in the selection of local crops and planting dates in different areas within the region [2]. In LMB countries, the wet season runs approximately from May to October, followed by a dry season from November to March. The Mekong River flow usually is at its lowest in April.

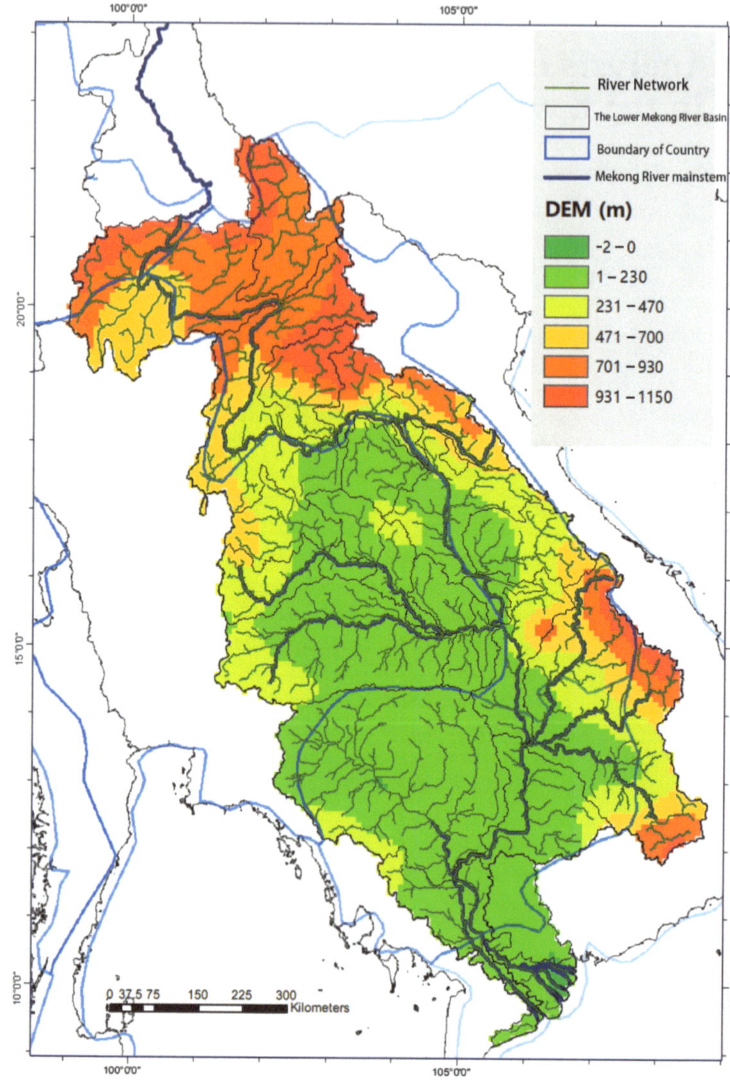

Figure 1. Mekong River topographic map.

Rural livelihoods in the LMB are highly dependent on agriculture [2]. This is particularly true for Cambodia, Myanmar and Laos PDR where agriculture related activities account for a significant share of GDP as shown in Table 1. Gross domestic product (GDP) is a monetary measure of the market value of all the final goods and services produced in a specific time period by a country. Irrigation plays an important role in securing monsoon crops and converting a single crop, namely rain-fed rice, to multiple cropping systems. Irrigation in the LMB relies on water diverted from the Mekong main stem and its tributaries. Irrigation is mostly carried out in the dry season when crop area is lower than that during the rainy season. Notably, the Mekong Delta is the most productive area in the LMB and a main irrigation water consumer. Crops in the LMB are heavily dependent on rainwater to grow, with rice being the dominant crop. The LMB produced 48.2 million tons, or around 51% of the 93.6 million tons of rice produced in the four LMB countries (Lao PDR, Cambodia, Thailand, and Vietnam) as a whole in 2013 [3].

Table 1. Key statistics in the LMB countries, 2021.

	Cambodia	Lao PDR	Thailand	Vietnam	Myanmar
GDP (10^6 USD)	26,961	18,827	505,947	366,137	65,091
Agriculture, forestry, and fishing, value added (% of GDP)	22.8	16.1	8.5	12.6	23.4
Population (10^3)	16,589	7425	71,601	97,468	53,798
GDP Growth (%)	3.0	2.5	1.5	2.6	−17.9
GDP per capita (USD)	1625	2535	7066	3756	1209
Urban population growth rate (%)	2.9	3.2	1.6	2.7	1.7
Poverty (%)	16.7	27.6	9.8	2.8	25.6

Source: https://data.worldbank.org/ (accessd on 15 March 2023).

In general, irrigation is seen as a crucial component of rural development and plays a significant role in boosting agricultural production and alleviating poverty in the LMB countries [4]. So far, irrigation systems are created expressly to produce rice, making it challenging for farmers to switch to other crops [4]. Nevertheless, the intensification of irrigated agriculture had led to farmers in Thailand and Vietnam to shift from rice to high-return activities such as aquaculture or cash crops. Nowadays, however, a considerable number of populations in the LMB still live in poverty, as shown in Table 1.

Water abstraction for agriculture makes up around 90% of all water diversions in the region (Cambodia: 94%, Thailand: 91%, Vietnam: 86% and Laos: 82%) [4]. However, the distribution of water resources in the Mekong River region is rather uneven, and there exist competing interests in water uses between the upper and lower reaches. Water is essential for sustainable agricultural production [5,6]. In agricultural water use, irrigation has the largest share, and irrigation efficiency and grain yield directly affect a country's agricultural development level [7,8].

Meanwhile, water, food and energy (WEF) are highly interconnected in the LMB. For example, water is used to produce food and energy (e.g., irrigation and hydropower) [9], and in return, energy is required to pump and distribute water and produce food [10,11]. In fact, irrigation uses large amounts of water, which poses a challenge to water resources management in the LMB, especially in the dry season. In this situation, it is useful to take a holistic perspective by considering water, food and energy simultaneously, or in other words, taking a WEF nexus perspective while analyzing irrigation in the LMB. Such a perspective can potentially enhance water, energy and food security by increasing efficiency, reducing trade-offs, building synergies and improving governance across the three inter-connected sectors [12]. Needless to say, WEF represents the most significant resources that support sustainable development of human society [13].

Most existing studies [14–17] related to irrigation in LMB focus on hydropower because of the close relationship between these two sectors. There are both synergies and tradeoffs between these two sectors. For instance, water stored in upstream hydropower reservoirs can augment irrigation water supply in the downstream during the dry season and in dry years, effectively relieving downstream drought. More enhanced cooperation between the riparian countries are expected to be an effective way in addressing water and food challenges in the LMB [18]. On this important and complex issue about irrigation in the LMB, despite a considerable number of existing studies conducted from various angles, there is generally a lack of systematic review of the development, status and future challenges of irrigation in the region, especially when it is related to the interactions and interdependencies between irrigation and other water-using sectors such as hydropower and ecosystem. To the best of our knowledge, an irrigation-focused review is yet to be done for the LMB. Based on a detailed investigation of available publications focusing on this topic through bibliometric analysis and specific theme-focused review, this paper aims to improve our understanding of key challenges and opportunities in irrigation management and development in the LMB.

2. Methodology

2.1. Data Collection

The bibliometric analysis and review in this paper rely on literature on LMB irrigation found from the Web of Science database [19]. We chose the Web of Science because it is the most well-established bibliometric dataset, and the inclusion criteria for journal titles are transparent and being regularly updated. The Web of Science database has synthetical and integrated paper records [20]; the high number of journals and articles in its collection permitted us to generate representative results.

Firstly, bibliometric records can be constructed using a keyword such as "irrigation Mekong" in the title, the abstract or the keyword of publications searching in Web of Science [21]. Secondly, we used key words such as land use, crop productivity and land cover in the search query associated with "irrigation Mekong". Thirdly, we used "Food, Water and Energy Nexus" and its variants (Water, Food and Energy Nexus; Energy, Food and Water Nexus, etc.) [22] to search for related publications in the Web of Science. In addition, we also use some search wildcards and similar keywords to increase the scope of the search. Overall, the summary of searching keywords includes "Irrigation, Mekong, land use, crop productivity, land cover, Food water and Energy Nexus". Lastly, after going through the search, we exported the text documents of those identified publications from the Web of Science and put them into the Citespace for bibliometric analysis.

2.2. Data Preprocessing

A screening of the collected articles from the Web of Science suggested that some publications not related to the topic were included, and they were subsequently removed from the collection after double-checking those publications from the original websites where they were published online, if applicable. We then extracted the information of each author's country of origin, affiliations, year of publication, publisher and citation numbers. For the purpose of thematic analysis, we extracted the main research content from the title and abstract of each paper for further analysis [23].

2.3. Bibliometric Analysis

Bibliometric analysis is effective in analyzing research conducted in a specific research field with a large amount of data. It helps discover the evolution of research activities in a specific field and can also help identify emerging areas in that field [24]. The bibliometric analysis was conducted using publications searched from the Web of Science database.

This paper is based on visual analytic functions executed in a bibliometric analysis software called Citespace [25]. Citespace is a java-based bibliometric analysis software tool to visualize and analyze temporal and structural patterns in scientific literature [26]. To obtain the required results, we set parameter thresholds in the software, following its instructions. The software has been constantly updated in recent years to accommodate various research objectives for visual analytic tasks [25]. In this study, we used Citespace 5.8.R3 (64 bit) [27].

2.4. Thematic Analysis

In addition to the bibliometric analysis, we identified several major research hotspots in irrigated agriculture in the LMB—for example, climate change, environmental issues, human alterations to flow regime due to hydropower generation, the impact of irrigation water diversion on downstream water uses and ecosystem health and so forth. The positive impacts of hydroelectric power reservoirs are also explored. In addition, we also evaluate transboundary water resource management across riparian countries from the perspective of the WEF nexus.

Figure 2 illustrates the organization of the study presented in this paper. First, we established a literature dataset. Second, we conducted a bibliometric analysis to identify, categorize and prioritize topics of high relevance and importance regarding irrigation in the LMB. It also categorizes the institutions and countries of the leading authors who wrote those articles. Third, with these findings, a thematic review was also carried out, focusing on each of the four identified areas. The bibliometric analysis identifies the patterns of the literature, whereas thematic review moves one step further to explore key issues under each selected theme. This review aims to tap into recent publications on irrigation in the LMB to produce a systematic review on key issues regarding irrigation in the basin.

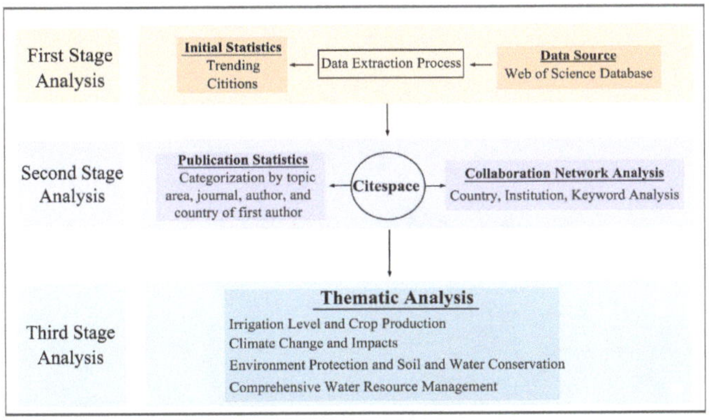

Figure 2. Overall organization of review of LMB irrigation.

3. Bibliometric Analysis Results

3.1. Literature Statistics

A total of 270 publications related to irrigation in the LMB were chosen from the body of literature extracted from the Web of Science database, after manually excluding publications that are not related to the research topic. The bar chart of Figure 3 shows the number of published articles per year, which has an increasing trend from 2000 to 2021. In 2019, there were over 35 publications related to irrigation in the LMB. The number of publications is still rather low, considering the importance of this topic. The line on Figure 2 shows citations of those publications, which has been rapidly increasing since 2006.

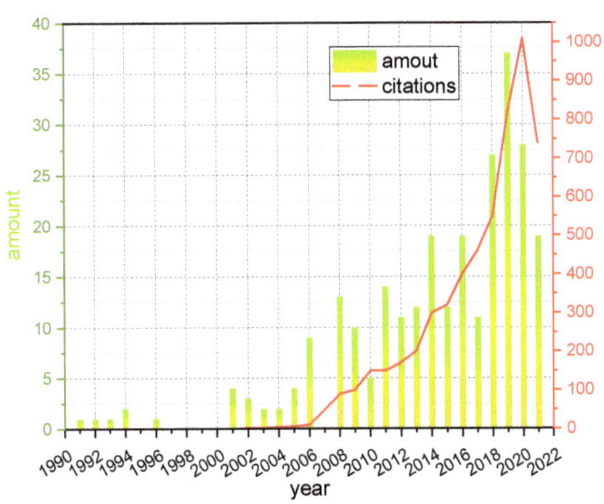

Figure 3. The number of publications and citations related to irrigation in the LMB.

Table 2 shows that most of the dataset's publications are centered on environmental sciences, ecology, agriculture, water resources and plant sciences. Most articles about irrigation in the LMB focus on ecological and environmental, agricultural and water resources aspects. The Mekong River's water crisis has long been a major regional issue [28,29] Furthermore, some articles [30–32] also analyzed irrigation from the regional economic point of view.

Table 2. Statistical results in research areas of the selected dataset.

Research Areas	Record Count	% of 270
Environmental Sciences Ecology	189	70.0%
Agriculture	167	61.9%
Water Resources	162	60.0%
Plant Sciences	100	37.0%
Meteorology Atmospheric Sciences	78	28.9%
Marine Freshwater Biology	76	28.2%
Biodiversity Conservation	68	25.2%
Geography	65	24.1%
Business Economics	61	22.6%
Science Technology Other Topics	61	22.6%
Engineering	59	21.9%
Mathematics	50	18.5%
Energy Fuels	41	15.2%
Geochemistry Geophysics	40	14.8%
Oceanography	40	14.8%
Physical Sciences Other Topics	37	13.7%
Geology	36	13.3%
Public Environmental Occupational Health	36	13.3%
Fisheries	30	11.1%
Food Science Technology	25	9.3%

Table 3 shows the distribution of those selected 270 articles in various academic journals. Most of the articles were published in agriculture-related journals even though they were widely distributed. Several journals, such as *Water*, *Agricultural Water Management*, *Field Crops Research*, *Advances in Global Research*, *Paddy and Water Environment*, and *Science of The Whole Environment* published more papers on LMB irrigation than other journals.

Table 3. Journal-wise distribution of the selected articles.

Journal	Impact Factor (2021)	H-Index (2021)	Average Number of Citations per Paper (2021)	Record Count	Share in Total Number of Selected Publications
Water	3.530	69	7.91	9	3.3%
Agricultural Water Management	6.611	139	13.67	7	2.6%
Field Crops Research	6.145	161	19.56	7	2.6%
Advances in Climate Change Research	4.746	30	5.93	6	2.2%
Paddy and Water Environment	1.554	38	9.24	6	2.2%
Science of The Total Environment	7.963	221	6.29	6	2.2%
Hydrological Processes	4.015	128	27.12	5	1.9%
International Journal of Remote Sensing	2.581	106	10.47	5	1.9%
Remote Sensing	5.076	161	25.19	5	1.9%
Water International	2.22	57	9.48	5	1.9%
Ambio	4.022	97	23.87	4	1.5%
Food Security	4.603	62	16.42	4	1.5%
Frontiers in Environmental Science	5.646	48	8.22	4	1.5%
Journal of Health and Pollution	2.249	10	4.82	4	1.5%
Journal of Hydrology	4.646	203	36.90	4	1.5%
Applied Geochemistry	3.621	121	24.86	3	1.1%

In addition, we also summarized countries of origin of the authors' affiliations (Table 4) and the highly cited papers, as shown in Table 5. Table 4 shows that 27.8% of the first authors of the selected publications are from Vietnam, 20.9% from the United States, 18.9 from Japan, 17.0% from Australia, 13.7% from China and 10.7% from Cambodia. This implies that irrigation in the LMB has drawn broad attention beyond riparian countries in the region.

Table 4. Countries of origin of authors.

Country	Paper Record Count	% of 270
Vietnam	75	27.8%
United States of America	59	21.9%
Japan	51	18.9%
Australia	46	17.0%
P.R. China	37	13.7%
Cambodia	29	10.7%
Laos	24	8.9%
Thailand	23	8.5%
Germany	18	6.7%
Netherlands	15	5.6%
England	10	3.7%

Table 5. Citations of the selected publications on LMB irrigation.

Publication	Average per Year	Total
Matthew L. Polizzotto [33]	30.6	429
Laurent Charlet [34]	13.4	214
Toshihiro Sakamoto [35]	12.4	198
Ingjerd Haddeland [36]	11.3	181
N.T. Sona [37]	16.8	168
Toshihiro Sakamoto [38]	14.0	168
Polya, DA [39]	9.8	167
Matti Kummu [40]	10.8	151
Benjamin D. Kocar [41]	9.6	135
F. Su [42]	18.3	110
Mauricio E. Arias [43]	13.1	105
Jory S. Hecht [44]	34.3	103
Nguyen Minh Dong [45]	9.9	99
Mauricio E. Arias [46]	9.3	93
Nguyen-Thanh Son [47]	10.8	86
Chandrashekhar M. Biradar [48]	7.3	80
Thi Thu Ha Nguyen [49]	7.9	79
Shawn G.Benner [50]	5.4	76
Gert-Jan Wilbers [51]	9.4	75
Toshihiro Sakamoto [52]	5.6	73

Table 5 lists the average number of citations per year and the total number of citations of highly cited publications on LMB irrigation, in descending order. The number of citations can explain the influence of an article in the field of the subject. At the same time, the number of citations has a close relationship with the quality of periodicals. The first few articles with a high number of citations focus on irrigation-related environmental issues in the Mekong River basin, the spatiotemporal distribution of rice phenology and cropping systems, the effects of irrigation, agriculture drought and so on.

3.2. Collaboration Analysis

Citation networks and co-citation networks have been long studied in information science and other disciplines [53]. Co-citation relations serve as a fundamental grouping mechanism [54]. We used Citespace to generate collaboration network graphs, as shown in Figure 4. In Figure 4a,b, we examined the distribution of research institutions and countries that the authors were affiliated with at the time those articles were published. The size of each node represents the number of citations to the corresponding article in this dataset. The theme of each cluster of cited articles is algorithmically labeled.

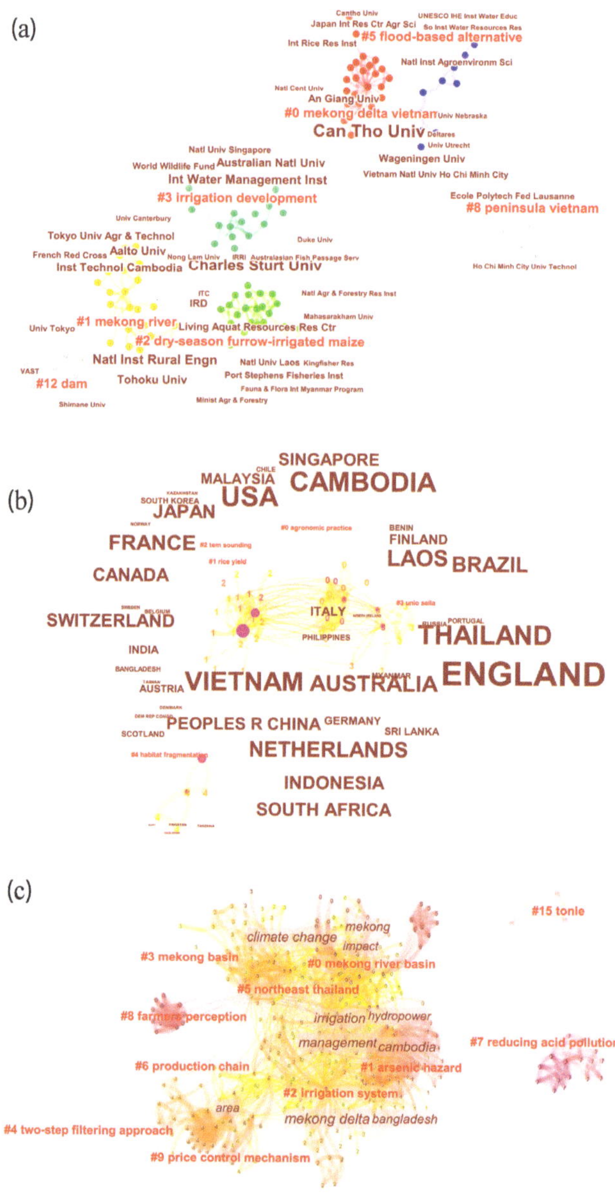

Figure 4. Bibliometric analysis results for cooperation network analysis: (**a**) institutions of authors; (**b**) countries of affiliated institutions; (**c**) co-citation analysis.

It gives a rough structure of the literature dataset generated using Citespace with the theme of institutions and countries of the authors. Citespace aggregates similar articles into a cluster with concentration, although the degree of concentration varies considerably between clusters. Each cluster generates a tag, which is assigned automatically by Citespace. The nodes in the figures are references cited by citing articles, and the distance between nodes represents the degree of association of keywords between articles. It is also reflected in the relative size of the different cluster tags. For example, "irrigation development" were studied by the National University of Singapore, the World Wildlife Fund, the Australian National University, and the International Water Management Institute. The labels of all nodes can be displayed in a picture. It is generally necessary to adjust the thresholds in the cluster to obtain a collaboration analysis graph that clearly illustrates the institutions. Some research themes deviate significantly from the clustered themes on the figure and, hence, are not displayed by the software.

Regarding the countries of authors' affiliations, Southeast Asian countries such as Vietnam, Laos and Cambodia have mainly focused on issues related to irrigation in the downstream areas. Research institutions in the United States, United Kingdom, Netherlands and China also made considerable contributions to those selected publications on LMB irrigation.

Figure 4c shows the co-citation results from the bibliometric analysis. Co-citation is used to determine the missing links between two relevant items using shared third items [55]. As mentioned above, each cluster has a label that represents the topic of the entire cluster. Figure 4c helps identify key literature for cross-disciplinary ideas. The label of clusters also shows a correlation between the articles that focus on Mekong irrigation issues, and environmental concerns, with a strong regional focus on the Mekong River Delta. Furthermore, there is a focus on issues between upstream hydropower development and irrigation of the LMB. Furthermore, climate change has a significant impact on the hydrological regimes in the Mekong River basin.

4. Thematic Analysis and Application of Remote Sensing

Figure 5 illustrates the matrix structure of the thematic analysis, which is organized based on initial analysis of the literature and on findings from the bibliometric review. According to modeling approaches, we categorized existing studies into land surface process modeling, systems modeling at river basin scale and reservoir operations. Considering irrigation, watershed management and hydropower as sub-systems in the LMB, land surface models nowadays incorporate irrigation module to analyze the effect of irrigation water withdrawal on downstream river flow; systems modeling at the river basin level, or river basin management modeling, address various complex issues related to water management in the basin; and reservoir operations modeling usually focuses on hydropower optimization and related downstream impacts. Within each sub-system, topics for review are organized into five aspects, including nature resources, socioeconomic, management, environment and infrastructure.

We selected several themes for further analysis, which are closely related to agricultural irrigation in the LMB: irrigation and crop production, water availability and land use, climate change and impact, environmental protection and soil and water conservation and integrated water resources management. Furthermore, remote sensing can be an effective tool for monitoring and managing irrigation in the Mekong region. Remote sensing can provide valuable insights into irrigation systems, crop growth and water availability in the Mekong region. By leveraging remote sensing technology, water authorities, farmers and policymakers can work together to create more sustainable and efficient irrigation systems that meet the needs of both people and the environment.

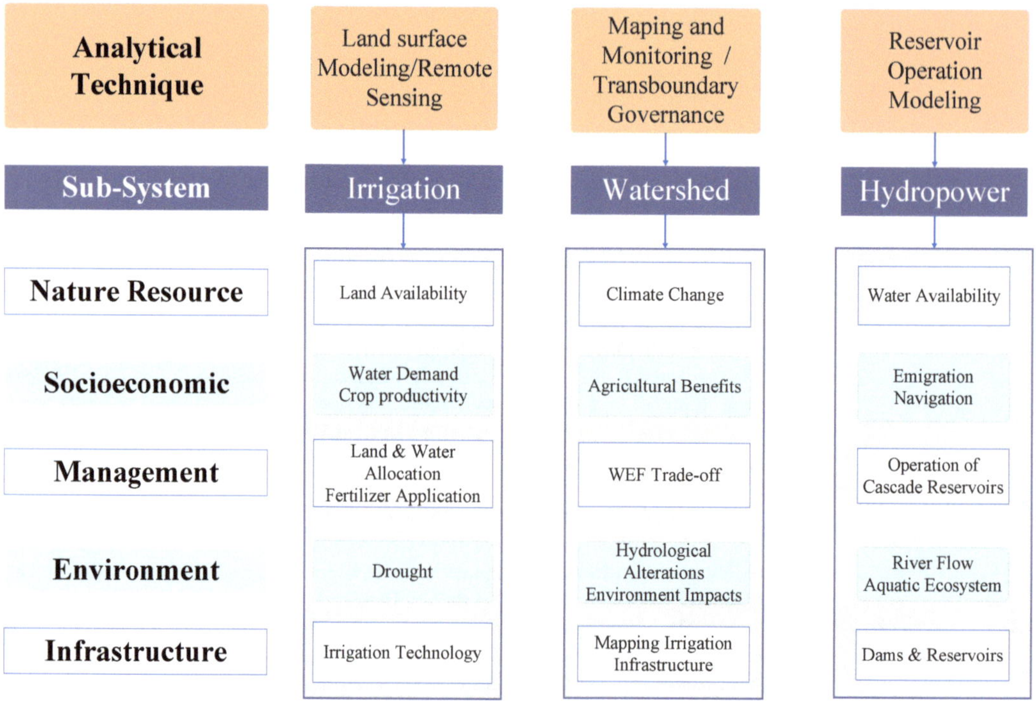

Figure 5. Organization of thematic review.

4.1. Irrigation and Crop Production

Irrigated crop production in the LMB plays an important role in regional economy. Moreover, Vietnam and Thailand are among the top rice exporters in the world. Thus, rice production in this region can have profound impacts on world rice prices. In the LMB, rice is the most important irrigated crop by irrigated crop area and by irrigation water consumption. Growing competition for water between irrigation and other economic sectors due to rapid population and economic growth poses a challenge to irrigation water management and requires close attention [56]. The Mekong River Basin features various hydraulic irrigation structures, such as dams, canals, weirs, pumping stations, embankments and water management systems. These structures play a crucial role in regulating water flow, distributing water for agriculture, managing floods and supporting crop cultivation. They contribute to enhanced agricultural productivity and livelihoods in the region.

Existing studies on irrigation estimated gridded irrigation water requirement in the LMB and other basins around the world [57]. For China, the impact of climate change was analyzed using the Community Land Model, version 4 [58]. Tatsumi et al. [59] conducted a detailed analysis of the impact of irrigation water abstraction on river flow using a simulation model to depict the changes in serval indicators within the LMB area. For the LMB, the CROPWAT model was also used for estimating irrigation water requirements [60]. In general, the literature suggests that more advanced irrigation technologies, appropriate crop varieties and enhanced agricultural extension are promising in improving irrigated agricultural production in the LMB.

In irrigation and crops, remote sensing can be used to map irrigation infrastructure, such as canals, reservoirs and water distribution systems. Landsat detected a significant decrease in the area of three types of rice crops as a way to judge the impact brought about [61]. This can help water authorities to identify areas where infrastructure upgrades or repairs are needed and to monitor changes in irrigation patterns and water use. Remote sensing has better applications in assessing water use and predicting crop water requirements. In crop health area, remote sensing can be used to assess crops health and to detect stress, disease or pest infestations [58]. This information can help farmers to take timely actions to mitigate crop losses and optimize water use and fertilizer application.

Remote sensing is a powerful technology that has found various applications in agriculture, with irrigation being a significant beneficiary. One of the critical aspects of effective irrigation management is determining the water requirements of crops, which is directly related to evapotranspiration (ET). Evapotranspiration is the combined process of water evaporation from the soil surface and transpiration from plant leaves. Remote sensing provides a non-intrusive and efficient method to monitor and estimate evapotranspiration over large agricultural areas. The process involves the use of satellite imagery, particularly the thermal infrared band and vegetation indices, to assess the temperature and health of crops. These data are integrated into surface energy balance models, which account for various factors affecting ET, such as solar radiation, sensible and latent heat fluxes.

4.2. Water Availability and Land Use

The country-wise water availability in the Mekong River Basin requires reassessment and a detailed description. The water availability in each country may fluctuate over time due to various factors, including changes in rainfall patterns, climate conditions, water usage and development projects such as dams and irrigation systems. A comprehensive reassessment would involve collecting and analyzing up-to-date data on water resources, including surface water and groundwater, in each country. It would also consider the impacts of existing and planned hydraulic structures, agricultural practices, industrial water usage and environmental considerations. There are significant variations in water availability in the Mekong region over time, including water levels in rivers, lakes and reservoirs. Water availability and land use are very important in the Mekong region, where water resources are scarce and unevenly distributed. Changes in land use patterns and flood conditions play an important role in the livelihood of local farmers, and remote sensing can be used to comprehensively monitor land use changes and flood conditions in the area and address the causes of land use changes [62].

Remote sensing information is used as a corresponding monitoring indicator to predict the timing and intensity of extreme events, including droughts and floods [63], as well as to allocate water resources more efficiently to manage them more effectively. Satellites are revolutionizing land use monitoring by providing unprecedented accuracy, resolution and coverage. Satellite-based land use monitoring has the capability to track land cover changes efficiently and effectively over a large range. This technology enables the detection of subtle changes in land use patterns [64], which can inform land management decisions and support sustainable development efforts. With its ability to capture frequent and consistent imagery, satellite-based land use monitoring has become an indispensable tool. Remote sensing can be used to identify areas that are currently being irrigated and to estimate the total area under irrigation. This information can be used to assess the overall water demand and to monitor changes in irrigation patterns over time.

4.3. Climate Change Impacts

Climate change is expected to affect the amount and timing of river flow in the LMB, and droughts and floods are likely to become more frequent [2]. On one hand, records already show that drought disaster is becoming more common in the region, resulting in reduced soil moisture, lower river flow and inadequate irrigation water supply [65–67]. On the other hand, floods may damage irrigation facilities, thereby restricting agricultural water use and even making irrigation impossible [68]. In addition, flood disasters can also cause water source pollution and water quality degradation, which have adverse effects on crop yield and quality [69]. Excessive soil moisture and high groundwater table also harm crop growth and increase the occurrence of pests and diseases of crops [70]. In this situation, the construction of flood prevention and drainage measures are essential. Faced with future climate-related stresses, crop profitability can be increased through water use efficiency, fertilizer management and crop management to improve the resilience of the local farming system [71].

Climate action is a crucial component in achieving the United Nations' Sustainable Development Goals, particularly in the context of sustainable irrigation practices. With the increasing impact of climate change on global water resources, it becomes imperative to implement effective measures that address both irrigation needs and environmental concerns. By embracing sustainable irrigation methods, we can address the challenges posed by climate change while striving for a more sustainable and food-secure future Through collaborative efforts between governments, stakeholders and communities, we can pave the way for a resilient and environmentally conscious approach to agriculture and water management.

Climate change impacts irrigation: rising temperatures, erratic weather, prolonged droughts and extreme events disrupt irrigation schedules and reduce water availability. Melting glaciers and reduced snowpack diminish natural water supply, while increased evapotranspiration intensifies water demand. Floods and storms damage irrigation infrastructure. Solutions include adopting innovative technologies like remote sensing and precision irrigation, implementing water-use efficiency incentives and promoting sustainable water management practices. Collaboration is crucial to secure food production and conserve water resources for the future.

Surface temperature monitoring using remote sensing is one of the key variables in climate forecasting and climate change research, and therefore, the analysis of its behavior is essential to assess climate variability. Remote sensing technology can be used to increase people's understanding of the overall climate system and its changing laws. It can continuously observe and analyze the earth's surface, ocean and atmosphere at different time and space scales, so as to observe the climate system and investigate climate-related processes or long-term and short-term phenomena [72,73].

4.4. Environment Impacts

Groundwater overdraft for irrigation causes land subsidence, which is a serious problem in the Mekong River Delta [74]. Sea water intrusion becomes more serious with land subsidence, sometimes making groundwater unusable [75,76]. In the Mekong Delta, salinity intrusion has a direct impact on agriculture and domestic water uses [75].

The development of irrigation systems has a direct impact on local aquatic resources [77], adversely affecting aquatic fauna, especially fish [78–81]. The shrinkage of the Tonle Sap Lake has been attributed to climate change [82–86]. The reduction in rainfall in the Mekong Basin area has led to a decline in runoff from the Mekong [84], and the flow regime of the Tonle Sap has been influenced by the Mekong's hydrological fluctuation and irrigation water alternations. The environmental changes that are taking place in the Mekong region are manifested in terms of land use and land cover types. The causes of this change are various, including river damming, an increase in agricultural land and frequent natural disasters [87]. In arid and semi-arid regions of the world where water resources are scarce, agricultural issues are also the focus of local water crises [88].

Remote sensing data can be used to evaluate the environmental changes of large-scale river systems in the study of land cover changes in Mekong River Basin and their impacts on rivers [89]. The establishment of eco-environmental status indicators relies heavily on remote sensing technology, which is of great significance to the monitoring and analysis of ecosystems and environment on a large scale or even on a global scale. Wu et al. studied the use of remote sensing technology to point out natural and human pressures, regional natural environmental conditions, ecosystem health, land cover changes and responses to pressures [90].

4.5. Comprehensive Water Resource Management

4.5.1. Rational Water Allocation

The overarching goal of water resources management in the Mekong region is to enable the maximization of water-using benefits through equitable and efficient water allocation across all countries, water-using sectors and stakeholders [91]. Given the complexity of water management in the region, negotiation is desirable and necessary in solving conflicts in water allocation and management [92]. Modeling techniques in this area can be used to estimate the gains and losses under different cooperation arrangements. For instance, the Water Management Group from HR Wallingford seeks to recognize Pareto-optimal water allocation solutions across various users [93]. Trade-off analysis is a persuasive tool to tackle interdependence and complexity. Elsewhere, the mixed-integer linear programming method was used to obtain maximum irrigation net benefits [94]. Further studies using mathematical models are desirable to look into the complex challenges raised by different patterns of cooperation, damming schemes and water requirements of riparian countries [95–98]. Remote sensing can be used to optimize irrigation scheduling by monitoring water use and crop water stress. This information can help farmers to avoid over- or under-irrigation, which can lead to water wastage or crop stress.

4.5.2. Transboundary Water Management

Effective irrigation practices in the Mekong River basin are essential to enhance agricultural productivity and mitigate the impacts of climate change on water availability. However, transboundary water management has become a complex issue due to the shared nature of the river among multiple countries.

To address these challenges, regional cooperation is crucial. Collaborative efforts between the Mekong River countries are necessary to develop a comprehensive and integrated approach to water management. This involves sharing data, coordinating water release from upstream dams and jointly planning for irrigation and hydropower development. By promoting dialogue and cooperation, the countries can work together to balance the water needs for irrigation, energy production and environmental sustainability.

Transboundary water management in the Mekong River requires sustained cooperation and commitment among the riparian countries to address the complex challenges posed by shared water resources effectively. Through effective management and collaborative efforts, the Mekong River basin can achieve a balance between economic development, water security and environmental conservation for the benefit of all involved nations and their people.

4.5.3. Reconciling Hydropower and Irrigation

Upstream reservoirs store and release river flow for power generation, while downstream agricultural production relies on water diverted from the river channel to irrigate crops. Hydropower does not really "consume" water; it relies on the hydraulic head created by the elevation difference between reservoir water level and tail water level, and flow running through the turbines. Thus, managing the competition for water uses between these two sectors is largely about reconciling the timing and magnitude of water release for hydropower generation and that of irrigation water withdrawal.

Hydropower development in the Mekong River basin has attracted wide attention regionally and globally [99] due to its potential impacts on local hydrological conditions, agricultural development and ecosystems. Such impacts are often uncertain, significant and sometimes irreversible [8]. For instance, studies found that the loss of fisheries due to dam construction in the basin may lead to significant additional land and water uses for food production in maintenance of food security in the region [100,101]. It is worth noting that upstream reservoir storage can augment downstream river flow during the dry season and in dry years, thus benefiting downstream irrigation. To explore effective transboundary water management strategies in the Mekong, based on a real case in the 2016 drought in the basin, Zhang et al. [102] analyzed upstream reservoir operation strategies that can effectively mitigate drought impacts in the downstream. In some cases, existing or emerging conflicts in transboundary water management may even provide a more productive and transformative way forward than purely cooperative arrangements [103]. Efforts must be made in all participating processes to reconcile competing and possibly conflicting interests [104].

Remote sensing can be used to assess the potential of hydropower resources by analyzing water flow rates and elevation changes. This information can be used to identify areas where hydropower potential is high and to plan hydropower development projects more effectively [105]. The safety of dams and other water infrastructures using remote sensing is vital. This includes structural health monitoring [106] and changes in water levels, detecting seepage or erosion and assessing the stability of the surrounding terrain.

4.5.4. WEF Nexus in the Mekong

A growing number of scholars refer to the social relationships and biophysical interflow linking water, food and energy networks as "nexus" to draw attention to important interactions and risks that have been overlooked [107–109]. Although interest in synergies and trade-offs in the provision of the water-energy-food nexus is growing rapidly, proposals and recommendations for technological interventions are increasing. However, it is unclear what value the adoption of a nexus perspective has in understanding public policy for water resource management [110]. Weitz [110] also pointed out that our understanding of the meaning of managing WEF relationships and the conditions under which it can or cannot work is still very limited, and more empirical research is warranted on this issue.

The broad scope of the topics of interest discussed above (i.e., hydropower generation, flow regime change and water supply for agriculture and ecosystems downstream) represent a complex water–energy–food nexus in the entire Mekong River Basin. The "water-energy-food" nexus was established primarily for increasing resource-using efficiency, reducing trade-offs, creating synergies and improving overall watershed management to enhance water, energy and food security. A systematic investigation of multisectoral and transboundary tradeoffs within this nexus can provide scientific guidance for the riparian countries to cooperate effectively.

The emergence of this interconnected relationship is related to the underlying drivers: the critical importance of water, energy and food for human survival and the need to ensure their security, as well as greater awareness of the social and economic risks posed by the increasing scarcity of natural resources [111]. The nexus approach may help to identify co-benefits and external costs at the international level associated with actions in different sectors. Studying transboundary river basin WEF coupling relationships from a historical perspective allows for a more rigorous institutional analysis of benefit sharing in different sectors.

Water is an active driving force in the WEF nexus, and together with food and energy, the three resources are indispensable for human society. Society and the economy are significantly impacted by the growing shortage of resources such as water. Therefore, water resource planning and management are decisive factors [111]. Existing WEF studies include the quantitative study of resource flow and the degree of dependence, evaluation of technical indicators and policies and quantitative study of the degree of performance of the entire system [112]. There are also review articles that illustrate the concept of WEF nexus [113–116], simulation tools [117,118] and nexus governance [110] or implementation [119].

The water–energy–food security nexus is particularly challenging in transboundary river basin management. The WEF nexus is found to be useful because more actors are recognizing the interdependence between water management and food and energy production. Through the dialogues at the World Environment Forum, many actors saw themselves as a reasonable and sectoral entry point in a compelling, new, multi-sectoral, interdisciplinary and cross-border discussion [120]. The concept of relationship has different manifestations according to the background, scale and geographic environment of the review [121].

In WEF nexus, remote sensing technology also plays an important role. Various remote sensing can be used to determine and analyze the technical feasibility of integrating urban agriculture, rainwater harvesting and photovoltaic systems [122]. Moreover, using the WEF framework based on GIS, stakeholders can also evaluate the utilization of resources to achieve sustainable productivity [123].

5. Conclusions

Enhanced awareness and understanding of water management in general and irrigation in the LMB are of great importance for cross-border cooperation on water resources management for achieving equitable, rational and sustainable water uses and for promoting sustainable economic prosperity and societal development in riparian areas. In this paper, we conducted a bibliometric analysis of irrigation in the LMB by tapping into a large volume of publications. Based on identified key thematic areas, we further synthesized key issues and findings from existing studies through a thematic review. The bibliometric analysis systematically summarized key statistics and features of the literature body concerning LMB irrigation, such as journals where those articles were published, authors' affiliations and countries of origin, and collaborator network. We found that LMB irrigation has drawn wide attention within and outside of the region, and the number of publications has been rapidly increasing in recent years. The thematic analysis is based on the results of bibliometric analysis, focusing on themes including irrigation management and crop production, transboundary water management, climate change, environmental impacts and comprehensive water resources management.

Remote sensing technology has played a great role in different fields of agricultural irrigation. This review of remote sensing applications in this paper is expected to advance socio-hydrological research and inform science-based management decisions and policy formulation for sustainable food-energy-water, livelihoods and ecosystem development in the Mekong region, providing a strong basis for decisions to promote sustainable ecological development in the Greater Mekong Subregion.

We advocate for heightened awareness and understanding of water management, specifically irrigation in the LMB. This will facilitate cross-border cooperation on water resources, promoting equitable and sustainable water usage and contributing to economic prosperity and societal development in riparian areas. It is strongly suggested to prioritize the adoption of remote sensing technology in agricultural irrigation for well-informed decision-making and sustainable development in the Mekong region. By incorporating research insights and fostering collaboration, we can ensure responsible water use amid evolving challenges, such as those posed by climate change impacts.

Author Contributions: Conceptualization, G.W. and T.Z.; methodology, G.W.; validation, G.W.; formal analysis, G.W. and T.Z.; investigation, G.W.; resources, T.Z.; data curation, G.W.; writing—original draft preparation, G.W.; writing—review and editing, G.W., S.B., T.Z., F.T. and M.A.O.; supervision, T.Z.; funding acquisition, T.Z. All authors have read and agreed to the published version of the manuscript.

Funding: This study was supported by the National Natural Science Foundation of China (Grant No. 51961125204).

Institutional Review Board Statement: Not applicable.

Informed Consent Statement: Not applicable.

Data Availability Statement: All data generated or analyzed are available from the corresponding author on reasonable request.

Acknowledgments: We are grateful to Citespace, which was used in this study.

Conflicts of Interest: The authors declare that they have no known competing financial interests or personal relationships that may influence the work reported in this paper.

References

1. World Bank Group. *World Bank Mekong Vision 3.0, Supporting Riparian Countries of the Mekong River Basin and Related Stakeholders to Align Their Development Goals and Natural Resources Management Activities More Closely towards Long-Term Sustainability in the Context of Emerging Chall*; World Bank: Washington, DC, USA, 2019; p. 8.
2. Yamauchi, K. Climate Change Impacts on Agriculture and Irrigation in the Lower Mekong Basin. *Paddy Water Environ.* **2014**, *12*, 227–240. [CrossRef]
3. Economic Dimension—State of the Basin Report 2018 Findings. Available online: https://www.mrcmekong.org/interactive-publications/state-of-the-basin-report-2018/economic-dimension.html (accessed on 1 March 2023).
4. Molle, F.; Foran, T.; Käkönen, M. Irrigation in the Lower Mekong Basin Countries: The Beginning of a New Era? In *Contested Waterscapes in the Mekong Region: Hydropower, Livelihoods and Governance*; Routledge: London, UK, 2012; pp. 165–194. ISBN 9781849770.
5. FAO. *Water for Sustainable Food and Agriculture*; Food and Agriculture Organization of the United Nations: Rome, Italy, 2017.
6. Sagara, J. *Surface Water Resources Improving Climate Resilience, Mekong Delta River Basin Groups: Assessment of the Tonle Sap and Productivity, and Sustainability*; ADB Briefs: Mandaluyong, Philippines, 2021; p. 20.
7. Boretti, A. Implications on Food Production of the Changing Water Cycle in the Vietnamese Mekong Delta. *Glob. Ecol. Conserv.* **2020**, *22*, e00989. [CrossRef]
8. Pokhrel, Y.; Burbano, M.; Roush, J.; Kang, H.; Sridhar, V.; Hyndman, D.W. A Review of the Integrated Effects of Changing Climate, Land Use, and Dams on Mekong River Hydrology. *Water* **2018**, *10*, 266. [CrossRef]
9. Smajgl, A.; Ward, J. *The Water-Food-Energy Nexus in the Mekong Region*; Springer: New York, NY, USA, 2013; pp. 971–978.
10. Liu, J.; Chen, S.; Wang, H.; Chen, X. Calculation of Carbon Footprints for Water Diversion and Desalination Projects. *Energy Procedia* **2015**, *75*, 2483–2494. [CrossRef]
11. Vora, N.; Shah, A.; Bilec, M.M.; Khanna, V. Food–Energy–Water Nexus: Quantifying Embodied Energy and GHG Emissions from Irrigation through Virtual Water Transfers in Food Trade. *ACS Sustain. Chem. Eng.* **2017**, *5*, 2119–2128. [CrossRef]
12. Holger, H. Clean Energy Solutions Center Understanding the Nexus. Background Paper for the Bonn 2011 Conference: The Water, Energy and Food Security Nexus. In Proceedings of the Bonn 2011 Conference, Bonn, Germany, 5 September 2011.
13. Liu, W.; Yang, H.; Tang, Q.; Liu, X. Understanding the Water-Food-Energy Nexus for Supporting Sustainable Food Production and Conserving Hydropower Potential in China. *Front. Environ. Sci.* **2019**, *7*, 1–10. [CrossRef]
14. Zhong, S.; Zhu, Y.; Zhao, J.; Shen, L. A Collaborative Framework for Hydropower Development and Sustainable Livelihood of Farmers in the Lancang-Mekong River Basin: A Review with the Perspective of Energy-Water-Food Nexus. *Water* **2022**, *14*, 499. [CrossRef]
15. Piman, T.; Lennaerts, T.; Southalack, P. Assessment of Hydrological Changes in the Lower Mekong Basin from Basin-Wide Development Scenarios. *Hydrol. Process.* **2013**, *27*, 2115–2125. [CrossRef]
16. Dombrowsky, I.; Hensengerth, O. Governing the Water-Energy-Food Nexus Related to Hydropower on Shared Rivers—The Role of Regional Organizations. *Front. Environ. Sci.* **2018**, *6*, 153. [CrossRef]
17. Zhang, X.; Li, H.-Y.; Deng, Z.D.; Ringler, C.; Gao, Y.; Hejazi, M.I.; Leung, L.R. Impacts of Climate Change, Policy and Water-Energy-Food Nexus on Hydropower Development. *Renew. Energy* **2018**, *116*, 827–834. [CrossRef]
18. Sabo, J.L.; Ruhi, A.; Holtgrieve, G.W.; Elliott, V.; Arias, M.E.; Ngor, P.B.; Räsänen, T.A.; Nam, S. Designing River Flows to Improve Food Security Futures in the Lower Mekong Basin. *Science* **2017**, *358*, 6368. [CrossRef] [PubMed]
19. Hamidov, A.; Helming, K. Sustainability Considerations in Water-Energy-Food Nexus Research in Irrigated Agriculture. *Sustainability* **2020**, *12*, 6274. [CrossRef]
20. Gao, W.; Guo, H.-C. Nitrogen Research at Watershed Scale: A Bibliometric Analysis during 1959–2011. *Scientometrics* **2014**, *99*, 737–753. [CrossRef]
21. Chen, C. Predictive Effects of Structural Variation on Citation Counts. *J. Am. Soc. Inf. Sci. Technol.* **2012**, *63*, 431–449. [CrossRef]
22. Opejin, A.K.; Aggarwal, R.M.; White, D.D.; Jones, J.L.; Maciejewski, R.; Mascaro, G.; Sarjoughian, H.S. A Bibliometric Analysis of Food-Energy-Water Nexus Literature. *Sustainability* **2020**, *12*, 1112. [CrossRef]
23. Chen, X.; Zou, D.; Xie, H. A Decade of Learning Analytics: Structural Topic Modeling Based Bibliometric Analysis. *Educ. Inf. Technol.* **2022**, *27*, 10517–10561. [CrossRef]
24. Donthu, N.; Kumar, S.; Mukherjee, D.; Pandey, N.; Lim, W.M. How to Conduct a Bibliometric Analysis: An Overview and Guidelines. *J. Bus. Res.* **2021**, *133*, 285–296. [CrossRef]
25. Chen, C. Science Mapping: A Systematic Review of the Literature. *J. Data Inf. Sci.* **2017**, *2*, 1–40. [CrossRef]
26. Chen, C. CiteSpace II: Detecting and Visualizing Emerging Trends and Transient Patterns in Scientific Literature. *J. Am. Soc. Inf. Sci. Technol.* **2006**, *57*, 359–377. [CrossRef]
27. Chen, C. *CiteSpace: A Practical Guide for Mapping Scientific Literature*; Nova Publishers: New York, NY, USA, 2016; ISBN 978-1-5361-0280-2.
28. Liu, J.; Chen, D.; Mao, G.; Irannezhad, M.; Pokhrel, Y. Past and Future Changes in Climate and Water Resources in the Lancang–Mekong River Basin: Current Understanding and Future Research Directions. *Engineering* **2022**, *13*, 144–152. [CrossRef]

29. Yamaguchi, T.; Tuan, L.M.; Minamikawa, K.; Yokoyama, S. Assessment of the Relationship between Adoption of a Knowledge-Intensive Water-Saving Technique and Irrigation Conditions in the Mekong Delta of Vietnam. *Agric. Water Manag.* **2019**, *212*, 162–171. [CrossRef]
30. Do, P.; Tian, F.; Zhu, T.; Zohidov, B.; Ni, G.; Lu, H.; Liu, H. Exploring Synergies in the Water-Food-Energy Nexus by Using an Integrated Hydro-Economic Optimization Model for the Lancang-Mekong River Basin. *Sci. Total Environ.* **2020**, *728*, 137996. [CrossRef] [PubMed]
31. Hoang, L.P.; van Vliet, M.T.H.H.; Kummu, M.; Lauri, H.; Koponen, J.; Supit, I.; Leemans, R.; Kabat, P.; Ludwig, F. The Mekong's Future Flows under Multiple Drivers: How Climate Change, Hydropower Developments and Irrigation Expansions Drive Hydrological Changes. *Sci. Total Environ.* **2019**, *649*, 601–609. [CrossRef]
32. Wang, K.; Liu, J.; Xia, J.; Wang, Z.; Meng, Y.; Chen, H.; Mao, G.; Ye, B. Understanding the Impacts of Climate Change and Socio-Economic Development through Food-Energy-Water Nexus: A Case Study of Mekong River Delta. *Resour. Conserv. Recycl.* **2021**, *167*, 105390. [CrossRef]
33. Polizzotto, M.L.; Kocar, B.D.; Benner, S.G.; Sampson, M.; Fendorf, S. Near-Surface Wetland Sediments as a Source of Arsenic Release to Ground Water in Asia. *Nature* **2008**, *454*, 505–508. [CrossRef]
34. Charlet, L.; Polya, D.A. Arsenic in Shallow, Reducing Groundwaters in Southern Asia: An Environmental Health Disaster. *Elements* **2006**, *2*, 91–96. [CrossRef]
35. Sakamoto, T.; Van Nguyen, N.; Ohno, H.; Ishitsuka, N.; Yokozawa, M. Spatio-Temporal Distribution of Rice Phenology and Cropping Systems in the Mekong Delta with Special Reference to the Seasonal Water Flow of the Mekong and Bassac Rivers. *Remote Sens. Environ.* **2006**, *100*, 1–16. [CrossRef]
36. Haddeland, I.; Lettenmaier, D.P.; Skaugen, T. Effects of Irrigation on the Water and Energy Balances of the Colorado and Mekong River Basins. *J. Hydrol.* **2006**, *324*, 210–223. [CrossRef]
37. Son, N.T.; Chen, C.F.; Chen, C.R.; Chang, L.Y.; Minh, V.Q. Monitoring Agricultural Drought in the Lower Mekong Basin Using MODIS NDVI and Land Surface Temperature Data. *Int. J. Appl. Earth Obs. Geoinf.* **2012**, *18*, 417–427. [CrossRef]
38. Sakamoto, T.; Wardlow, B.D.; Gitelson, A.A.; Verma, S.B.; Suyker, A.E.; Arkebauer, T.J. A Two-Step Filtering Approach for Detecting Maize and Soybean Phenology with Time-Series MODIS Data. *Remote Sens. Environ.* **2010**, *114*, 2146–2159. [CrossRef]
39. Polya, D.A.; Gault, A.G.; Diebe, N.; Feldman, P.; Rosenboom, J.W.; Gilligan, E.; Fredericks, D.; Milton, A.H.; Sampson, M.; Rowland, H.a.L.; et al. Arsenic Hazard in Shallow Cambodian Groundwaters. *Mineral. Mag.* **2005**, *69*, 807–823. [CrossRef]
40. Kummu, M.; Sarkkula, J. Impact of the Mekong River Flow Alteration on the Tonle Sap Flood Pulse. *Ambio* **2008**, *37*, 185–192. [CrossRef] [PubMed]
41. Kocar, B.D.; Polizzotto, M.L.; Benner, S.G.; Ying, S.C.; Ung, M.; Ouch, K.; Samreth, S.; Suy, B.; Phan, K.; Sampson, M.; et al. Integrated Biogeochemical and Hydrologic Processes Driving Arsenic Release from Shallow Sediments to Groundwaters of the Mekong Delta. *Appl. Geochem.* **2008**, *23*, 3059–3071. [CrossRef]
42. Su, F.; Zhang, L.; Ou, T.; Chen, D.; Yao, T.; Tong, K.; Qi, Y. Hydrological Response to Future Climate Changes for the Major Upstream River Basins in the Tibetan Plateau. *Glob. Planet. Chang.* **2016**, *136*, 82–95. [CrossRef]
43. Arias, M.E.; Cochrane, T.A.; Kummu, M.; Lauri, H.; Holtgrieve, G.W.; Koponen, J.; Piman, T. Impacts of Hydropower and Climate Change on Drivers of Ecological Productivity of Southeast Asia's Most Important Wetland. *Ecol. Model.* **2014**, *272*, 252–263. [CrossRef]
44. Hecht, J.S.; Lacombe, G.; Arias, M.E.; Dang, T.D.; Piman, T. Hydropower Dams of the Mekong River Basin: A Review of Their Hydrological Impacts. *J. Hydrol.* **2019**, *568*, 285–300. [CrossRef]
45. Dong, N.M.; Brandt, K.K.; Sørensen, J.; Hung, N.N.; Hach, C.V.; Tan, P.S.; Dalsgaard, T. Effects of Alternating Wetting and Drying versus Continuous Flooding on Fertilizer Nitrogen Fate in Rice Fields in the Mekong Delta, Vietnam. *Soil. Biol. Biochem.* **2021**, *47*, 166–174. [CrossRef]
46. Arias, M.E.; Cochrane, T.A.; Piman, T.; Kummu, M.; Caruso, B.S.; Killeen, T.J. Quantifying Changes in Flooding and Habitats in the Tonle Sap Lake (Cambodia) Caused by Water Infrastructure Development and Climate Change in the Mekong Basin. *J. Environ. Manag.* **2012**, *112*, 53–66. [CrossRef]
47. Son, N.-T.; Chen, C.-F.; Chen, C.-R.; Duc, H.-N.; Chang, L.-Y. A Phenology-Based Classification of Time-Series MODIS Data for Rice Crop Monitoring in Mekong Delta, Vietnam. *Remote Sens.* **2014**, *6*, 135–156. [CrossRef]
48. Biradar, C.M.; Xiao, X. Quantifying the Area and Spatial Distribution of Double- and Triple-Cropping Croplands in India with Multi-Temporal MODIS Imagery in 2005. *Int. J. Remote Sens.* **2011**, *32*, 367–386. [CrossRef]
49. Nguyen, T.T.H.; De Bie, C.A.J.M.; Ali, A.; Smaling, E.M.A.; Chu, T.H. Mapping the Irrigated Rice Cropping Patterns of the Mekong Delta, Vietnam, through Hyper-Temporal SPOT NDVI Image Analysis. *Int. J. Remote Sens.* **2012**, *33*, 415–434. [CrossRef]
50. Benner, S.G.; Polizzotto, M.L.; Kocar, B.D.; Ganguly, S.; Phan, K.; Ouch, K.; Sampson, M.; Fendorf, S. Groundwater Flow in an Arsenic-Contaminated Aquifer, Mekong Delta, Cambodia. *Appl. Geochem.* **2008**, *23*, 3072–3087. [CrossRef]
51. Wilbers, G.-J.; Becker, M.; Nga, L.T.; Sebesvari, Z.; Renaud, F.G. Spatial and Temporal Variability of Surface Water Pollution in the Mekong Delta, Vietnam. *Sci. Total Environ.* **2014**, *485*, 653–665. [CrossRef]
52. Sakamoto, T.; Van Phung, C.; Kotera, A.; Nguyen, K.D.; Yokozawa, M. Analysis of Rapid Expansion of Inland Aquaculture and Triple Rice-Cropping Areas in a Coastal Area of the Vietnamese Mekong Delta Using MODIS Time-Series Imagery. *Landsc. Urban. Plan.* **2009**, *92*, 34–46. [CrossRef]

53. Eto, M. Extended Co-Citation Search: Graph-Based Document Retrieval on a Co-Citation Network Containing Citation Context Information. *Inf. Process. Manag.* **2019**, *56*, 102046. [CrossRef]
54. Surwase, G.; Sagar, A.; Kademani, B.S.; Bhanumurthy, K. Co-Citation Analysis: An Overview. In *Beyond Librarianship: Creativity, Innovation and Discovery*; BOSLA: Mumbai, India, 2011.
55. Yun, J. Generalization of Bibliographic Coupling and Co-Citation Using the Node Split Network. *J. Informetr.* **2022**, *16*, 101291. [CrossRef]
56. Feng, J. Optimal Allocation of Regional Water Resources Based on Multi-Objective Dynamic Equilibrium Strategy. *Appl. Math Model.* **2021**, *90*, 1183–1203. [CrossRef]
57. Döll, P.; Siebert, S. Global Modeling of Irrigation Water Requirements. *Water Resour. Res.* **2002**, *38*, 8–10. [CrossRef]
58. Leng, G.; Tang, Q. Modeling the Impacts of Future Climate Change on Irrigation over China: Sensitivity to Adjusted Projections. *J. Hydrometeorol.* **2014**, *15*, 2085–2103. [CrossRef]
59. Yamashiki, Y. Effect of Irrigation Water Withdrawals on Water and Energy Balance in the Mekong River Basin Using an Improved VIC Land Surface Model with Fewer Calibration Parameters. *Agric. Water Manag.* **2015**, *159*, 92–106. [CrossRef]
60. Le, A.T. Water Balance for Agriculture Production in the Dry Seasons of the Mekong River Delta in Vietnam. *Vietnam. J. Sci Technol. Eng.* **2020**, *62*, 56–61. [CrossRef]
61. Hui, T.R.; Park, E.; Loc, H.H.; Tien, P.D. Long-Term Hydrological Alterations and the Agricultural Landscapes in the Mekong Delta: Insights from Remote Sensing and National Statistics. *Environ. Chall.* **2022**, *7*, 100454. [CrossRef]
62. Vu, H.T.D.; Tran, D.D.; Schenk, A.; Nguyen, C.P.; Vu, H.L.; Oberle, P.; Trinh, V.C.; Nestmann, F. Land Use Change in the Vietnamese Mekong Delta: New Evidence from Remote Sensing. *Sci. Total Environ.* **2022**, *813*, 151918. [CrossRef] [PubMed]
63. Mohammed, I.N.; Bolten, J.D.; Srinivasan, R.; Lakshmi, V. Satellite Observations and Modeling to Understand the Lower Mekong River Basin Streamflow Variability. *J. Hydrol.* **2018**, *564*, 559–573. [CrossRef]
64. Fonji, S.F.; Taff, G.N. Using Satellite Data to Monitor Land-Use Land-Cover Change in North-Eastern Latvia. *SpringerPlus* **2014**, *3*, 61. [CrossRef]
65. Kang, H.; Sridhar, V.; Mainuddin, M.; Trung, L.D. Future Rice Farming Threatened by Drought in the Lower Mekong Basin. *Sci. Rep.* **2021**, *11*, 9383. [CrossRef]
66. Zhang, L.; Song, W.; Song, W. Assessment of Agricultural Drought Risk in the Lancang-Mekong Region, South East Asia. *Int. J. Environ. Res. Public Health* **2020**, *17*, 6153. [CrossRef]
67. Zhang, B.; Zhang, L.; Guo, H.; Leinenkugel, P.; Zhou, Y.; Li, L.; Shen, Q. Drought Impact on Vegetation Productivity in the Lower Mekong Basin. *Int. J. Remote Sens.* **2014**, *35*, 2835–2856. [CrossRef]
68. Chen, X.; Liu, H.; Baiyinbaoligao; Mu, X. Summary of Flood and Drought in Mekong River Basin. In *Flood Prevention and Drought Relief in Mekong River Basin*; Liu, H., Ed.; Springer Tracts in Civil Engineering; Springer: Cham, Switzerland, 2020; pp. 27–54. ISBN 9789811520068.
69. Tilleard, S.; Turral, H.; Ketelsen, T.; Whiting, L. Climate Change, Water Scarcity and Agriculture: Lessons from the Countries of the Lower Mekong. In *Climate Risks to Water Security: Framing Effective Response in Asia and the Pacific*; Ojha, H., Schofield, N., Camkin, J., Eds.; Palgrave Studies in Climate Resilient Societies; Springer International Publishing: Cham, Switzerland, 2023; pp. 215–238. ISBN 978-3-031-16648-8.
70. Adeola, O.M.; Ramoelo, A.; Mantlana, B.; Mokotedi, O.; Silwana, W.; Tsele, P. Review of Publications on the Water-Energy-Food Nexus and Climate Change Adaptation Using Bibliometric Analysis: A Case Study of Africa. *Sustainability* **2022**, *14*, 13672. [CrossRef]
71. Nhẫn, P.P.; Hòa, L.V.; Quí, C.N.; Huy, N.X.; Hữu, T.P.; Macdonald, B.C.T.; Tường, T.P. Increasing Profitability and Water Use Efficiency of Triple Rice Crop Production in the Mekong Delta, Vietnam. *J. Agric. Sci.* **2016**, *154*, 1015–1025. [CrossRef]
72. Wang, X. Remote Sensing Applications to Climate Change. *Remote Sens.* **2023**, *15*, 747. [CrossRef]
73. Yang, J.; Gong, P.; Fu, R.; Zhang, M.; Chen, J.; Liang, S.; Xu, B.; Shi, J.; Dickinson, R. The Role of Satellite Remote Sensing in Climate Change Studies. *Nat. Clim. Chang.* **2013**, *3*, 875–883. [CrossRef]
74. Galloway, D.L.; Erkens, G.; Kuniansky, E.L.; Rowland, J.C. Preface: Land Subsidence Processes. *Hydrogeol. J.* **2016**, *24*, 547–550. [CrossRef]
75. Toan, T.Q. 9-Climate Change and Sea Level Rise in the Mekong Delta: Flood, Tidal Inundation, Salinity Intrusion, and Irrigation Adaptation Methods. In *Coastal Disasters and Climate Change in Vietnam*; Thao, N.D., Takagi, H., Esteban, M., Eds.; Elsevier: Amsterdam, The Netherlands, 2014; pp. 199–218. ISBN 978-0-12-800007-6.
76. Khang, N.D.; Kotera, A.; Sakamoto, T.; Yokozawa, M. Sensitivity of Salinity Intrusion to Sea Level Rise and River Flow Change in Vietnamese Mekong Delta-Impacts on Availability of Irrigation Water for Rice Cropping. *J. Agric. Meteorol.* **2008**, *64*, 167–176. [CrossRef]
77. Nilsson, C.; Reidy, C.A.; Dynesius, M.; Revenga, C. Fragmentation and Flow Regulation of the World's Large River Systems. *Science* **2005**, *308*, 405–408. [CrossRef]
78. Anderson, D.; Moggridge, H.; Warren, P.; Shucksmith, J. The Impacts of 'Run-of-River' Hydropower on the Physical and Ecological Condition of Rivers. *Water Environ. J.* **2015**, *29*, 268–276. [CrossRef]
79. Benejam, L.; Saura-Mas, S.; Bardina, M.; Solà, C.; Munné, A.; García-Berthou, E. Ecological Impacts of Small Hydropower Plants on Headwater Stream Fish: From Individual to Community Effects. *Ecol. Freshw. Fish.* **2016**, *25*, 295–306. [CrossRef]

80. Mbaka, J.G.; Wanjiru Mwaniki, M. A Global Review of the Downstream Effects of Small Impoundments on Stream Habitat Conditions and Macroinvertebrates. *Environ. Rev.* **2015**, *23*, 257–262. [CrossRef]
81. Mueller, M.; Pander, J.; Geist, J. The Effects of Weirs on Structural Stream Habitat and Biological Communities. *J. Appl. Ecol.* **2011**, *48*, 1450–1461. [CrossRef]
82. Arias, M.E.; Cochrane, T.A. *Impacts of Hydrological Alterations to the Tonle Sap Ecosystem of the Mekong River Basin*; American Geophysical Union: Washington, DC, USA, 2013.
83. Cochrane, T.A.; Arias, M.E.; Piman, T. Historical Impact of Water Infrastructure on Water Levels of the Mekong River and the Tonle Sap System. *Hydrol. Earth Syst. Sci.* **2014**, *18*, 4529–4541. [CrossRef]
84. Frappart, F.; Biancamaria, S.; Normandin, C.; Blarel, F.; Bourrel, L.; Aumont, M.; Azemar, P.; Vu, P.-L.; Le Toan, T.; Lubac, B.; et al. Influence of Recent Climatic Events on the Surface Water Storage of the Tonle Sap Lake. *Sci. Total Environ.* **2018**, *636*, 1520–1533. [CrossRef] [PubMed]
85. Wang, Y.; Feng, L.; Liu, J.; Hou, X.; Chen, D. Changes of Inundation Area and Water Turbidity of Tonle Sap Lake: Responses to Climate Changes or Upstream Dam Construction? *Environ. Res. Lett.* **2020**, *15*, 0940a1. [CrossRef]
86. Chen, X.; Zheng, Y.; Xu, B.; Wang, L.; Han, F.; Zhang, C. Balancing Competing Interests in the Mekong River Basin via the Operation of Cascade Hydropower Reservoirs in China: Insights from System Modeling. *J. Clean. Prod.* **2020**, *254*, 119967. [CrossRef]
87. Tran, H.; Tran, T.; Kervyn, M. Dynamics of Land Cover/Land Use Changes in the Mekong Delta, 1973–2011: A Remote Sensing Analysis of the Tran Van Thoi District, Ca Mau Province, Vietnam. *Remote Sens.* **2015**, *7*, 2899–2925. [CrossRef]
88. Rosegrant, M.W.; Ringler, C.; Zhu, T. Water for Agriculture: Maintaining Food Security under Growing Scarcity. *Annu. Rev. Environ. Resour.* **2009**, *34*, 205–222. [CrossRef]
89. Evers, J.; Pathirana, A. Adaptation to Climate Change in the Mekong River Basin: Introduction to the Special Issue. *Clim. Change* **2018**, *149*, 1–11. [CrossRef]
90. Wu, J.; Wang, X.; Zhong, B.; Yang, A.; Jue, K.; Wu, J.; Zhang, L.; Xu, W.; Wu, S.; Zhang, N.; et al. Ecological Environment Assessment for Greater Mekong Subregion Based on Pressure-State-Response Framework by Remote Sensing. *Ecol. Indic.* **2020**, *117*, 106521. [CrossRef]
91. Yuan, L.; He, W.; Liao, Z.; Degefu, D.M.; An, M.; Zhang, Z.; Wu, X. Allocating Water in the Mekong River Basin during the Dry Season. *Water* **2019**, *11*, 400. [CrossRef]
92. Degefu, D.M.; He, W. Allocating Water under Bankruptcy Scenario. *Water Resour. Manag.* **2016**, *30*, 3949–3964. [CrossRef]
93. Hurford, A.P.; Harou, J.J. Balancing Ecosystem Services with Energy and Food Security—Assessing Trade-Offs from Reservoir Operation and Irrigation Investments in Kenya's Tana Basin. *Hydrol. Earth Syst. Sci.* **2014**, *18*, 3259–3277. [CrossRef]
94. Tingsanchali, T.; Singh, P.R. Optimum Water Resources Allocation for Mekong-Chi-Mun Transbasin Irrigation Project, Northeast Thailand. *Water Int.* **1996**, *21*, 20–29. [CrossRef]
95. Johnston, R.; Kummu, M. Water Resource Models in the Mekong Basin: A Review. *Water Resour. Manag.* **2012**, *26*, 429–455. [CrossRef]
96. Ringler, C. ZEF Discussion Papers on Development Policy. In *Optimal Water Allocation in the Mekong River Basin*; ZEF Center for Development Research: Bonn, Germany, 2016; p. 41.
97. Wang, L.Z.; Fang, L.; Hipel, K.W. Water Resources Allocation: A Cooperative Game Theoretic Approach. *J. Environ. Inform.* **2003**, *2*, 11–22. [CrossRef]
98. Ziv, G.; Baran, E.; Nam, S.; Rodríguez-Iturbe, I.; Levin, S.A. Trading-off Fish Biodiversity, Food Security, and Hydropower in the Mekong River Basin. *Proc. Natl. Acad. Sci. USA* **2012**, *109*, 5609–5614. [CrossRef] [PubMed]
99. Ruida, Z.; Tongtiegang, Z.; Xiaohong, C. Evaluating the Tradeoff between Hydropower Benefit and Ecological Interest under Climate Change: How Will the Water-Energy-Ecosystem Nexus Evolve in the Upper Mekong Basin? *Energy* **2021**, *237*, 121518.
100. Stone, R. Dam-Building Threatens Mekong Fisheries. *Science* **2016**, *354*, 1084–1085. [CrossRef]
101. Orr, S.; Pittock, J.; Chapagain, A.; Dumaresq, D. Dams on the Mekong River: Lost Fish Protein and the Implications for Land and Water Resources. *Glob. Environ. Chang.* **2012**, *22*, 925–932. [CrossRef]
102. Yu, Y.; Zhao, J.; Li, D.; Wang, Z. Effects of Hydrologic Conditions and Reservoir Operation on Transboundary Cooperation in the Lancang–Mekong River Basin. *J. Water Resour. Plan. Manag.* **2019**, *145*, 04019020. [CrossRef]
103. Keskinen, M.; Salminen, E.; Haapala, J. Water Diplomacy Paths—An Approach to Recognise Water Diplomacy Actions in Shared Waters. *J. Hydrol.* **2021**, *602*, 126737. [CrossRef]
104. Middleton, C.; Devlaeminck, D.J. Reciprocity in Practice: The Hydropolitics of Equitable and Reasonable Utilization in the Lancang-Mekong Basin. *Int. Environ. Agreem. Politics Law. Econ.* **2021**, *21*, 235–253. [CrossRef]
105. Lin, Z.; Qi, J. A New Remote Sensing Approach to Enrich Hydropower Dams' Information and Assess Their Impact Distances: A Case Study in the Mekong River Basin. *Remote Sens.* **2019**, *11*, 3016. [CrossRef]
106. Zhu, Y.; Tang, H. Automatic Damage Detection and Diagnosis for Hydraulic Structures Using Drones and Artificial Intelligence Techniques. *Remote Sens.* **2023**, *15*, 615. [CrossRef]
107. Covarrubias, M. The Nexus between Water, Energy and Food in Cities: Towards Conceptualizing Socio-Material Interconnections. *Sustain. Sci.* **2019**, *14*, 277–287. [CrossRef]
108. Leck, H.; Conway, D.; Bradshaw, M.; Rees, J. Tracing the Water–Energy–Food Nexus: Description, Theory and Practice. *Geogr. Compass* **2015**, *9*, 445–460. [CrossRef]

109. Smajgl, A.; Ward, J.; Pluschke, L. The Water–Food–Energy Nexus–Realising a New Paradigm. *J. Hydrol.* **2016**, *533*, 533–540 [CrossRef]
110. Weitz, N.; Strambo, C.; Kemp-Benedict, E.; Nilsson, M. Closing the Governance Gaps in the Water-Energy-Food Nexus: Insights from Integrative Governance. *Glob. Environ. Chang.* **2017**, *45*, 165–173. [CrossRef]
111. Keskinen, M.; Varis, O. Water-Energy-Food Nexus in Large Asian River Basins. *Water* **2016**, *8*, 446. [CrossRef]
112. Chi, Z.; Xiaoxian, C.; Yu, L.; Wei, D.; Guangtao, F. Water-Energy-Food Nexus: Concepts, Questions and Methodologies. *J. Clean. Prod.* **2021**, *2018*, 625–639. [CrossRef]
113. Biggs, E.M.; Bruce, E.; Boruff, B.; Duncan, J.M.A.; Horsley, J.; Pauli, N.; McNeill, K.; Neef, A.; Van Ogtrop, F.; Curnow, J.; et al. Sustainable Development and the Water–Energy–Food Nexus: A Perspective on Livelihoods. *Environ. Sci. Policy* **2015**, *54*, 389–397. [CrossRef]
114. Cai, X.; Wallington, K.; Shafiee-Jood, M.; Marston, L. Understanding and Managing the Food-Energy-Water Nexus—Opportunities for Water Resources Research. *Adv. Water Resour.* **2018**, *111*, 259–273. [CrossRef]
115. De Grenade, R.; House-Peters, L.; Scott, C.A.; Thapa, B.; Mills-Novoa, M.; Gerlak, A.; Verbist, K. The Nexus: Reconsidering Environmental Security and Adaptive Capacity. *Curr. Opin. Environ. Sustain.* **2016**, *21*, 15–21. [CrossRef]
116. Wichelns, D. The Water-Energy-Food Nexus: Is the Increasing Attention Warranted, from Either a Research or Policy Perspective? *Environ. Sci. Policy* **2017**, *69*, 113–123. [CrossRef]
117. Dai, J.; Wu, S.; Han, G.; Weinberg, J.; Xie, X.; Wu, X.; Song, X.; Jia, B.; Xue, W.; Yang, Q. Water-Energy Nexus: A Review of Methods and Tools for Macro-Assessment. *Appl. Energy* **2018**, *210*, 393–408. [CrossRef]
118. Kaddoura, S.; El Khatib, S. Review of Water-Energy-Food Nexus Tools to Improve the Nexus Modelling Approach for Integrated Policy Making. *Environ. Sci. Policy* **2017**, *77*, 114–121. [CrossRef]
119. Vakilifard, N.; Anda, M.; Bahri, P.A.; Ho, G. The Role of Water-Energy Nexus in Optimising Water Supply Systems—Review of Techniques and Approaches. *Renew. Sustain. Energy Rev.* **2018**, *82*, 1424–1432. [CrossRef]
120. Dore, J.; Lebel, L.; Molle, F. A Framework for Analysing Transboundary Water Governance Complexes, Illustrated in the Mekong Region. *J. Hydrol.* **2012**, *466–467*, 23–36. [CrossRef]
121. Matthews, N.; Motta, S. Chinese State-Owned Enterprise Investment in Mekong Hydropower: Political and Economic Drivers and Their Implications across the Water, Energy, Food Nexus. *Water* **2015**, *7*, 6269–6284. [CrossRef]
122. Zambrano-Prado, P.; Muñoz-Liesa, J.; Josa, A.; Rieradevall, J.; Alamús, R.; Gasso-Domingo, S.; Gabarrell, X. Assessment of the food-water-energy nexus suitability of rooftops. A methodological remote sensing approach in an urban Mediterranean area. *Sustain. Cities Soc.* **2021**, *75*, 103287. [CrossRef]
123. Azzam, A.; Samy, G.; Hagras, M.A.; ElKholy, R. Geographic Information Systems-Based Framework for Water–Energy–Food Nexus Assessments. *Ain Shams Eng. J.* **2023**, 102224. [CrossRef]

remote sensing

MDPI

Article

Impacts of Shape Assumptions on Z–R Relationship and Satellite Remote Sensing Clouds Based on Model Simulations and GPM Observations

Liting Mai [1], Shuping Yang [1], Yu Wang [1] and Rui Li [1,2,3,*]

1 School of Earth and Space Sciences, University of Science and Technology of China, Hefei 230026, China
2 Deep Space Exploration Laboratory, Hefei 230026, China
3 State Key Laboratory of Fire Science, University of Science and Technology of China, Hefei 230026, China
* Correspondence: rli7@ustc.edu.cn; Tel.: +86-183-2619-5630

Abstract: In this study, the spherical particle model and ten nonspherical particle models describing the scattering properties of snow are evaluated for potential use in precipitation estimation from spaceborne dual-frequency precipitation radar. The single scattering properties of nonspherical snow particles are computed using discrete dipole approximation (DDA), while those of spherical particles are determined using Mie theory. The precipitation profiles from WRF output are then input to a forward radiative transfer model to simulate the radar reflectivity at Ka-band and Ku-band. The results are validated with Global Precipitation Mission Dual-Frequency Precipitation Radar measurements. Greater consistency between the simulated and observed reflectivity is obtained when using the sector- and dendrite-shape assumptions. For the case in this study, when using the spherical-shape assumption, radar underestimates the error of the cloud's top by about 300 m and underestimates the error of the cloud's area by about 15%. As snowflake shapes change with temperature, we use the range between −40 °C and −5 °C to define three temperature layers. The relationships between reflectivity (Z) and precipitation rate (R) are fitted separately for the three layers, resulting in $Z = 134.59 \cdot R^{1.184}$ (sector) and $Z = 127.35 \cdot R^{1.221}$ (dendrite) below −40 °C.

Keywords: shape of snowflakes; radiative transfer; Z–R relationship; DPR; detection threshold

Citation: Mai, L.; Yang, S.; Wang, Y.; Li, R. Impacts of Shape Assumptions on Z–R Relationship and Satellite Remote Sensing Clouds Based on Model Simulations and GPM Observations. *Remote Sens.* **2023**, *15*, 1556. https://doi.org/10.3390/rs15061556

Academic Editors: Haobo Li, Suelynn Choy, Yuriy Kuleshov, Mayra I. Oyola-Merced and Xiaoming Wang

Received: 14 January 2023
Revised: 6 March 2023
Accepted: 8 March 2023
Published: 12 March 2023

1. Introduction

Precipitation is one of the most crucial processes in the global water cycle and energy balance, not only because water is a fundamental need for all life around the world but also because significant energy will be released to the atmosphere accompanied by the phase-changing process at different altitudes from the surface to the cloud top [1–3]. Such energy, ending up as latent heat (LH), is the primary driving source of atmospheric circulation and acts to transfer a significant proportion (about 23%, [1]) of solar energy to atmospheric kinetic energy. Therefore, knowing the vertical distribution of precipitation rates is essential to understand the thermodynamics inside storms.

To measure the vertical structure of precipitation at a global scale, a satellite precipitation radar device was sent to space in 1997. The Tropical Rainfall Measuring Mission [4–6] and the Global Precipitation Mission [7–9] are the only two in history that carried single- and dual-frequency precipitation radar (hereafter DPR) working at 13.6 GHz and 35.5 GHz to directly measure the backscattering echo from precipitating particles at a vertical resolution of 250 m and 125 m, respectively. Many formulas describing the radar reflectivity–precipitation rate (hereafter Z–R relationship) have been published to retrieve precipitation rates from the radar measurements.

A challenging task in Z–R relationship parameterization is treating solid phase particles. Traditionally, the spherical-shape assumption was used to calculate the optical properties, including the attenuation coefficient, scattering coefficient, and phase function, based on Mie theory [10,11]. However, it is known that solid particles such as cloud ice snow, and graupels in the real atmosphere are not spherical. Instead, their shapes can be complicated both as single crystals and as aggregates. Many studies have been published to show the nonspherical effects in the microwave spectral region [12–18]. Some studies have applications in precipitation radar retrieval [19,20].

Will treating particles as nonspherical necessarily be better than using the sphere assumption? The answer may not be as simple as expected. Since each radar bin contains many particles with different shapes in different size distributions, the measurement of radar reflectivity is a synergetic effort that involves accounting for backscattered energy from different shapes at different scattering angles. Moreover, the shape of solid phase particles depends on temperature [21–23]. For example, in temperatures colder than −40 °C, simple columns, plates, or plate-like polycrystal-shaped ice dominate in the atmosphere [24,25] due to the lack of water vapor [26]. In temperatures from −40 to −20 °C, particles grow by further deposition, aggregation, and collection and show much more complicated shapes, including bullet-rosette, dendrite, and ice aggregates [24,27–29] With higher temperature, the ice phase particles partly melt, starting from the periphery, especially the corner of the particles. This results in liquid–solid mixed-phase particles with particular shapes, such as the water-coated ice ball shape [30–32]. Therefore, the parameterization of the shape of ice phase particles in radiative transfer calculation is complicated.

There are few validation studies of radiative transfer modeling of nonspherical effects in microwave regions from real satellite observations. Kulie et al. [33] simulated the brightness temperature between 6.9 GHz and 157 GHz for precipitation observed simultaneously by CloudSat's Cloud Profiling Radar (CPR), the Advanced Microwave Scanning Radiometer for Earth Observing System (AMSR-E), and the Microwave Humidity Sounder (MHS). They found that a few ice particle models demonstrate low bias among 25 tested, especially the long hex column, sector snowflake, 3-bullet rosette, short hex column derived from Liu [13], and aggregate derived from Hong [34]. Leinonen et al. [35] and Kulie et al. [36] compared the observed reflectivity derived from the Wakasa Bay field campaign with the simulated reflectivity under different ice particle shape assumptions through the relationship between two dual-frequency ratios: $DFR_{Ku/Ka}$ and $DFR_{Ka/W}$. They found that nonspherical-shape assumptions effectively interpreted the observed data. Olson et al. [19] interpreted radar observations from the ER-2 airborne High-Altitude Imaging Wind and Rain Airborne Profiler (HIWRAP) and then simulated upwelling microwave radiances for channels in which the Conical Scanning Millimeter-Wave Imaging Radiometer (CosMIR) operates. They concluded that the nonspherical crystal/aggregate snow particle model limited the discrepancies between the simulated and observed CosMIR radiances at 89 and 165.5 GHz to less than 4 K, and the discrepancies were larger than 8 K when using homogeneous ice–air spheres.

In this study, we focus on simulations and satellite validations of radar reflectivity from solid-phase precipitating particles at Ku and Ka band, and we attempt to answer the following questions: (1) What are the performances of simulations using different shape assumptions and using GPM DPR observations as a reference? (2) What are the effects of the temperature-dependent shape assumption? (3) What are the associated retrieval biases in the Z–R relationship, precipitation top height, and rain area when using different shape assumptions?

The remainder of this paper is organized as follows. Section 2 introduces basic information on the precipitation case, as well as the setting of WRF and the microwave radiative transfer model. Section 3 details the analyzed results for the case, including the verification of the simulations, the fitted Z–R relationships, and the effect of shape assumptions on detectable errors of the case. Finally, Section 4 summarizes the method and the findings of this study.

2. Data and Method

A snowfall event in East China at about 16:30 UTC on 6 January 2018 is used in this study. The flow chart of this study is shown in Figure 1. Given the lack of airborne radar observation data and the limited precision of radar observation, WRF output is utilized as an input for radiative transfer calculation, in which water content, the particle size distribution of five types of hydrometers (hereafter hydrometer profiles), and the atmospheric environment (temperature, pressure, and relative humidity) are needed. When calculating radar reflectivity for GPM DPR at Ka-band (35.5 GHz) and Ku-band (13.6 GHz), a sphere assumption, ten other nonspherical-shape assumptions, and a temperature-dependent shape assumption are considered. Then, by comparing the simulated reflectivity products and DPR observations, the accuracy of simulations and performances with different shape assumptions can be revealed. Beyond that, the associated retrieval bias in Z–R relationships, precipitation top height, and rain area are analyzed by combining the density of solid water simulated by WRF and reflectivity simulated by the radiative transfer model.

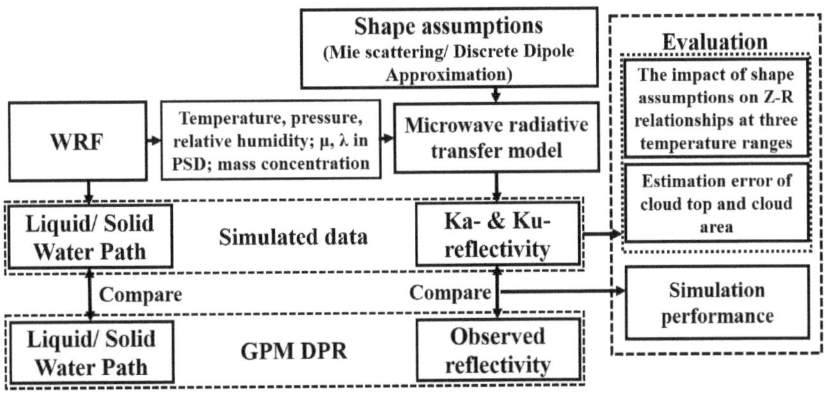

Figure 1. Framework of this study.

WRF V4.0 was used in this study. The model simulation used two nested domains, as shown in Figure 2. The inner and outer spatial resolutions were 4 km (convective permitting resolution) and 12 km, respectively. The cumulus parametrization scheme was turned on only for the outer domain. Other detailed simulation parameter settings are shown in Table 1. In the WRF output, the hydrometer profiles and the atmospheric environment were needed for radiative transfer simulations, while the hydrometer profiles are also used when analyzing Z–R relationships and cloud detection errors. To evaluate the results of the WRF simulation, the DPR products were compared with the precipitation water path (PrecipWP, includes snow, graupel, and rain), solid water path (SWP, includes snow and graupel), and liquid water path (LWP, includes rain). The reason why the water path is used instead of precipitation rate is that to compare the difference between solid and liquid water, only the water path distinguishes the phase among DPR products.

The radar simulator used in this study is the University of Science and Technology of China (USTC) Space-Borne Equivalent Radar Simulator (USERS) [37], which simulates the radiative transfer process of five hydrometers (cloud water, cloud ice, rain, snow, graupel) separately. To that end, it makes full use of the outputs of WRF that are given according to the type of hydrometer. During the simulation, all liquid water drops are assumed to be spherical, and their optical properties are calculated based on Mie theory. Solid-phase precipitating particles are assumed to have eleven shapes [13], including sphere, four of the five column/plate shapes designed in the model, four rosettes with different bullets, sector, and dendrite, as shown in Figure 3. Aggregations are also simulated. However, the simulated reflectivity is not closer to the observation than that under a single crystals shape assumption, and it is similar to that which short column performs (shown in Supplementary

Materials). Considering aggregations may be a collection of different single crystals, we consider single crystals first. The optical characteristics of these shape assumptions were simulated using discrete dipole approximation (DDA) [38] and called using a lookup table mode in USERS.

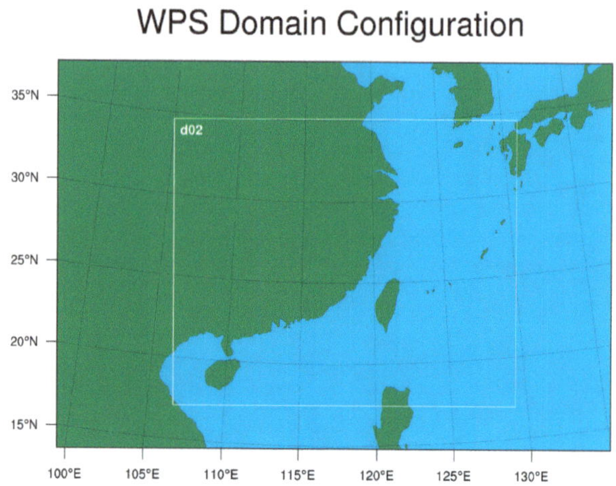

Figure 2. Domain setting of WRF.

Table 1. Setting of the WRF simulation.

Domain ID	01	02
Lateral/initial data	CFSv2 0.5° × 0.5° 6 hourly	
MP physics	Morrison	
CU physics	Modified Tiedtke scheme	None
Boundary layer physics	Mellor–Yamada–Janjic TKE scheme	
Surface layer physics	Monin–Obukhov (Janjic) scheme	
Land surface physics	Unified Noah land-surface model	
Longwave radiation physics	RRTMG scheme	
Shortwave radiation physics	RRTMG scheme	
Time step	60 s	20 s
Spatial resolution	12 km	4 km
Time range	1 January 2018–8 January 2018	
Output interval	None	30 min
Feedback	False	

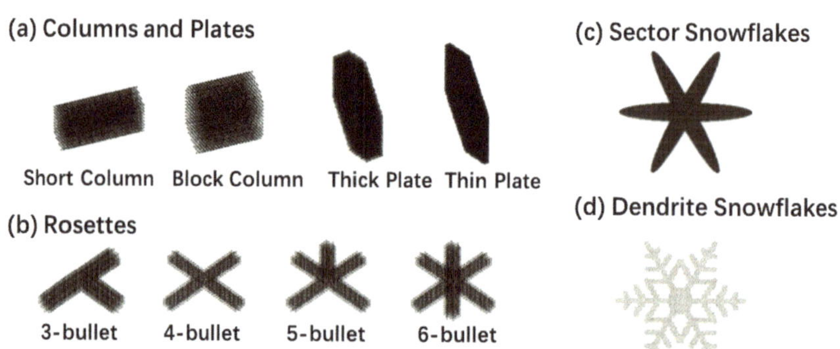

Figure 3. Ten nonspherical-shape assumptions of snowflakes used in radiative transfer simulation (from Liu GS, 2008).

In this study, the two parameters are set in WRF microphysical schemes as:

$$N(D) = N_0 D^{\mu} e^{-\lambda D}. \quad (1)$$

where D is the diameter, N_0 (unit: $1/m^4$), μ and λ (unit: 1/m) are the intercept, spectral and slope parameter of the PSD. In Morrison microphysics scheme used in this study, μ is specified and the slope parameter λ is calculated by:

$$\lambda = \left[\frac{cN\Gamma(\mu + d + 1)}{q\Gamma(\mu + 1)}\right]^{1/d}. \quad (2)$$

Here, number density N (unit: 1/kg) and q (unit: kg/kg) are obtained from WRF directly, and $c = \rho\frac{\pi}{6}$ (kg/m^3) and d are two parameters.

In the assumed mass–diameter relationship $m = cD^d$, ρs (the density of the particles) are set to be 977, 977, 500, 100, and 900 kg/m^3 for cloud water, rain, cloud ice, snow, and graupe, respectively. The coefficients c and d are determined to be associated. Therefore, the mass-weighted diameter Dm = (μ + 2)/λ is determined.

To illustrate the accuracy of simulated radar reflectivity, it is compared with the GPM DPR product. Only the pixels identified as precipitation pixels by the DPR retrieval algorithm (in which the precipitation rate product of DPR is not the default value) are considered in this study, which means the signals are free from surface clutter and nonprecipitation noise. After excluding the influence of noise, the spatiotemporal assimilation of data needs to be considered. WRF simulated the atmospheric conditions at 16:30 UTC on 6 January 2018, while DPR observed the region from 16:24 UTC to 16:29 UTC (5100–5400 scan) during the day. However, this time difference is a tiny gap compared to the time scale of a mesoscale cloud system. Though any method will cause differences between simulated and observed signals (Sun and Fu, 2021), spatial assimilation is needed before comparing them. In the vertical direction, the linear interpolation method is used to interpolate the simulated radar reflectivity to the vertical resolution of DPR; this method uses 176 layers from the ground (0 km) to 22 km. In the horizontal direction, the closest simulated profile is used to present the DPR profile. Considering that, in reality, the solid particles in a cloud system exist in various shapes, we assume that the proportion of simple-shaped solid particles in the total solid particles is a function of temperature, and this proportion increases when temperature decreases.

Hydrometer profiles and simulated radar reflectivity are used to fit Z–R relationships for eleven shapes of snowflakes, which are later exploited to illustrate the effects of using theoretical Z–R relationships in the snowfall retrieval algorithm. The fundamental reason for the effect on the Z–R relationship of rain types and weather conditions is the difference in microphysical characteristics and microphysical processes, such as the phases of particles at different temperature layers [39]. Thus, in this study, −40 °C and −5 °C are used to distinguish three temperature layers in order to minimize the effect of these issues. This temperature-dependent approach differs from the DPR retrieval algorithm, which uses radar reflectivity as an index [4]. However, from a physical point of view, it is temperature and precipitation rate (R) that determine radar reflectivity, which means that using temperature as the index is more appropriate than using reflectivity.

3. Results

A single case at 16:30 UTC on 6 January 2018, in East China, was used in this study.

3.1. Correctness of Simulation

Figure 4d shows that WRF simulated two cloud systems, which we labeled cloud system 1 at (116E, 35N) and cloud system 2 at (121E, 26N). Cloud system 1 does not exist in the DPR product, which is mainly a snowfall system. However, the simulated position, horizontal distribution, LWP, and SWP for cloud system 2 are very consistent with the DPR observed values. Therefore, we focus on cloud system 2 when analyzing the consistency between simulated radar reflectivity and observed radar reflectivity (after spatiotemporal assimilation, only the data in the black box are used). Ground-based observations (dataset named Weather of Now at 17:00 UTC is used, which represents the weather phenomenon in the past one hour, as shown in Supplementary Materials Figure S1) illustrate that there was a large area of snow and rain near the simulated moment in this area. Note that the minima of detectable radar reflectivity are 5 dBZ (Ka, but about 17 dBZ in the inner swath, and 12 dBZ (Ku), which means weak precipitation is unobservable. Slight snowfall may be ignored in the DPR retrieval algorithm (mentioned later), so, understandably, LWP and SWP in the DPR product are slightly smaller than the simulation results of WRF. Furthermore, there are some differences between WRF simulation and reality due to the selection of boundary conditions and parameterization schemes, but this is not the focus of this experiment.

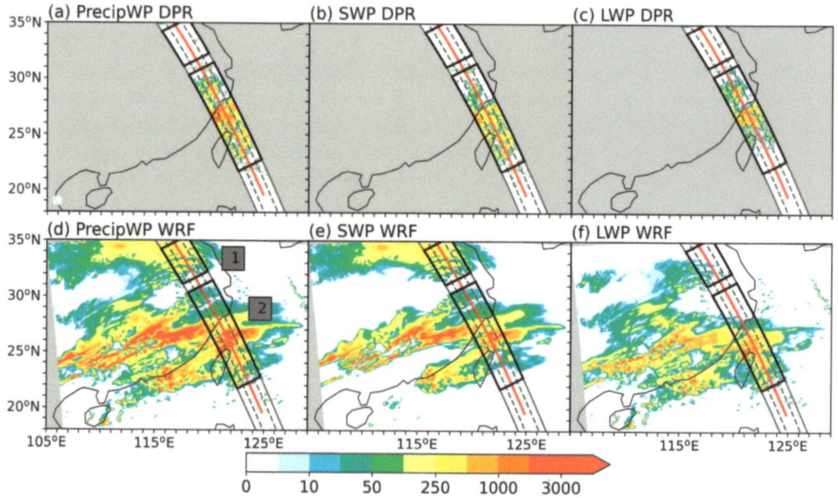

Figure 4. Satellite observations and model simulations of precipitation at 16:30 UTC on 6 January 2018. (**a–c**) GPM DPR observed precipitation water path (PrecipWP), solid water path (SWP), and liquid water path (LWP); (**d–f**) WRF simulated PrecipWP, SWP, and LWP. Unit: g/m^2.

Figure 5 shows cross-sections of radar reflectivity along the red line in Figure 4. Similar cross-sections of reflectivity along the north gray dashed line are shown in the Supplementary Materials (Figure S2). It can be easily determined from Figure 5b that WRF simulates the shape of cloud system 2 well, although the height is slightly overestimated; the magnitude of simulated radar reflectivity and that of observed radar reflectivity is quite consistent. As mentioned before, cloud system 1 is dominated by snowfall, and cloud system 2 by snow at high altitudes and rain at low altitudes. Therefore, to compare the effects of shape assumption on radar reflectivity simulation, attention should be paid to the upper level of cloud system 2 and to cloud system 1. Comparing Figure 5b,f to Figure 5a,e, it is obvious that when using the spherical-shape assumption, the simulated radar reflectivity at Ka-band is close to the observed radar reflectivity, while the simulation and the observation are pretty different at Ku-band. When using the simple-shape assumptions (Figure 5d,h) for simulation, the situation is just the opposite. This means that the snow and cloud ice particles in the precipitation cloud are neither spheres nor simple shapes. However, when using the complex-shape assumption (Figure 5c,g), the simulated radar signal is similar to the observed signal in both Ku-band and Ka-band, which means snow and cloud ice particles in the real world are likely to exist in complex shapes. There is a significant feature of bright-band at about 4 km height observed by GPM DPR, as shown in Figure 5a,e. This is a strong indicator of a melting layer and the stratiform precipitation type (Houze 1997). It was found that different shape assumptions resulted in different simulation performances of this feature. The dendrite assumption led to a good simulation of clear bright-band, while simulations with the sphere assumption captured this characteristic but with a relatively weaker signature. In contrast, the short column assumption completely missed bright-band in the simulation.

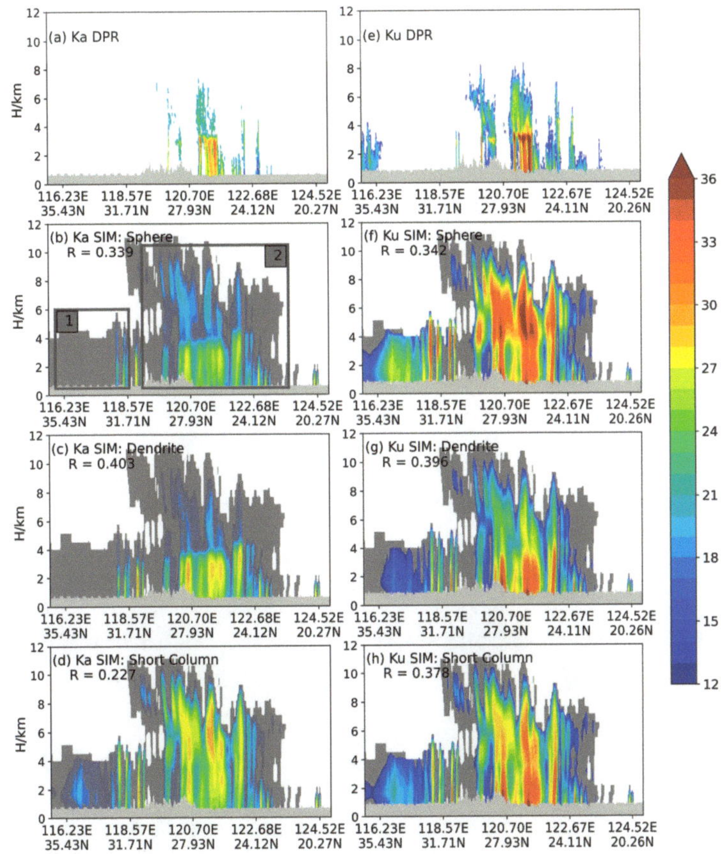

Figure 5. Cross-sections of reflectivity along the red line in Figure 4. (**a–h**) The first column is for Ka-band, and the second is for Ku-band. The first row represents GPM DPR observations. From the second to the fourth rows show simulations using sphere, dendrite, and short column assumptions, respectively. The lighter gray areas show surface echo, and the darker gray areas show the reflectivity under the detectable threshold (17 dBZ for Ka-band, 12 dBZ for Ku-band). R refers to the correlation coefficient between simulation and observation for the cross-section. Only the three most representative assumptions are shown here, but all eleven shape assumptions mentioned in the Method section are simulated; the complete results are shown in Supplementary Materials Figures S5 and S6.

The contoured frequency by altitude diagrams (CFADs) of Ka-band reflectivity observations and those of simulated reflectivity show that, compared to simple-shape assumptions (Figure 6b–i), the complex-shape assumptions (Figure 6j,k) or sphere assumption (Figure 6l) being the CFADs of reflectivity simulation closer to those of observations. More specifically, the maximum reflectivity of CFAD (Figure 6a) changes with height as follows: the maximum reflectivity value increases smoothly from ~6 km to ~3 km, and increases sharply between ~3 km and ~2.5 km, then decreases slightly toward lower levels. A similar trend can be seen in the CFADs of reflectivity simulation under the sphere assumption and complex-shape assumptions. Nevertheless, such a trend cannot be seen in simulation under the simple-shape assumption.

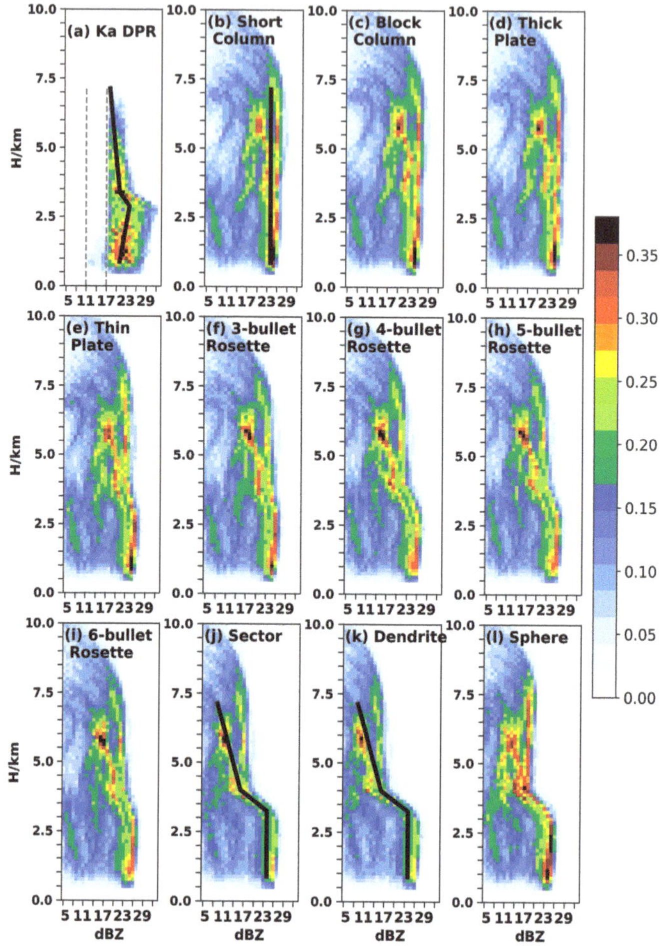

Figure 6. Contoured frequency by altitude diagrams (CFADs) illustrating the frequency of occurrence of values of reflectivity at Ka-band at different heights for the sample of cloud system 2. Data are binned at 1 dBZ intervals at each level and then normalized by the total number of samples in all levels. (**a**) CFAD of observation, (**b–l**) CFADs of simulations using eleven shape assumptions: short column, block column, thick column, thin column, 3-bullet rosette, 4-bullet rosette, 5-bullet rosette, 6-bullet rosette, sector, dendrite, and sphere. Black curves in (**a**,**b**,**j**,**k**) show the maximum occurrence probability of dBZ in each layer.

As for CFADs in Ku-band (Figure 7), CFADs of reflectivity simulation are also closer to the observation under complex-shape assumptions than others, as the maximum dBZ in the high altitude (3–6 km) appears at a lower value (~21 dBZ). For the radar reflectivity simulation under nonspherical-shape assumptions, the maximum value of CFAD appears at ~25 dBZ (simple shapes) or even higher (the spherical-shape assumption, ~30 dBZ).

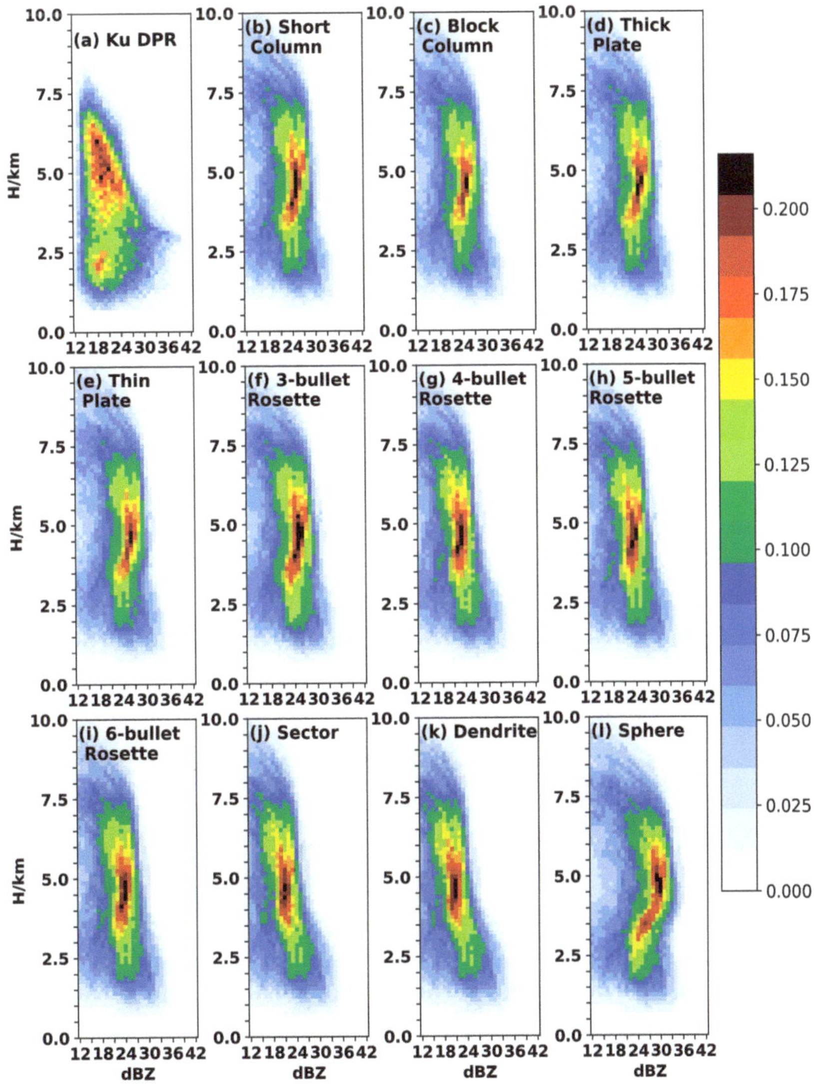

Figure 7. (**a–l**) same as Figure 6, but at Ku-band.

The dynamic and thermodynamic have a great impact on the precipitation profile and thus on the reflectivity profile. Therefore, a classification methodology is required. Reflectivity at low altitude is considered a great indicator as the higher the reflectivity, the greater the precipitation, and more vigorous convection also often relates to greater precipitation. Then, in order to avoid the interference of surface echo, reflectivity at the height of 2 km (hereafter dBZ (2 km)) was used. Figure 8a,b show that simulated profiles significantly differ from the observed reflectivity profile when dBZ (2 km) is lower than 20 dBZ. The number of effectively observed reflectivity grids (the grids where dBZ (2 km) has an exact value) is far less than the simulation when dBZ (2 km) is small, so it is too arbitrary to conclude directly that the simulation was incorrect. For example, the number of effective dBZ (2 km) falling in the range of 5–17 dBZ with the DPR product is 53, which is far less than that of simulated reflectivity (the number of effective dBZ (2 km) falling in the range of 5–17 dBZ), which revealed about 1398–1909, so we hypothesize that the neglect of slight precipitation by the DPR retrieval algorithm or the difference between the minimum detectable signals in the inner and outer swaths (17 dBZ and 5 dBZ) is responsible for the difference. Nevertheless, the numbers of observed data points within other classes are nearly the same, and the number of simulated data points is comparable to that of observed data points (shown in the Supplementary Materials Table S1). In these classes, almost all simulated reflectivity is also smaller than the observed reflectivity at high altitudes while they are equivalent at low altitudes (rainfall grids). This means that the USERS may underestimate the reflectivity on snow and cloud ice at Ka-band; otherwise, the actual snowflake shape is not among the eleven shapes assumed by the USERS. No matter what when the dBZ (2 km) is large, the simulated reflectivity under the spherical shape or two complex-shape assumptions (sector and dendrite) shows the same phenomenon of a steep increase from high altitudes to low altitudes. This similarity indicates that these three shapes may be close to the real shapes of snowflakes.

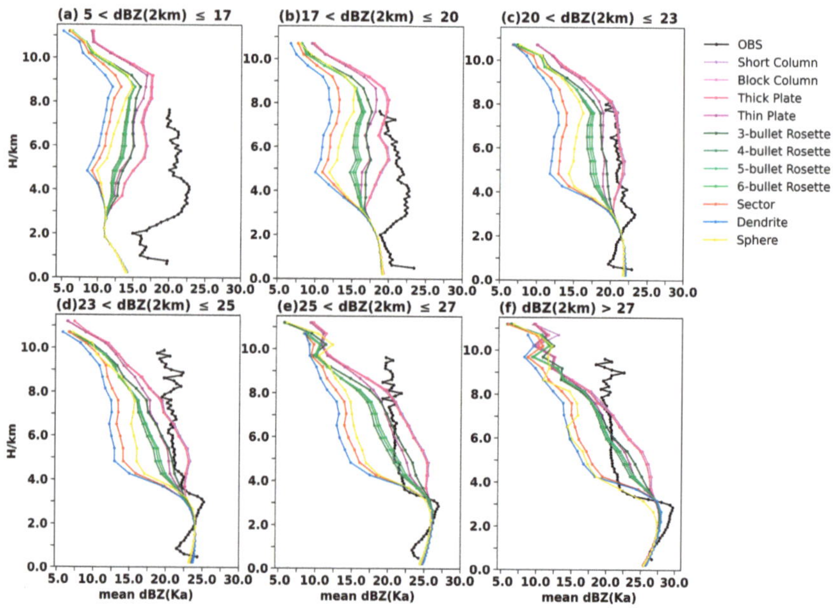

Figure 8. Mean Ka-band reflectivity profile of GPM DPR and of simulations under eleven shape assumptions for the sample of cloud system 2, with given reflectivity at the height of 2 km (dBZ (2 km)) in ranges of (**a**) 5–17 dBZ, (**b**) 17–20 dBZ, (**c**) 20–23 dBZ, (**d**) 23–25 dBZ, (**e**) 25–27 dBZ, (**f**) >27 dBZ.

Figure 9 shows that the probability distribution functions (PDFs) of simulated reflectivity shifted to the smaller end comparing to those of GPM-observed reflectivity. This is due to the limited detection sensitivity of GPM DPR. It illustrates that our guess about the effect of the DPR retrieval algorithm neglecting slight precipitation, mentioned in the previous paragraph, is probably right. This conclusion is consistent with that of previous studies comparing DPR-retrieved snowfall with other satellite or ground-based observation data [7,40–42]. Beyond that, in the radar reflectivity simulation, under all shape assumptions, the peak values of PDFs between 20 and 30 dBZ were well displayed; the sources of these signals were primarily liquid precipitation, which indicated that the USERS simulation of liquid water was more accurate than that of solid water.

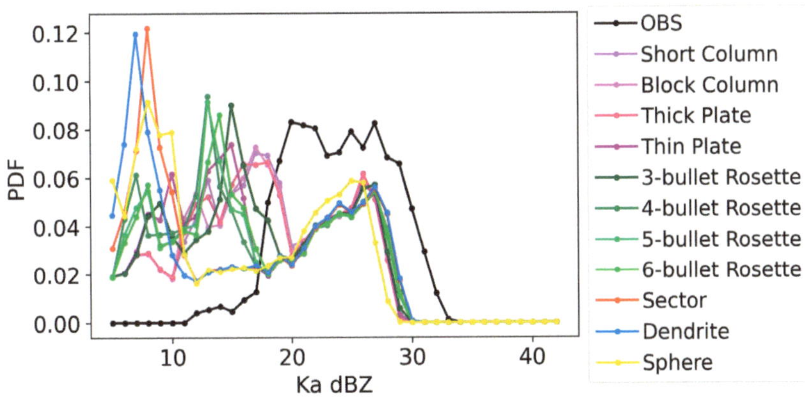

Figure 9. Probability distribution functions (PDF) of observed and simulated Ka-band dBZ (2 km) with eleven shape assumptions for cloud system 2.

In the real atmosphere, solid water cannot exist in just one shape, so a mixture of simple shapes (here, only short column is used) and complex shapes (dendrite) was considered in this study. Considering that the snowflakes are primarily of simple shape at high altitude and are of complex shape at low altitude because of condensation and collision effects during falling, the percentage of snowflakes in the short column to the total reflectivity (C) was assumed to be a function of temperature:

$$C = 100.0 * \left(\frac{T}{40}\right)^2 \quad (3)$$

Here, T is the temperature (unit: °C) from the WRF simulation. However, when T is lower than −40 °C, C is set to 1. Figure 10 shows the cross-section of C along the red line in Figure 4.

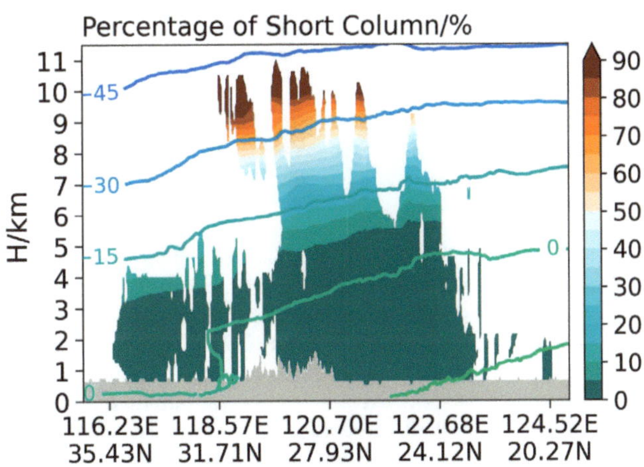

Figure 10. Percentage of snowflakes of short column shape to total reflectivity: $C = 100.0 * \left(\frac{T}{40}\right)^2$, where T is temperature in °C, which the contours represent.

Figure 11 shows that the correlation coefficient (R) between the observed and simulated reflectivity of cloud system 2 under any shape assumption is between 0.310 and 0.42, close to the correlation coefficient of the WRF-simulated precipitation rate and DPR precipitation rate product [43]. When comparing R under eleven single-shape assumptions (relative to aggregate-shape assumptions), we found that complex-shape assumptions performed better than any other shape assumption at both Ku-band and Ka-band. The spherical-shape assumption was acceptable in Ka-band simulation but performed worse at Ku-band than simple-shape assumptions. When considering the mixed-shape assumption, R was 0.412 in Ka-band, which was slightly higher than that under any simple-shape assumptions (short column: 0.206, dendrite: 0.411, as the mixed shape is a mixture of short column and dendrite), and R was 0.358 in Ku-band, which was superior to all simple-shape assumptions, although slightly lower than two complex-shape assumptions (short column: 0.300, dendrite: 0.357). This is because the bright-band was clearly shown in the Ka-band reflectivity cross-section when simulated under the mixed-shape assumption (see Supplementary Materials Figure S3), just like the signal under complex-shape assumptions. In addition, the reflectivity for each band in the upper atmosphere was closer to the measured one. Other temperature-dependent functions were tested, including the third, fourth, fifth, and sixth power functions of temperature, and the best denominators of temperature (40.0 for the square function), which exhibited the best fitting effects, were found for each function. However, the correlation coefficients calculated under these mixed-shape assumptions were similar to those of the square function, only ranging from 0.409 to 0.412 (Ka-band) and from 0.352 to 0.358 (Ku-band).

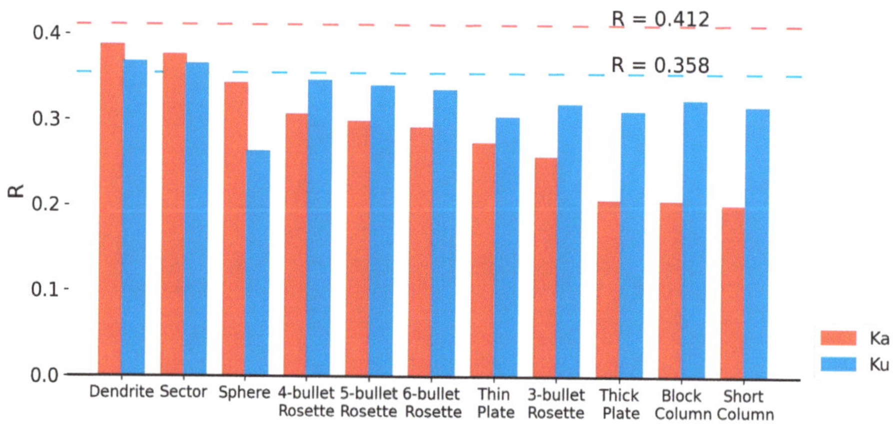

Figure 11. Correlation coefficient (R) between observed and simulated reflectivity under eleven shape assumptions for cloud system 2. The red/blue dotted lines represent R at Ka/Ku bands when simulated reflectivity is determined under the mixed-shape assumption (more simple shapes at higher altitudes).

In this case, the simulations of reflectivity in both Ka-band and Ku-band were consistent with DPR observations and thus could be used in the following study. In addition, we found: (1) The snow shape assumption greatly influenced radar reflectivity. (2) The difference between the simulation of reflectivity and observation was significant when using simple-shape assumptions; sector and dendrite provided a more reasonable scheme for studying snow's real shape and for radiative transfer process calculation. (3) In a real atmosphere, there were likely to be more snowflakes with simple shapes at the upper altitude and more with complex shapes at the lower altitude; the mixed-shape assumption depending on temperature provided a new perspective for studying snowflake shapes.

3.2. The Impact of Shape Assumptions on Z–R Relationships in Three Temperature Ranges

The Z–R relationship is the basis of precipitation retrieval from radar measurements. The shape assumption has significant impacts on this relationship and thus can lead to different retrieval bias, which has not previously been quantified. In this section, Z–R relationship is fitted from the model simulations with multiple shape assumptions and compared to selected relationships in literature. In addition, the associated retrieval error and bias are investigated to evaluate the impacts in different temperature layers.

The relationship between Z in Ku-band and the rain rate (R in mm/h), $Z = a{\cdot}R^b$ is often used to calculate R from Z directly in active radar retrieval algorithms [16,19,44]. Marshall and Palmer fitted the parameters a and b using observations [45], but the Z–R relationship is only for precipitation between 1 and 23 mm/h because of the limited instrument detection accuracy. Some researchers derived Z–R relationships based on the Rayleigh assumption [46,47], but that introduces errors into the estimation of reflectivity since both of the sizes and and the real shapes of snowflakes differ from that under sphere shape assumption. Therefore, it is necessary to study Z–R relationships for snowflakes with different shapes. Note that R is the short form for precipitation rate in this study, as there is also rainfall in this case.

As it is common to use dBZ as the unit of reflectivity (Z) in observation, and because we wanted to connect reflectivity observation and precipitation directly, hereafter, dBZ is used:

$$\mathrm{dBZ} = 10\mathrm{lg}Z \tag{4}$$

Moreover, the dBZ–R relationship is fitted:

$$\mathrm{dBZ} = 10\mathrm{lg}(Z) = 10\mathrm{lg}\left(a{\cdot}R^b\right) = A + B{\cdot}\mathrm{lg}R, \tag{5}$$

Here, A and B are two parameters to be fitted. MSE is used for the regression score:

$$\mathrm{MSE} = \frac{\sum_i \left(\mathrm{dBZ}_i - \mathrm{d\hat{B}Z}_i\right)^2}{N} \tag{6}$$

Here, N is the data volume, and the superscript indicates the fitting value. The MSEs for Ku-band reflectivity of snowflakes with six shapes in three temperature layers are reported in Table 2. The fitting effects are good for the particles in the atmosphere under −5 °C, as most MSEs are less than 1. However, MSEs for the reflectivity of spherical snowflakes existing between −40 °C and −5 °C are relatively high (>5), which results from the dispersion of the fitted data points and leads to bias that stems from using a fixed Z–R relationship in the retrieval algorithm. The MSEs are usually high for the particles existing between −5 °C and 0 °C because within this region, the scattering properties vary as precipitation particles tend to exist as a mixture of solid and liquid. Furthermore, the reflectivity partly results from cloud water, which implies strong effects from the microphysical properties of cloud water.

Table 2. Mean squared error (MSE) for Ku-band reflectivity of snowflakes with six shapes in three temperature layers.

Temperature	Sphere	Short Column	Thin Plate	6-Bullet Rosette	Sector	Dendrite
$T \leq -40$ °C	0.18	0.04	0.03	0.12	0.03	0.05
$-40 < T \leq -5$ °C	4.14	0.88	0.74	0.78	0.27	0.28
$-5 < T \leq 0$ °C	22.72	11.19	12.82	8.07	4.53	4.25

Figure 12 shows dBZ–R relationships for snowflakes with six shapes in three temperature layers, and the fitted parameters are listed in Table 3. The shape assumptions and temperature stratification clearly affected the fitting results of the dBZ–R relationships, so it was necessary to consider them in the radar retrieval algorithm.

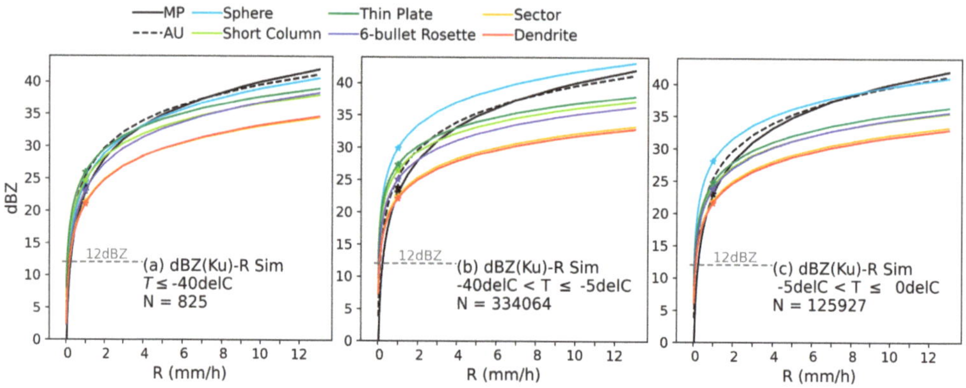

Figure 12. Simulated Z–R relationships of snowflakes with six shapes in three temperature layers (<40 °C, from −40 to −5 °C, from −5 to 0 °C), and the theoretical Z–R relationships (MP and AU relationships). (**a**–**c**) relationships for each temperature layers seperately.

Table 3. Parameters in dBZ–R relationships (dBZ = A + B·lgR) of snowflakes with six shapes in three temperature layers.

Parameters	Temperature/°C	Sphere	Short Column	Thin Plate	6-Bullet Rosette	Sector	Dendrite
A	$T \leq -40$	24.43	24.68	25.96	23.08	21.29	21.05
	$-40 < T \leq -5$	29.88	26.43	27.33	25.03	22.41	22.07
	$-5 < T \leq 0$	28.07	24.10	24.76	23.74	21.88	21.60
B	$T \leq -40$	14.51	11.95	11.72	13.68	11.84	12.21
	$-40 < T \leq -5$	11.85	9.61	9.48	10.06	9.74	9.67
	$-5 < T \leq 0$	11.57	10.21	10.39	10.65	10.27	10.17

Next, we compared the fitted relationships with $Z = 190 \cdot R^{1.72}$ derived by Marshell and Palmer (hereafter MP relationship, [46]), and $Z = 366 \cdot R^{1.42}$ derived by Atlas and Ulbrich (hereafter AU relationship, [47]), which are dBZ = 22.79 + 17.2·lgR and dBZ = 25.36 + 14.2·lgR (Figure 12). Below −40 °C, dBZ–R relationships of spherical snowfall and small (precipitation) snowfall in simple shapes were close to the theoretical relationships; between −40 °C and −5 °C, dBZ–R relationships of snowfall in complex shapes were close to theoretical relationships; above −5 °C, when snowfall was small (<1 mm/h), dBZ–R relationships of snowflakes in complex shapes were closer to theoretical relationships, but when snowfall was large (>6 mm/h), dBZ–R relationships of spherical snowflakes were closer. Despite all this, when theoretical dBZ–R relationships (MP or AU) were used in the retrieval algorithm, the following deviations arose: (1) The method underestimated snowfall of simple-shaped snowflakes when R was large and overestimated snowfall of simple-shaped snowflakes from −40 °C to −5 °C when R was small. (2) The method significantly underestimated snowfall of sector/dendrite-shaped snowflakes when R was large (>1 mm/h). (3) As for snowfall of spherical snowflakes, snowfall from −40 °C to −5 °C and light snowfalls (<9 mm/h) above −5 °C were significantly overestimated, but large snowfalls (>9 mm/h) above −5 °C were underestimated.

Table 3 shows that for the two shapes, sector and dendrite, that made our reflectivity simulation the most consistent with the observation in this experiment, the dBZ–R relationships below −40 °C were dBZ = 21.29 + 11.84·lgR and dBZ = 21.05 + 12.21·lgR, which were $Z = 134.59 \cdot R^{1.184}$ (sector) and $Z = 127.35 \cdot R^{1.221}$ (dendrite), respectively.

3.3. Estimation Error of Cloud Top and Cloud Area

On the one hand, the detection sensitivity of radar affects the detectable three-dimensional structure of the cloud. On the other hand, the detection capabilities are different for snowflakes in different shapes, so the estimation of detection error based on the spherical-shape assumption differs from reality. In this study, the difference of top of cloud (DTOC) and the difference in area of cloud (DAOC) are two indexes expressing cloud system detection accuracy. As shown in Figure 13, the real cloud boundary refers to where the total density of hydrometers equals 0.001 g/m^3, and the detectable cloud boundary is determined according to the radar detection threshold. $DTOC_i$ is defined as the difference between the real cloud top and the detectable cloud top in grid i in the horizontal direction. $DAOC_j$ is defined as the ratio of AOC detection error (the area where precipitation exists but is undetectable) and real AOC at height j. In addition, the DTOC and DAOC are illustrated for DPR deployed on GPM and FY3E, whose minimum detectable signals are assumed to be 5 dBZ and 10.4 dBZ at Ka-band, and 12 dBZ and 14.0 dBZ at Ku-band, respectively.

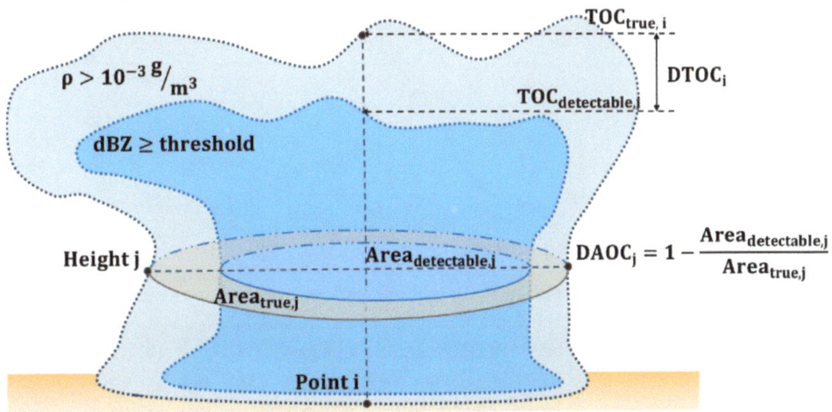

Figure 13. Sketch map of the difference of top of cloud (DTOC) and the difference in area of cloud (DAOC). Here, i represents a horizontal position, and j represents height. A pixel where the total density of hydrometers (ρ) is larger than 10^{-3} g/m^3 is considered a precipitation pixel, and pixels with radar reflectivity higher than the threshold are detectable. $TOC_{true,i}/TOC_{detectable,i}$ is the top of the cloud that is precipitation/detectable at the position i, and $Area_{true,j}/Area_{detectable,j}$ is the area of cloud that is precipitation/detectable at the height j; thus, $DTOC_i = TOC_{true,i} - TOC_{detectable,i}$ represents the DTOC at position i, and $DAOC_j = 1 - \frac{Area_{detectable,j}}{Area_{true,j}}$ represents the DAOC at height j.

Figure 14 shows that the detected TOC and AOC error would be close to zero if the detection threshold of Ka-band and Ku-band could be limited to −15 dBZ. However, the real threshold was larger than this value; thus, the DTOCs of DPR for snowflakes in dendrite shape exceeded 1750 m (Ka, use the minimum detectable signal of the whole swath hereafter) and 1250 m (Ku), and the DAOCs were about 60% for two bands at the height of 2 km or 8 km. Moreover, the curves of DAOC variation with detection thresholds at the height of 8 km were different from those at 2 km, showing a steep increase at ~5 dBZ (Figure 14c,d). This is because cloud system 1 was a low snowfall cloud whose reflectivity was always at a low value, so if the detection threshold was higher than ~5 dBZ, the radar would ignore cloud system 1. In addition, the DTOC of cloud system 2 with snowflakes of simple shapes was smaller than that with snowflakes of complex shapes (Figure 14a,b), which further confirmed that the cloud top detected by radar was mainly composed of simple-shaped snowflakes. Typically, the detection error (DTOC and DAOC) for the cloud system with spherical snowflakes was lower than for sector or dendrite snowflakes but higher than for simple particle-shaped snowflakes (short column, thin plate). However, the previous retrieval algorithm considers the real nonspherical particles as spherical particles, which means that it will overestimate the detection ability of solid precipitation with complex shapes and underestimate the detection ability of solid precipitation with simple shapes. Under DPR's detection threshold, the DTOC (DAOC) of complex-shape solid precipitation was 200–400 m (15%), larger than that of spherical solid precipitation.

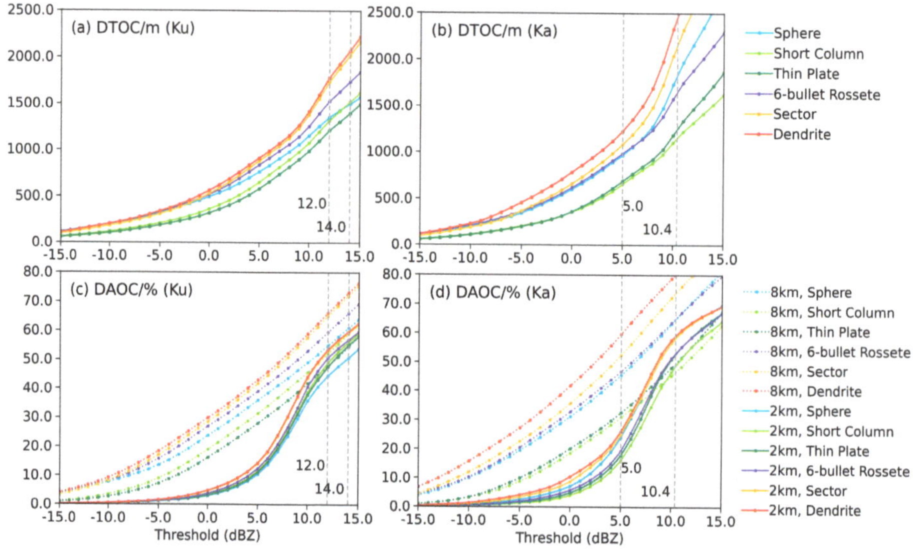

Figure 14. Under different shape assumptions, quantitative relationships between DTOC and DAOC at heights of 2 km/8 km and the radar detection thresholds of Ka-band and Ku-band. (**a**) Relationship between DTOC and the threshold of Ku-band, (**b**) relationship between DTOC and the threshold of Ka-band, (**c**) relationship between DAOC and the threshold of Ka-band, and (**d**) relationship between DAOC and the threshold of Ku-band. The gray auxiliary lines in (**a**,**c**) are thresholds of 12.0 dBZ (DPR) and 14.0 dBZ (FYPR) in Ku-band, and the gray auxiliary lines in (**b**,**d**) are thresholds of 5.0 dBZ (DPR) and 10.4 dBZ (FYPR) in Ka-band, respectively.

It was easily determined that large DTOC values often appeared at the edge of the cloud system (Figure 15a,b,e). By comparing the DTOC of Ka-band and Ku-band for DPR (Figure 15a,b), we found that Ka-band was better than Ku-band at detecting cloud tops, while for designed thresholds of FYPR, the detection capabilities of these two bands were similar. However, it should be mentioned that the channels of DPR (13.6 GHz/35.5 GHz) were slightly different from those of FYPR (13.35/35.55 GHz), and the detection thresholds of DPR in the actual use stage were lower than in the test stage; thus, the actual detection capability of FYPR may have been higher than its design value. Furthermore, Figure 15c–f show that, compared to the dendrite assumption, the spherical-shape assumption often underestimated DTOC (as we found in Figure 14), but it did not uniformly underestimate DTOC horizontally. Specifically, it underestimated DTOC in the main part of the cloud and overestimated DTOC in areas with low water paths. The reflectivity of dendrite snowflakes is lower than that of spherical snowflakes. Therefore, for cloud system 1, nearly all the reflectivity was under the threshold, which made it inconsequential in the DTOC calculation. For cloud system 2, the reflectivity at high altitudes was lower, which made the DTOC larger.

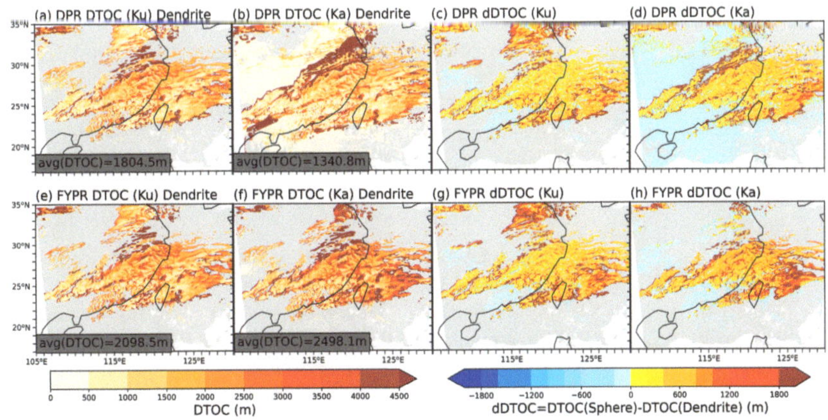

Figure 15. (**a–d**) Horizontal distribution of DTOC under the dendrite assumption and the difference of DTOC calculated under the dendrite and spherical-shape assumptions (dDTOC). (**a**,**b**) DTOC under the dendrite assumption for Ku-band and Ka-band, respectively. (**c**,**d**) dDAOC for Ku-band and Ka-band, respectively. (**e–h**) are the same as (**a–d**) under the design detection thresholds of FYPR. The gray part shows the area where PrecipWP > 0.001 g/m^2, but no hydrometer has a density greater than 1^{-5} g/m^3 in the atmospheric column.

4. Discussion

In addition, we hypothesize that the actual detection threshold of DPR may be higher than 5 dBZ or that the DPR retrieval algorithm ignores small snowfall. In summary, the shape assumption of snowflakes is very important in the radar radiative transfer process simulation. Using an inappropriate shape assumption affects the simulation of the Z–R relationship, thus affecting precipitation retrieval and estimation of detection error.

In this study, to focus on a case study in detail, only one snow case in East China was used. It remains necessary to conduct simulation research on different regions to develop a Z–R relationship that can be applied to the active radar retrieval algorithm, in addition to carefully considering shape assumptions in the vertical stratification.

As suggested by one anonymous reviewer, the mass-weighted diameter, Dm (fourth over third moment of the distribution), and the particle number concentration in particle size distribution (PSD) are very important factors affecting the Ku/Ka radar reflectivity of snow and deserve further in-depth study. It is worth noting that, in this study, we only focused on the impacts of particle shape assumption. The two parameters just mentioned were fixed for different shape assumptions. Further studies on the relative importance of the above three aspects (mass-weighted diameter, particle number concentration, and particle shape assumption) will be valuable for fully understanding the impacts of particle microphysical properties on radar reflectivity.

It is challenging to directly compare a storm instantaneously observed using a low-orbit satellite, such as GPM, to a simulated one, generated by a CRM or MWRT model. Both simulation errors by the CRM and the radiative transfer calculations may lead to significant discrepancies. In terms of this study, the horizontal pattern, size, and location of the simulated storms showed good consistency with the satellite observation (Figure 1). In the vertical cross-section (Figure 5), the GPM-observed convection cores between 120.70E and 122.68E were well simulated by the model using dendrite assumption in terms of location, bright-band, detectable height, etc. Therefore, although there is great uncertainty in the model, this case is valuable for informing future model sensitivity studies.

Furthermore, the dual-frequency ratio (DFR) is an important parameter when considering the mixing ratio, particle size, or mass densities.

5. Conclusions

The combination of WRF and a radiative transfer model such as the USERS is an effective way to test and verify the radiative transfer model and radar retrieval algorithm. This study used a precipitation case in East China at 16:30 UTC on 6 January 2018. This article has discussed the following issues: (1) The effect of shape assumptions (including a temperature-dependent shape assumption) on the performance of simulations. (2) The associated bias in retrieval results when using theoretical Z–R relationships. (3) The detectable errors of precipitation top height, and the rain area. Here are the main conclusions:

1. Compared with the simple-shape assumptions, our complex-shape assumptions (sector and dendrite) performed better in both Ka-band and Ku-band reflectivity simulations. This was shown by the higher correlation coefficients between the simulated and observed reflectivity and smaller differences between their reflectivity profiles. Therefore, snowflakes in the real atmosphere might be closer to sector and dendrite than sphere. The Z–R relationships for these shape assumptions under -40 °C are $Z = 134.59 \cdot R^{1.184}$ (sector) and $Z = 127.35 \cdot R^{1.221}$ (dendrite). However, snowflakes tend to exist in simple shapes when temperature is low and in complex shapes when temperature is high. The temperature-dependent assumption performs well, especially at Ka-band, but the operational method still needs further study.
2. In most conditions, the theoretical Z–R relationships (MP/AU relationships) differed from the fitted Z–R relationships of snowflakes, regardless of their shape. Furthermore, the differences led to estimation errors that stemmed from using a theoretical relationship in the retrieval algorithm. The errors were to underestimate large snowfalls with simple-shaped snowflakes below -40 °C or with complex shapes, and to overestimate snowfalls with spherical snowflakes or small snowfalls with simple-shaped snowflakes below -40 °C.
3. Under the existing detection sensitivity, the DTOCs of DPR for this case were 1804.5 m (Ka) and 1340.8 m (Ku), and the DAOCs reached 50% and 20% at heights of 8 km and 2 km for Ka-band. If the detection threshold of spaceborne dual frequency radar could reach 5 dBZ (Ku)/0 dBZ (Ka), its detection capability for snowfall in eastern China would be greatly improved.
4. An inappropriate shape assumption affected the estimation of detection error: the DTOC of a complex-shape assumption was 200–400 m larger than that of the spherical-shape assumption, while the DAOC was ~15% larger.

Supplementary Materials: The following supporting information can be downloaded at: https://www.mdpi.com/article/10.3390/rs15061556/s1, Figure S1. The same as Figure 3, and the scatters show ground-based observed weather type; Figure S2. Same as Figure 5, but along the gray dashed line on the north side in Figure S1, and the darker gray area show reflectivity between 5 dBZ and 12 dBZ only for Ku-band; Figure S3. Same as Figure 5, but simulations are under the aggregation and the mixed-shape assumptions; Figure S4. The cross sections of Ka-band reflectivity along the red line in Figure 3. (a) GPM DPR observation, (b–l) simulations using eleven shape assumptions separately, as their titles show. R means the correlation coefficient between simulation and observation for this cross section; Figure S5. Same as Figure S4, but at Ku-band; Table S1. Numbers of dBZ (2 km) within six ranges for observations and simulations under eleven shape assumptions.

Author Contributions: Conceptualization, R.L. and L.M.; methodology, R.L.; investigation, L.M. and S.Y.; data curation, L.M.; writing—original draft preparation, L.M. and S.Y.; writing—review and editing, R.L. and Y.W.; visualization, L.M.; supervision, R.L.; project administration, R.L.; funding acquisition, R.L. and Y.W. All authors have read and agreed to the published version of the manuscript.

Funding: This work was supported by the National Natural Science Foundation of China (NSFC) (grant nos. 41830104, 42275139, 41675022, 42075124), the National Key Research and Development Program of China (grant no. 2021YFC3000300), and Fengyun Satellite Application Pilot Plan (2022) and Innovation Center for Fengyun Meteorological Satellite Special Project (grant no. FY-APP-ZX-2022.02XX).

Data Availability Statement: The dataset related to this article is available on request.

Acknowledgments: We gratefully thank the three reviewers for their helpful suggestions and insightful opinions. They were all great help to improve our manuscript.

Conflicts of Interest: The authors declare no conflict of interest.

References

1. Trenberth, K.E.; Fasullo, J.T.; Kiehl, J. Earth's global energy budget. *Bull. Am. Meteorol. Soc.* **2009**, *90*, 311–324. [CrossRef]
2. Stephens, G.L.; Li, J.; Wild, M.; Clayson, C.A.; Loeb, N.; Kato, S.; L'ecuyer, T.; Stackhouse, P.W.; Lebsock, M.; Andrews, T. An update on Earth's energy balance in light of the latest global observations. *Nat. Geosci.* **2012**, *5*, 691–696. [CrossRef]
3. Budyko, M.I. The heat balance of the earth's surface. *Sov. Geogr.* **1961**, *2*, 3–13. [CrossRef]
4. Iguchi, T.; Kozu, T.; Meneghini, R.; Awaka, J.; Okamoto, K.I. Rain-Profiling algorithm for the TRMM precipitation radar. *J. Appl. Meteorol.* **2000**, *39*, 2038–2052. [CrossRef]
5. Iguchi, T.; Kozu, T.; Kwiatkowski, J.; Meneghini, R.; Awaka, J.; Okamoto, K.I. Uncertainties in the rain profiling algorithm for the TRMM precipitation radar. *J. Meteorol. Soc. Japan. Ser. II* **2009**, *87*, 1–30. [CrossRef]
6. Kummerow, C.; Barnes, W.; Kozu, T.; Shiue, J.; Simpson, J. The tropical rainfall measuring mission (TRMM) sensor package. *J. Atmos. Ocean. Technol.* **1998**, *15*, 809–817. [CrossRef]
7. Skofronick-Jackson, G.; Kulie, M.; Milani, L.; Munchak, S.J.; Wood, N.B.; Levizzani, V. Satellite Estimation of Falling Snow: A Global Precipitation Measurement (GPM) Core Observatory Perspective. *J. Appl. Meteorol. Climatol.* **2019**, *58*, 1429–1448. [CrossRef]
8. Liao, L.; Meneghini, R. GPM DPR Retrievals: Algorithm, Evaluation, and Validation. *Remote Sens.* **2022**, *14*, 843. [CrossRef]
9. Neeck, S.P.; Kakar, R.K.; Azarbarzin, A.A.; Hou, A.Y. Global Precipitation Measurement (GPM) launch, commissioning, and early operations. In Proceedings of the Sensors, Systems, and Next-Generation Satellites XVIII, Amsterdam, The Netherlands, 22–25 September 2014; pp. 31–44.
10. Lock, J.A.; Gouesbet, G. Generalized Lorenz–Mie theory and applications. *J. Quant. Spectrosc. Radiat. Transf.* **2009**, *110*, 800–807. [CrossRef]
11. Wriedt, T. Mie theory: A review. In *The Mie Theory*; Springer: Berlin/Heidelberg, Germany, 2012; pp. 53–71.
12. Liu, G. Approximation of single scattering properties of ice and snow particles for high microwave frequencies. *J. Atmos. Sci.* **2004**, *61*, 2441–2456. [CrossRef]
13. Liu, G. A Database of Microwave Single-Scattering Properties for Nonspherical Ice Particles. *Bull. Am. Meteorol. Soc.* **2008**, *89*, 1563–1570. [CrossRef]
14. Yang, J.; Weng, F. Analysis of Microwave scattering properties of non-spherical ice particles using discrete dipole approximation. In Proceedings of the IGARSS 2020-2020 IEEE International Geoscience and Remote Sensing Symposium, Waikoloa, HI, USA, 26 September–2 October 2020; pp. 5309–5312.
15. Xu, X.; Zheng, G.; Wang, Z.; Wang, Q. Experimental and theoretical analyses on the microwave backscattering ability and differential radar reflectivity of nonspherical hydrometeors. *J. Quant. Spectrosc. Radiat. Transf.* **2005**, *92*, 61–72.
16. Kuo, K.-S.; Olson, W.S.; Johnson, B.T.; Grecu, M.; Tian, L.; Clune, T.L.; van Aartsen, B.H.; Heymsfield, A.J.; Liao, L.; Meneghini, R. The microwave radiative properties of falling snow derived from nonspherical ice particle models. Part I: An extensive database of simulated pristine crystals and aggregate particles, and their scattering properties. *J. Appl. Meteorol. Climatol.* **2016**, *55*, 691–708. [CrossRef]
17. Lu, Y.; Jiang, Z.; Aydin, K.; Verlinde, J.; Clothiaux, E.E.; Botta, G. A polarimetric scattering database for nonspherical ice particles at microwave wavelengths. *Atmos. Meas. Tech.* **2016**, *9*, 5119–5134. [CrossRef]
18. Petty, G.W.; Huang, W. Microwave backscatter and extinction by soft ice spheres and complex snow aggregates. *J. Atmos. Sci.* **2010**, *67*, 769–787. [CrossRef]
19. Olson, W.S.; Tian, L.; Grecu, M.; Kuo, K.-S.; Johnson, B.T.; Heymsfield, A.J.; Bansemer, A.; Heymsfield, G.M.; Wang, J.R.; Meneghini, R. The microwave radiative properties of falling snow derived from nonspherical ice particle models. Part II: Initial testing using radar, radiometer and in situ observations. *J. Appl. Meteorol. Climatol.* **2016**, *55*, 709–722. [CrossRef]
20. Johnson, B.; Skofronick-Jackson, G. The influence of non-spherical particles and land surface emissivity on combined radar/radiometer precipitation retrievals. In Proceedings of the EGU General Assembly Conference Abstracts, Vienna, Austria, 19–24 April 2009; p. 6151.
21. Baran, A.J. A review of the light scattering properties of cirrus. *J. Quant. Spectrosc. Radiat. Transf.* **2009**, *110*, 1239–1260.
22. Baran, A.J. From the single-scattering properties of ice crystals to climate prediction: A way forward. *Atmos. Res.* **2012**, *112*, 45–69. [CrossRef]
23. Field, P.; Hogan, R.; Brown, P.; Illingworth, A.; Choularton, T.; Cotton, R. Parametrization of ice-particle size distributions for mid-latitude stratiform cloud. *Q. J. R. Meteorol. Soc. J. Atmos. Sci. Appl. Meteorol. Phys. Oceanogr.* **2005**, *131*, 1997–2017.
24. Heymsfield, A.J.; Platt, C. A parameterization of the particle size spectrum of ice clouds in terms of the ambient temperature and the ice water content. *J. Atmos. Sci.* **1984**, *41*, 846–855. [CrossRef]
25. Connolly, P.; Saunders, C.; Gallagher, M.; Bower, K.; Flynn, M.; Choularton, T.; Whiteway, J.; Lawson, R. Aircraft observations of the influence of electric fields on the aggregation of ice crystals. *Q. J. R. Meteorol. Soc. J. Atmos. Sci. Appl. Meteorol. Phys. Oceanogr.* **2005**, *131*, 1695–1712. [CrossRef]

26. Bailey, M.; Hallett, J. Growth rates and habits of ice crystals between −20° and−70 °C. *J. Atmos. Sci.* **2004**, *61*, 514–544. [CrossRef]
27. Saha, S.; Niranjan Kumar, K.; Sharma, S.; Kumar, P.; Joshi, V. Can Quasi-Periodic Gravity Waves Influence the Shape of Ice Crystals in Cirrus Clouds? *Geophys. Res. Lett.* **2020**, *47*, e2020GL087909. [CrossRef]
28. Heymsfield, A.J.; Miloshevich, L.M. Parameterizations for the cross-sectional area and extinction of cirrus and stratiform ice cloud particles. *J. Atmos. Sci.* **2003**, *60*, 936–956. [CrossRef]
29. Gallagher, M.W.; Connolly, P.; Whiteway, J.; Figueras-Nieto, D.; Flynn, M.; Choularton, T.; Bower, K.; Cook, C.; Busen, R.; Hacker, J. An overview of the microphysical structure of cirrus clouds observed during EMERALD-1. *Q. J. R. Meteorol. Soc. J. Atmos. Sci. Appl. Meteorol. Phys. Oceanogr.* **2005**, *131*, 1143–1169. [CrossRef]
30. Kintea, D.M.; Hauk, T.; Roisman, I.V.; Tropea, C. Shape evolution of a melting nonspherical particle. *Phys. Rev. E* **2015**, *92*, 033012. [CrossRef]
31. Pan, D.; Liu, L.-M.; Slater, B.; Michaelides, A.; Wang, E. Melting the ice: On the relation between melting temperature and size for nanoscale ice crystals. *ACS Nano* **2011**, *5*, 4562–4569. [CrossRef]
32. Johnson, B.; Olson, W.; Skofronick-Jackson, G. The microwave properties of simulated melting precipitation particles: Sensitivity to initial melting. *Atmos. Meas. Tech.* **2016**, *9*, 9–21. [CrossRef]
33. Kulie, M.S.; Bennartz, R.; Greenwald, T.J.; Chen, Y.; Weng, F. Uncertainties in microwave properties of frozen precipitation: Implications for remote sensing and data assimilation. *J. Atmos. Sci.* **2010**, *67*, 3471–3487. [CrossRef]
34. Hong, G. Parameterization of scattering and absorption properties of nonspherical ice crystals at microwave frequencies. *J. Geophys. Res. Atmos.* **2007**, *112*. [CrossRef]
35. Leinonen, J.; Kneifel, S.; Moisseev, D.; Tyynelä, J.; Tanelli, S.; Nousiainen, T. Evidence of nonspheroidal behavior in millimeter-wavelength radar observations of snowfall. *J. Geophys. Res. Atmos.* **2012**, *117*. [CrossRef]
36. Kulie, M.S.; Hiley, M.J.; Bennartz, R.; Kneifel, S.; Tanelli, S. Triple-frequency radar reflectivity signatures of snow: Observations and comparisons with theoretical ice particle scattering models. *J. Appl. Meteorol. Climatol.* **2014**, *53*, 1080–1098. [CrossRef]
37. Wang, Y.; Han, T.; Guo, J.; Jiang, K.; Li, R.; Shao, W.; Liu, G. Simulation of three-dimensional structure of cloud and rain detected by Ku, Ka and W radar. *Sci. Bull.* **2019**, *64*, 430–443.
38. Draine, B.T.; Flatau, P.J. User Guide for the Discrete Dipole Approximation Code DDSCAT, Version 5a10. Available online: https://arxiv.org/abs/astro-ph/0008151 (accessed on 7 March 2023).
39. Tokay, A.; Short, D.A. Evidence from tropical raindrop spectra of the origin of rain from stratiform versus convective clouds. *J. Appl. Meteorol. Climatol.* **1996**, *35*, 355–371. [CrossRef]
40. Casella, D.; Panegrossi, G.; Sanò, P.; Marra, A.C.; Dietrich, S.; Johnson, B.T.; Kulie, M.S. Evaluation of the GPM-DPR snowfall detection capability: Comparison with CloudSat-CPR. *Atmos. Res.* **2017**, *197*, 64–75. [CrossRef]
41. Heymsfield, A.; Bansemer, A.; Wood, N.B.; Liu, G.; Tanelli, S.; Sy, O.O.; Poellot, M.; Liu, C. Toward improving ice water content and snow-rate retrievals from radars. Part II: Results from three wavelength radar–collocated in situ measurements and CloudSat–GPM–TRMM radar data. *J. Appl. Meteorol. Climatol.* **2018**, *57*, 365–389. [CrossRef]
42. Mroz, K.; Montopoli, M.; Battaglia, A.; Panegrossi, G.; Kirstetter, P.; Baldini, L. Cross validation of active and passive microwave snowfall products over the continental United States. *J. Hydrometeorol.* **2021**, *22*, 1297–1315. [CrossRef]
43. Zhou, R.; Yan, T.; Yang, S.; Fu, Y.; Huang, C.; Zhu, H.; Li, R. Characteristics of clouds, precipitation, and latent heat in midlatitude frontal system mixed with dust storm from GPM satellite observations and WRF simulations. *JUSTC* **2022**, *52*, 3-1–3-15. [CrossRef]
44. Iguchi, T.; Meneghini, R. Intercomparison of single-frequency methods for retrieving a vertical rain profile from airborne or spaceborne radar data. *J. Atmos. Ocean. Technol.* **1994**, *11*, 1507–1516. [CrossRef]
45. Marshall, J.; Palmer, W.M. Shorter Contributions. *J. Meteorol.* **1948**, *5*, 166.
46. Ulbrich, C.W.; Atlas, D. A method for measuring precipitation parameters using radar reflectivity and optical extinction. *Ann. Des. Télécommunications* **1977**, 415–421. [CrossRef]
47. Uijlenhoet, R. Raindrop size distributions and radar reflectivity–rain rate relationships for radar hydrology. *Hydrol. Earth Syst. Sci.* **2001**, *5*, 615–628. [CrossRef]

remote sensing

Technical Note

Comprehensive Evaluation of Near-Real-Time Satellite-Based Precipitation: PDIR-Now over Saudi Arabia

Raied Saad Alharbi [1,*], Vu Dao [2], Claudia Jimenez Arellano [2] and Phu Nguyen [2]

[1] Department of Civil Engineering, College of Engineering, King Saud University, P.O. Box 800, Riyadh 11421, Saudi Arabia
[2] Center for Hydrometeorology and Remote Sensing (CHRS), Department of Civil and Environmental Engineering, University of California, Irvine, CA 90095, USA; vndao1@uci.edu (V.D.); claudij@uci.edu (C.J.A.); ndphu@uci.edu (P.N.)
* Correspondence: rsalharbi@ksu.edu.sa

Abstract: In the past decade, Saudi Arabia has witnessed a surge in flash floods, resulting in significant losses of lives and property. This raises a need for accurate near-real-time precipitation estimates. Satellite products offer precipitation data with high spatial and temporal resolutions. Among these, the Precipitation Estimation from Remotely Sensed Information using Artificial Neural Networks–Dynamic Infrared Rain Rate near-real-time (PDIR-Now) stands out as a novel, global, and long-term resource. In this study, a rigorous comparative analysis was conducted from 2017 to 2022, contrasting PDIR-Now with rain gauge data. This analysis employs six metrics to assess the accuracy of PDIR-Now across various daily rainfall rates and four yearly extreme precipitation indices. The findings reveal that PDIR-Now slightly underestimates light precipitation but significantly underestimates heavy precipitation. Challenges arise in regions characterized by orographic rainfall patterns in the southwestern area of Saudi Arabia, emphasizing the importance of spatial resolution and topographical considerations. While PDIR-Now successfully captures annual maximum 1-day and 5-day precipitation measurements across rain gauge locations, it exhibits limitations in the length of wet and dry spells. This research highlights the potential of PDIR-Now as a valuable tool for precipitation estimation, offering valuable insights for hydrological, climatological, and water resource management studies.

Keywords: PDIR-Now; satellite precipitation products (SPPs); extreme precipitation; Saudi Arabia

Citation: Alharbi, R.S.; Dao, V.; Jimenez Arellano, C.; Nguyen, P. Comprehensive Evaluation of Near-Real-Time Satellite-Based Precipitation: PDIR-Now over Saudi Arabia. *Remote Sens.* **2024**, *16*, 703. https://doi.org/10.3390/rs16040703

Academic Editors: Haobo Li, Suelynn Choy, Yuriy Kuleshov, Mayra I. Oyola-Merced and Xiaoming Wang

Received: 13 January 2024
Revised: 14 February 2024
Accepted: 15 February 2024
Published: 17 February 2024

1. Introduction

In the last decade, the Kingdom of Saudi Arabia has faced a high number of flash floods, which resulted in many losses of lives and property. Specifically, more than 13,000 people were impacted by these floods between 1993 and 2013, which occurred approximately nine times throughout that time period, causing an economic loss of over 20 million USD [1]. Due to the increasing frequency of these flood events, it is crucial to characterize and forecast them for the purposes of flood control and warning [2–4]. Furthermore, Saudi Arabia also suffers from severe drought, which can have severe impacts on the water resources of the region [5]. Thus, it is of extreme importance to have access to accurate precipitation estimates.

Even though there are many factors that impact flooding, precipitation is one of the key influencing factors, and the accuracy of the precipitation estimates has a substantial impact on flood predictions and mitigation strategies [4]. Rain gauges effectively and reliably measure precipitation, but a sparse and uneven gauge network's lack of representativeness is a major concern [6–8]. Moreover, a low-density gauge network also makes it difficult to provide accurate rainfall measurements over a large area [9,10].

Over the past three decades, numerous satellite precipitation products (SPPs) have been developed as alternative sources of precipitation estimates [11–13]. These SPPs are able to provide information about the spatial variability of precipitation, unlike rainfall gauges, and can be available at fine temporal and spatial resolutions. Hydrological and climatological studies benefit substantially from SPPs as they are capable of providing rainfall estimates over areas with few to none monitoring stations [13–15]. Numerous SPPs with high spatial and temporal resolutions have been produced and are available for free online [8,15]. Some SPPs include: the Tropical Rainfall Measuring Mission (TRMM) [16] by the National Aeronautics and Space Administration (NASA), the National Ocenanic and Atmospheric Administration's (NOAA) Climate Prediction Center Morphing Technique (CMORPH) [17], the Climate Hazards Group InfraRed Precipitation with Station data (CHIRPS) [18] by the Climate Hazards Center at the University of California (UC) Santa Barbara, Precipitation Estimation from Remotely Sensed Information using Artificial Neural Networks (PERSIANN) [19], PERSIANN-Cloud Classification System (PERSIANN-CCS) [20], and PERSIANN- Climate Data Record (PERSIANN-CDR) [21] all by the Center for Hydrometeorology and Remote Sensing (CHRS) at UC Irvine.

Due to its precise spatial and temporal properties, SPPs have been extensively used to analyze the life cycle of extreme precipitation events [22]. However, due to retrieval procedures, indirect measurements, and numerical model parameters, SPPs have errors and biases [23–25]. As a result, it is vital to analyze and compare the performance of various SPPs before their implementation [26].

The accuracy of SPPs can also be influenced by the complex interplay between topography and precipitation. Mountains and hills can significantly influence precipitation patterns by directing prevailing winds over their slopes, leading to varied precipitation levels in different regions. In the context of southwestern Saudi Arabia, the study by Al-Ahmadi titled "Spatiotemporal variations in rainfall–topographic relationships in southwestern Saudi Arabia" sheds light on the localized dynamics [27]. The research emphasizes the importance of local topographic factors, including topography, altitude, slope, proximity to ridge, and proximity to the Red Sea, in shaping annual and seasonal rainfall. Specifically, the Asir Mountains emerge as key players, with higher altitudes, more mountainous areas, steeper slopes, and areas closer to ridges correlating with increased rainfall. Notably, the concentration of gauge stations along a coastal strip emphasizes the need for more examination of topography's influence on precipitation variations. Despite advancements in precipitation estimation algorithms, the complex terrain in southwestern Saudi Arabia poses challenges in obtaining reliable quantitative precipitation estimation (QPE), given the rapid changes in precipitation characteristics due to orographic enhancement. As we navigate these complexities, continued research and methodological refinements are crucial for a comprehensive understanding of the relationship between topography and precipitation in such complex terrains.

Among the various SPPs available, PDIR-Now [28] is one of the most recently used satellite-based gridded datasets in hydrological models [11,29,30]. It is one of the latest PERSIANN products developed by the Center for Hydrometeorology and Remote Sensing (CHRS) at the University of California, Irvine (UCI). PDIR-Now is a quasi-global near-real-time precipitation product that provides a long record of precipitation estimates, spanning from the year 2000 to the present at a 0.04° × 0.04° spatial resolution. Precipitation is better detected by PDIR-Now than by other near-real-time products by the CHRS because it links brightness temperature and rain rate to remedy errors using climatological data. PDIR-Now's high-frequency infrared image sampling allows it to deliver accurate rainfall estimates quickly after precipitation begins. This algorithm is based on the framework of the PERSIANN-CCS algorithm but uses dynamical shifting of the cloud-top temperature and rainfall rate curves using climatology data [28]. Furthermore, the precipitation threshold was lowered from 273 K to 263 K to improve the detection of warm precipitation.

There has been a limited number of studies analyzing SPPs over Saudi Arabia. Kheimi and Gutub [31] evaluated TRMM 3B42, CMORPH, Global Satellite Mapping of Precipitation Microwave-IR Combined Product (GSMap-MVK), and PERSIANN against rain gauges. The correlation coefficients (CC) of TRMM 3B42, CMORPH, GSMap-MVK, and PER-

SIANN were 0.44, 0.44, 0.45, and 0.11, respectively. Furthermore, the mean errors (ME) of TRMM 3B42, CMORPH, GSMap-MVK, and PERSIANN were 0.26 mm/day, 0.62 mm/day, 0.65 mm/day, and 0.30 mm/day. Their probability of detection (POD) was 0.39, 0.52, 0.53, and 0.24 for TRMM 3B42, CMORPH, GSMap-MVK, and PERSIANN, respectively. This study concluded that even though all products could detect precipitation accurately, they overestimated the amount of rainfall over the study area. Mahmoud et al. [32] assessed the capability of IMERG Early, IMERG Late, and IMERG Final to capture precipitation over Saudi Arabia using gauge data as the ground truth. The main finding from this study is that all products exhibit increased accuracy, with the exception of some regions in the northern part of the study area. Furthermore, they found that the estimates improved from the Early run to the Final run. Sultana and Nasrollahi [33] evaluated the performance of PERSIANN, PERSIANN-CDR, TRMM-RT, TRMM-3B42, and CMORPH over Saudi Arabia using gauge data as the ground truth. The main conclusion in the study was that PERSIANN-CDR, TRMM-3B42, and CMORPH performed best over the study area. Specifically, PERSIANN-CDR, TRMM-3B42, and CMORPH had PODs of 0.322, 0.424, and 0.654 and CCs of 0.171, 0.42, and 0.471. Helmi and Abdelhamed [34] evaluated CMORPH, PERSIANN-CDR, CHIRPS V2.0, TRMM 3B42 V7, and IMERG V6 against rain gauge data from 2000 to 2012. This study found that all these products performed best at an altitude of 500–700 m in the central and northern parts of the country and that CMORPH performed best in their monthly assessment.

Due to the critical need for accurate near-real-time estimates in flood-prone regions and the product's capacity to accurately depict the spatial variation of precipitation, evaluating this product is crucial. Furthermore, this product has never been evaluated over Saudi Arabia. Thus, the first objective in this study is to evaluate the accuracy of PDIR-Now using Saudi Arabian daily gauge precipitation data. The second objective is to assess the capability of the SPP to identify extreme precipitation events. The results of this study can provide an in-depth understanding of the capabilities of this near-real-time product and can lead to more informed water management decisions.

2. Materials and Methods

2.1. Study Area

Saudi Arabia is located between 15°N and 35°N and 35°E and 57°E, as shown in Figure 1. According to the Köppen climate classification system [35,36], the majority of Saudi Arabia is a hot and arid desert; however, the southwest region of Saudi is considered semi-arid [37,38]. Precipitation is scant and infrequent in most regions of Saudi Arabia, with the wet season occurring from October to April [37,39]. During dry months, hardly any precipitation occurs, with the exception of the southwest area of the country [39,40].

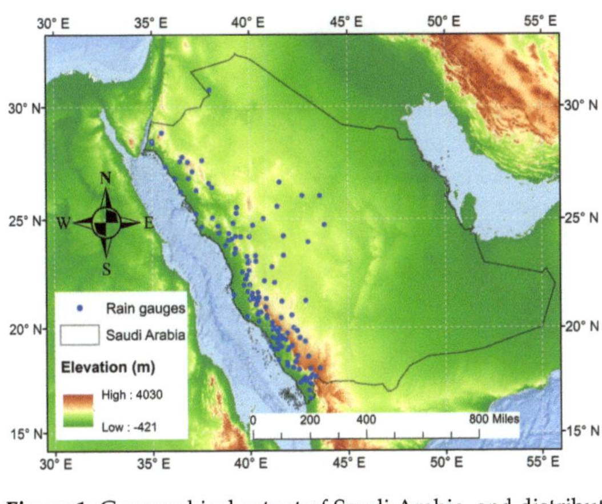

Figure 1. Geographical extent of Saudi Arabia, and distribution of rainfall gauges over the region of study.

2.2. Datasets

2.2.1. Rain Gauges

In this study, the daily rain gauge dataset obtained from the Ministry of Environment, Water, and Agriculture (MEWA) for 130 meteorological observatories across Saudi Arabia (Figure 1) from 2017 to 2022 is used. To ensure the robustness of our analysis, a data quality control process was conducted. This involved an inspection of the dataset to identify and address missing or Not a Number (NaN) values, ensuring consistency throughout the entire study period.

2.2.2. PERSIANN Dynamic Infrared (PDIR-Now)

PDIR-Now is a near-real-time global high-resolution SPP developed by the University of California, Irvine's Center for Hydrometeorology and Remote Sensing (CHRS). The PDIR-Now algorithm considers several factors beyond brightness temperature to estimate precipitation intensity. It utilizes a catalog of cloud types that is created by training self-organizing feature maps (SOFMs) with passive microwave (PMW) data. Then, the rain rate estimation is based on the brightness temperature, size, and texture features of the cloud patches at temperatures below 263 K, utilizing the IMERG PMW precipitation dataset for model training. Furthermore, this algorithm involves the dynamical shifting of cloud-top brightness temperatures–rain rate (Tb-R) curves using rainfall climatology data. The spatiotemporal resolution of 0.04° on a monthly basis is employed to adjust the position of the curves, aiming to correct biases and produce more accurate precipitation estimates. The adjustment is intended to account for regional variations in wetness, generating more precipitation in areas with a wetter climatology and vice versa. Compared to other SPPs, the characteristics of the PDIR-Now algorithm include being a real-time, high-resolution precipitation product with a short delay time (15 min–1 h) [28]. From 2000 forward, PDIR-Now offers quasi-global coverage (60°S–60°N) and high spatiotemporal resolution (almost 0.04° and hourly data) precipitation data. The CHRS database of daily PDIR-Now readings was acquired from the CHRS data portal (http://chrsdata.eng.uci.edu/ (accessed on 1 December 2023)).

2.3. Methods

Daily satellite precipitation data for each rain gauge were extracted from the nearest grid point in the satellite-based precipitation products, matching the locations of the rain gauges used in this study. The evaluation in this study is divided into two categories: rainfall intensity and extreme rainfall assessments. Statistical metrics were calculated for PDIR-Now using the gauge data as the ground truth. This method is widely adopted for assessing the accuracy of SPPs.

2.3.1. Evaluation Metrics

First, six widely used statistical metrics, shown in Table 1, were applied to assess the accuracy of PDIR-Now across various rainfall rates when compared to the rain gauge observations, with the goal of better understanding the PDIR-Now performance in terms of precipitation amount and occurrence. The correlation coefficient (CC), mean bias (MB), and the root-mean-squared error (RMSE) were used to evaluate the PDIR-Now performance in estimating the amount of rainfall, whereas the probability of detection (POD), the false alarm ratio (FAR), and the critical success index (CSI) were used to evaluate the performance of PDIR-Now in rainfall detection.

Table 1. Statistical metrics employed to gauge the performance of satellite-based precipitation products.

Index	Equation	Optimal Value
CC	$CC = \frac{\sum_{i=1}^{n}(E_i - \bar{E})(O_i - \bar{O})}{\sqrt{\sum_{i=1}^{n}(E_i - \bar{E})^2}\sqrt{\sum_{i=1}^{n}(O_i - \bar{O})^2}}$	1
MB	$MB = \sum_{i=1}^{n}(E_i - O_i)$	0
RMSE	$RMSE = \sqrt{\frac{1}{n}\sum_{i=1}^{n}(E_i - O_i)^2}$	0
POD	$POD = \frac{TP}{TP + FN}$	1
FAR	$FAR = \frac{FP}{TP + FP}$	0
CSI	$CSI = \frac{TP}{TP + FP + FN}$	1

Where O_i represents the reference rain gauge data; $\bar{O}$ represents the mean of the reference; E_i represents the PDIR-Now estimation; $\bar{E}$ represents the mean of the estimation; and n refers to the number of samples. In the case of the last three indices, *TP* represents the number of precipitation events detected within the rain gauges and PDIR-Now synchronously; *FN* represents the number of precipitation events observed by rain gauges but not PDIR-Now; *FP* refers to the number of precipitation events detected by PDIR-Now that were not observed by rain gauges. The threshold for defining precipitation occurrence was 0.5 mm/day, as specified by MEWA.

CC quantifies the linear relationship strength between PDIR-Now and the rain gauge variables. MB and RMSE highlight discrepancies between PDIR-Now and the rain gauge measurements, with lower values indicating smaller differences. Additionally, POD assesses PDIR-Now's reliability in detecting precipitation events, FAR measures its tendency to identify unobserved precipitation, and CSI evaluates its proficiency in recognizing precipitation events relative to the rain gauge data. To assess PDIR-Now's capability to detect precipitation, especially intense precipitation, various rain thresholds were used. Table 2 presents the defined rainfall intensity classes within the scope of the study, along with their corresponding daily rainfall thresholds.

Given that the study area encompasses arid and semi-arid regions where rainfall is often minimal or absent, we first calculate the metrics for all precipitation throughout the study period and then categorize rainfall data into four classes, including "No Rain", "Light Rain", "Moderate Rain", and "Heavy Rain". The subjective selection of thresholds is carefully considered after analyzing the distribution of the daily data and reviewing different thresholds and classification standards in similar regions. The "No Rain" category covers instances of no rain and instances where the rainfall rate is equal to or less than 0.5 mm/day, implying negligible rainfall. The "Light Rain" category covers instances where the rainfall rate is between 0.5 and 2 mm/day, implying relatively minimal rainfall. The third category, "Moderate Rain", covers a range of rainfall rates between 2 and 10 mm/day, showing a moderate level of precipitation, and the last category, "Heavy Rain", is determined by rainfall rates exceeding 10 mm/day, representing extreme and impactful rainfall events.

Table 2. Rainfall intensity classes defined in the study with their respective daily rainfall thresholds in mm/day.

Index	Equation
No Rain	Rainfall Rate $\leq$ 0.5
Light Rain	0.5 $<$ Rainfall Rate $\leq$ 2
Moderate Rain	2 $<$ Rainfall Rate $\leq$10
Heavy Rain	Rainfall Rate $>$ 10

2.3.2. Extreme Precipitation Analysis

The second part of the study consisted of calculating four common standard extreme precipitation indices to assess PDIR-Now's performance in capturing extreme precipitation events using gauge data as the baseline. These indices, shown in Table 3, were defined by the World Climate Research Programme (WRCP) and have been used all around the globe [41–43]. The four indices were initially calculated for each year within the 6-year period at each rain gauge, then at each station, the mean of the annual extreme index values was computed. Furthermore, the 130 rain gauge stations were classified into three altitude ranges, specifically: <500 m, 500 m–1000 m, and >1000 m. This was implemented to assess the performance of PDIR-Now in capturing extremes at different elevations.

Table 3. The four extreme indices selected for this study with their respective definitions and units.

Index	Descriptive Name	Definition	Units
RX1day	Maximum 1-day precipitation	Annual maximum 1-day rainfall	mm
RX5day	Maximum 5-day precipitation	Annual maximum consecutive 5-day rainfall	mm
CWD	Consecutive wet days	Annual maximum consecutive rainy days	days
CDD	Consecutive dry days	Annual maximum consecutive dry days	days

3. Results

3.1. Analysis of Rainfall Estimation Errors

The evaluation of PDIR-Now against the gauge data was performed at a daily scale, involving the calculation of several metrics, including the CC, MB, RMSE, POD, FAR, and CSI. To assess the results, the mean of each of the metrics over the study area was calculated for each rainfall category, and the spatial distribution of the metrics was visually assessed. The results reveal a CC of 0.33, a MB of 0.07 mm/day, and a RMSE of 3.04 mm/day (Figure 2). Figure 3 shows the CC, MB, and RMSE results for the three rainfall categories, "Light Rain", "Moderate Rain", and "Heavy Rain". PDIR-Now exhibits the largest mean absolute CC in the "Heavy Rain" category, 0.29, whereas the "Moderate Rain" and "Light Rain" categories had lower mean absolute CCs, 0.17 and 0.16, respectively. This is as expected given that smaller precipitation rates are more susceptible to errors. The lower CC values during moderate rainfall events could be linked to the orographic effects of the prominent mountainous region in Saudi Arabia [27,44]. In arid regions like Saudi Arabia, the interaction between topography and precipitation is a crucial factor influencing the spatial distribution of rainfall [39]. Mountains can act as barriers to moist air masses, leading to orographic lifting on the windward side and subsequently enhancing rainfall in those areas. Conversely, on the leeward side of the mountains, a rain shadow effect may occur, resulting in reduced rainfall. PDIR-Now exhibits the lowest MB in the "Light Rain" category across all areas evaluated, with a mean of 1.14 mm/day. As expected, the MB increases as the rainfall threshold increases, thus, the "Moderate Rain" and "Heavy Rain" categories show larger, but negative, MB values with a mean of −1.76 mm/day and −14.55 mm/day, respectively. The negative sign of the MB results represents that PDIR-Now underestimates precipitation in the two heavier precipitation categories. Similarly, the RMSE values in the "Light Rain" category are the lowest compared to the heavier precipitation categories, with a mean value of 4.45 mm/day. Just as in the case of the MB, the RMSE increases as the precipitation threshold increases, leading the "Moderate Rain" and "Heavy Rain" categories to have increasingly higher values, with means of 6.65 mm/day and 21.88 mm/day, respectively. In the case of the "Light Rain" and "Moderate Rain" categories, the southwest region of the study area along the coast shows higher RMSE values, once again, due to the orographic effect over this mountainous area. Finally, the "Heavy Rain" category does not show one specific area with higher or lower RMSE values than the rest, and instead, these values vary greatly from one gauge to another.

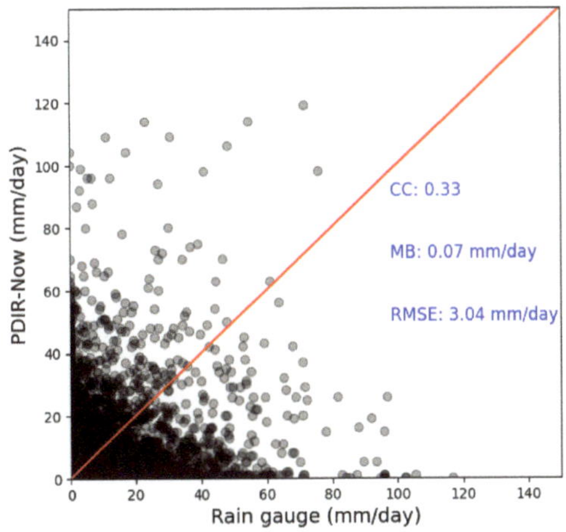

Figure 2. Scatter plot of daily PDIR-Now results compared to the rain gauge observation with statistical indices results. The red line represents the best fit.

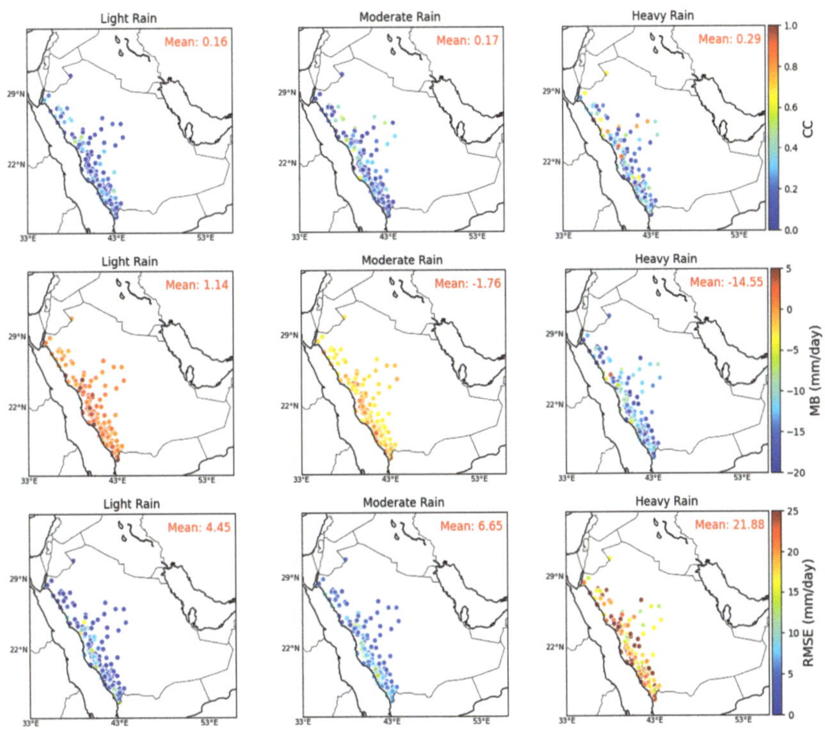

Figure 3. Spatial distribution of CC (upper row), MB (middle row), and RMSE (lower row) of daily PDIR-Now against rain gauge observations for different rain categories with respective mean values.

3.2. Performance Indicator Based on Events

After assessing the results of the rainfall estimation errors, it is necessary to assess the detection capability of PDIR-Now at different rainfall rate thresholds. The result of the rainfall detection in Figure 4 shows that PDIR-Now has a POD of 0.73, a FAR of 0.80, and a CSI of 0.18. While rain gauges along the coast do not achieve high detection performance compared to inland ones, it is evident that the southwestern coast exhibits lower false alarms (first and second columns). This pattern is reflected in the detection accuracy, as seen in the CSI plot. It is important to note that a threshold of 0.5 mm/day is applied to differentiate between rain and no rain instances for this calculation. More detailed results for rainfall categories such as "Light Rain", "Moderate Rain", and "Heavy Rain" are discussed below.

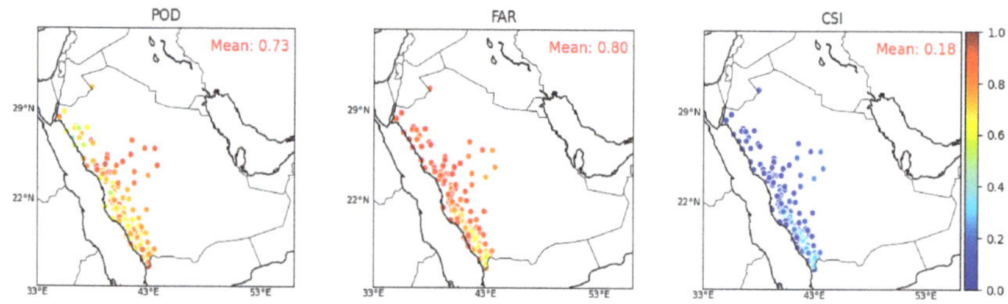

Figure 4. Spatial distribution of POD, FAR, and CSI of daily PDIR-Now against rain gauge observations in rainfall detection.

Figure 5 offers a nuanced insight into the capacity of the PDIR-Now dataset to accurately capture rain events across different rainfall rate categories. The left column of this figure exhibits the great capability of PDIR-Now to capture precipitation in the "Light Rain" category, with mean POD, FAR, and CSI values of 0.47, 0.94, and 0.05, respectively. PDIR-Now performs similarly in all regions of the study area for this category. The POD of PDIR-Now is lower, but FAR decreases with higher rainfall rates, as shown by the middle and right columns of Figure 5. In the case of the "Moderate Rain" category, the POD is substantially lower than in the "Light Rain" category, with a mean value of 0.20. The mean FAR and CSI values obtained were 0.85 and 0.09, respectively. These values indicate that PDIR-Now's accuracy in detecting moderate precipitation events is higher than in the case of light precipitation detection. Finally, PDIR-Now's performance slightly increases when detecting heavy precipitation events, as shown in the right column of Figure 5. Despite showing a lower mean POD than in the "Moderate Rain" category, 0.18 specifically, the FAR decreased, and the CSI increased, with values of 0.75 and 0.11, respectively. This indicates that PDIR-Now's detection of events is slightly lower but more accurate for heavy precipitation events than in the "Moderate Rain" and "Light Rain" categories. The middle and right columns of Figure 5 also reveal that PDIR-Now is more accurate in the middle region of Saudi Arabia than along the coast, as was also depicted by the RMSE previously.

An observation made from the detection results of the rain/ no rain analysis is that the detection is more accurate along the southwestern coast, whereas this is not the case for the different rainfall categories. The discrepancy can be attributed to the more rigorous thresholds set for the categories (Table 2). In this analysis, PDIR-Now is required to meet the criteria for "Light Rain", "Moderate Rain", and "Heavy Rain" based on gauge readings, as opposed to a simpler rain/no rain detection.

3.3. *Analysis of Rainfall Extremes and Climatic Patterns*

The last part of the study was to assess the capability of PDIR-Now at capturing extreme precipitation events. Figure 6 provides a comprehensive overview of several critical climatic and hydrological parameters across all the meteorological stations employed in this study. The top panel shows the results for the mean of the annual maximum daily precipitation amount (RX1day) in mm. The results depict that the RX1day index is similarly captured by PDIR-Now for most of the rainfall gauges, with some minor underestimation. This similitude accompanied by underestimation can be confirmed by the leftmost panel of Figure 7, which shows a high mean CC of 0.50 but a mean bias of −2.83 mm. The mean of the annual maximum five consecutive day precipitation amount, or RX5day index, exhibits similar results, indicating a slight underestimation by PDIR-Now. In regards to the RX5day, the CC and RMSE were 0.53 and −3.10 mm, respectively. The results displayed on the top panels of Figure 6 and the leftmost panels of Figure 7 indicate that the precipitation distribution of PDIR-Now across gauge locations is in agreement with the precipitation amount captured by the rainfall gauge network.

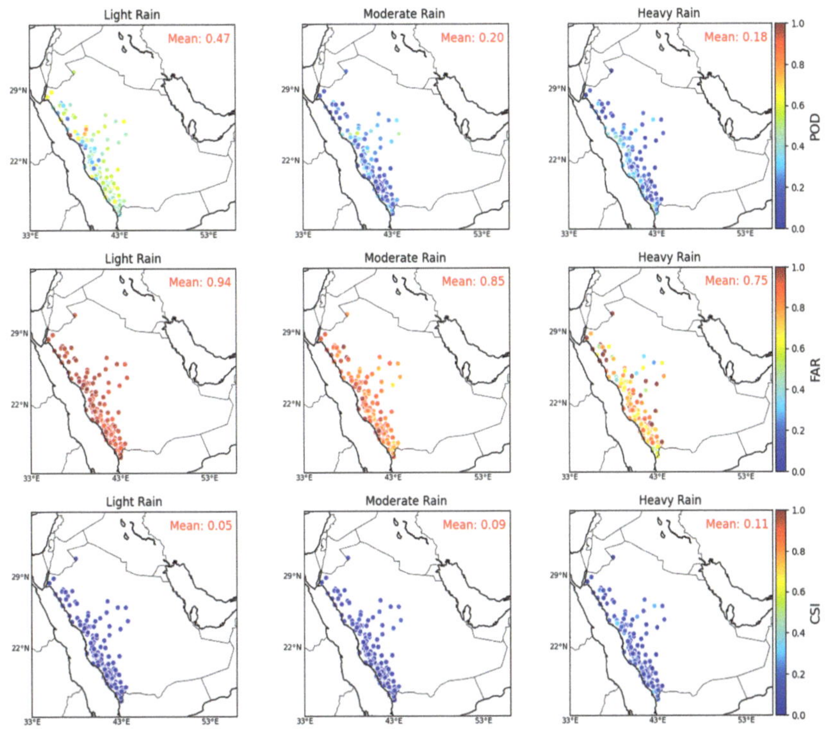

Figure 5. Spatial distribution of POD (upper row), FAR (middle row), and CSI (lower row) of daily PDIR-Now against rain gauge observations for different rain categories with respective mean values.

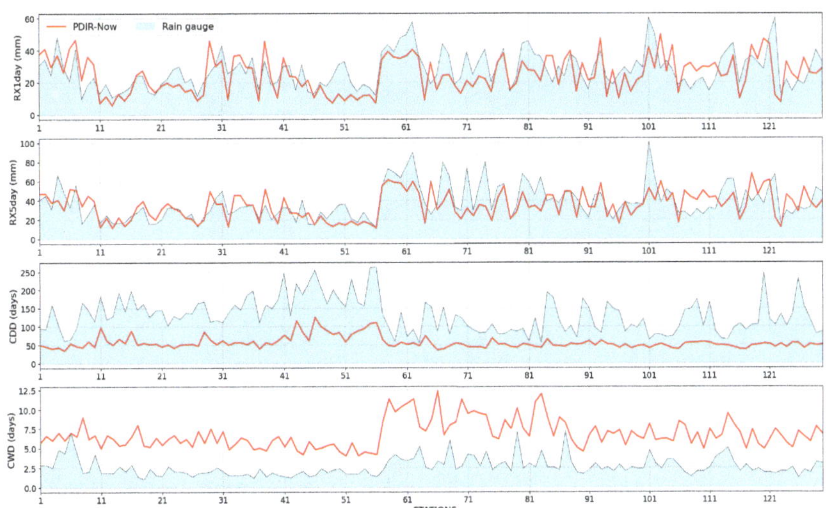

Figure 6. Extreme precipitation indices (RX1day, RX5day, CDD, and CWD) of PDIR-Now and rain gauge data at observed gauge stations.

In regards to the final two indices, PDIR-Now exhibits an underestimation of the CDD and an overestimation of CWD. This indicates that PDIR-Now captures longer wet spells and shorter dry spells than the rainfall gauges do. Figure 7 supports this by showing a large negative bias in the CDD, −74.25 days, and a positive mean bias in the CWD of 4.22 days. The difference in the values of the bias in the CDD and CWD corresponds to the fact that wet spells are usually a lot shorter than dry spells, especially in arid and semi-arid regions such as Saudi Arabia.

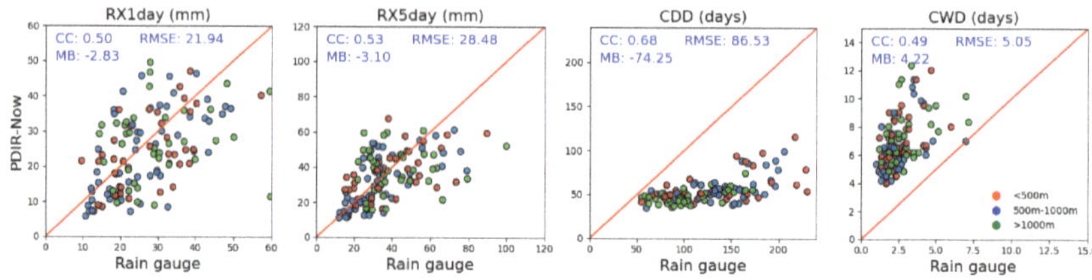

Figure 7. Scatter plots of PDIR-Now and rain gauge data for RX1day, RX5day, CDD, and CWD indices at different elevation thresholds with statistics of mean CC, MB, and RMSE.

Figure 8 illustrates the spatial pattern of extreme indices derived from both gauge data and PDIR-Now. The figure reveals that PDIR-Now demonstrates a similar spatial distribution in the RX1day and RX5day indices compared to the gauge data. However, the accurate representation of the spatial distribution of dry and wet spell durations is not achieved by PDIR-Now, as evident in the four rightmost panels of Figure 8. This observation aligns with expectations, given the superior performance demonstrated by PDIR-Now in the RX1day and RX5day indices, as depicted in Figure 6.

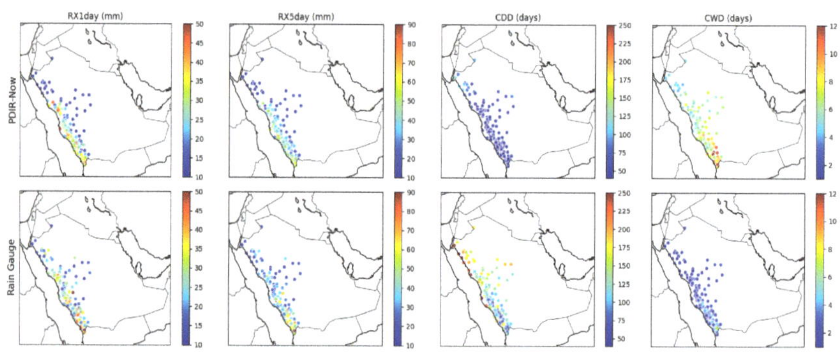

Figure 8. Spatial distribution of extreme precipitation indices (RX1day, RX5day, CDD, and CWD) of PDIR-Now and rain gauge data at observed gauge stations.

In addition, 130 rain gauge stations were classified into different altitude ranges: <500 m, 500–1000 m, and >1000 m. This categorization is employed to assess the performance of PDIR-Now across RX1day, RX5day, CWD, and CDD indices, considering various topographies. Stations below 500 m are represented by the color red, those between 500 m and 1000 m by blue, and those above 1000 m by green in Figure 7. A comprehensive statistical evaluation of PDIR-Now in comparison to the rain gauge data for RX1day, RX5day, CDD, and CWD indices across various topographies is presented in Table 4. The higher CC and lower RMSE and MB observed for stations with elevations <500 m and between 500 m to 1000 m confirm that PDIR-Now exhibits greater accuracy at elevations below 1000 m. As the altitude increases and the orographic effect of the southwestern mountain range comes into play, PDIR-Now exhibits poor performance above 1000 m, with a low CC of 0.18 for RX1day and 0.17 for RX5day, along with higher RMSE and MB compared to the stations below 1000 m.

Table 4. Performance of PDIR-Now compared to the rain gauge for RX1day, RX5day, CDD, and CWD indices across different elevation thresholds.

	RX1day (mm)	RX5day (mm)	CDD (Days)	CWD (Days)
Elevation < 500 m				
CC	0.60	0.60	0.58	0.57
RMSE	20.05	28.76	85.91	5.08
MB	−2.89	−3.98	−75.53	4.30
500–1000 m				
CC	0.64	0.66	0.76	0.42
RMSE	21.56	25.51	98.97	4.86
MB	−2.23	−1.62	−84.72	4.05
>1000 m				
CC	0.18	0.17	0.53	0.42
RMSE	24.31	32.23	70.24	5.28
MB	−3.58	−4.24	−60.73	4.38

4. Discussion

In this study, an evaluation of PDIR-Now's performance was conducted, specifically focusing on its comparison against ground-based rain gauges. This analysis is implemented in the context of various geographical regions across Saudi Arabia, allowing for insights into the national effectiveness of PDIR-Now.

There are noticeable patterns of overestimation and underestimation within PDIR-Now's rainfall estimations, particularly in connection to various rainfall rate categories. These patterns highlight the challenges faced by satellite-based precipitation products when estimating precipitation across a wide range of rainfall rates. It is crucial to carefully examine the topographical and geological variations within the research area. Factors such as topography, altitude, orographic effects, localized wind patterns, and microclimatic differences significantly affect the accuracy of remote sensing-based precipitation estimation products [45–47]. The algorithms employed in satellite-based precipitation products not only add an extra layer of complexity but also demonstrate their sensitivity to climatic conditions and precipitation event features [46,48].

There are several factors that can affect the accuracy of satellite precipitation products when evaluated against gauge data, such as orographic precipitation, spatial and temporal matching, and algorithm characteristics. Orographic precipitation usually falls into the category of warm precipitation, given that orographic lift is not conducive to producing frozen hydrometeors [49]. Moreover, warm precipitation is described as rainfall that occurs when there are no frozen hydrometeors present and the cloud-top temperature is above 273 K. IR-based algorithms, such as PDIR-Now, use a cloud-top temperature threshold that is then related to precipitation rate. Thus, if the cloud-top temperature is higher than the threshold, the precipitation will be underestimated by the product. The results of this paper confirm the limitation of IR algorithms in areas where orographic precipitation is present, as the accuracy of PDIR-Now diminished in areas prone to this type of precipitation. In terms of geospatial and temporal matching, each PDIR-Now pixel of 0.04° × 0.04° is aligned with the corresponding rain gauge data point based on their coordinates. This can cause differences between the products, because a point measurement is compared to an estimate of an area of approximately 16 km^2. Additionally, the daily PDIR-Now, aggregated from half-hourly estimates, is synchronized with the daily readings from the rain gauge [28]. Systematic and aggregation errors during these processes could also contribute to the divergence between PDIR-Now and the rain gauges.

Furthermore, the length and quality of the accessible data are important in determining efficiency indicators in this study. The study's 6-year duration enables an evaluation of PDIR-Now's ability to accurately represent rainfall intensities compared to rain gauge data. Nevertheless, it is important to acknowledge the limitations of the existing ground truth dataset with an uneven distribution of rain gauges across the region. Given the limited availability of observation data, this research is fundamentally comparative over a 6-year study period and should not be considered a conclusive basis for assessing the overall performance of PDIR-Now. A more thorough evaluation using over two decades of rainfall data is recommended.

As mentioned in the introduction, the past studies have assessed other SPPs over the country. However, the results between those studies and this one are not comparable. In each of these studies, the assessed periods are different. In the case of this study, the period between 2017 through 2022 is being assessed, whereas in other studies, such as Helmi and Abdelhamed [34], the period from 2000 to 2012 was used. Furthermore, in Kheimi and Gutub [31], January 2003 through November 2010 was utilized. Furthermore, in the past literature, there has not been a study that only assesses near-real-time products, which is the only type of product that PDIR-Now could be objectively compared to, due to the difference in purpose of near-real-time products and climate data records. Mahmoud et al. [32] assessed IMERG Early, NASA's near-real-time product, but the length of the study period, October 2015 to April 2016, does not match this study's. Additionally, the rain gauge network is different. Thus, the results from this study cannot be objectively compared to past studies.

This application of PDIR-Now is a good starting point for the assessment of this SPP over the study area. However, there is still plenty of space for improvement, particularly in arid and semiarid regions. An appropriate technique for future developments is minimizing the existing biases in PDIR-Now. This may be accomplished by using topographical data, considering other climatic features, and applying machine learning techniques [46,50]. It is essential to assess the accuracy of satellite products, such as PDIR-Now, and detect any hidden biases through comprehensive evaluations across different spatial and temporal scales. More validation findings for remote sensing precipitation products will hold significance and offer benefits across a wide range of applications, including hydrology, agriculture, and water management.

5. Conclusions

This study was the first one of its kind to test the capabilities of the latest near-real-time satellite product by the CHRS, PDIR-Now, over Saudi Arabia. This analysis involved a thorough approach of matching gauge-station coordinates with satellite precipitation coordinates, which enabled the evaluation of the SPP. By evaluating PDIR-Now against the rain gauge data, the study aimed to discern the product's strengths and limitations. The evaluation can be separated into two sections. First, six well-established statistical metrics were employed to assess the accuracy of PDIR-Now to detect rainfall events and their precipitation amounts at a daily scale. The metrics utilized to assess the accuracy of PDIR-Now in capturing precipitation amounts were the CC, the MB, and the RMSE, whereas the metrics for assessing the correctness of the rainfall detection were the POD, FAR, and CSI. Moreover, four annual extreme precipitation indices were used to analyze the capability of PDIR-Now when capturing extreme precipitation events. The extreme precipitation indices included in the study were RX1day, RX5day, CDD, and CWD.

Accurate estimates of precipitation are extremely valuable in arid and semi-arid areas. All of the tools used to measure or estimate precipitation have advantages and disadvantages. Even though the fine spatial and temporal resolutions of satellite products are a great advantage, these products also have limitations. SPP evaluations are essential and have great implications in assessing the validity of SPPs over Saudi Arabia. This is because given that these products can provide information about the spatial distribution of precipitation, they are very often used in studies related to hydrology and climatology. Thus, assessing their reliability is crucial. Additionally, having accurate near-real-time estimates of precipitation over the study area, including spatial distribution information, is key for water budget studies, thus, directly affecting water resources management. Furthermore, having a product that accurately depicts extreme events is useful for post-disaster assessment, enabling prompt response in mitigating the impacts of these heavy precipitation events.

This analysis provided a depiction of PDIR-Now's performance over Saudi Arabia. PDIR-Now showed the ability to capture precipitation with a low MB of 0.07 mm/day and a CC of 0.33. In the case of the detection of precipitation, PDIR-Now showed a better capability in the southwestern part of the study area. In the categories analysis, PDIR-Now performed best in the detection of "Heavy Rain". Even though PDIR-Now shows higher POD in the "Light Rain" category, the accuracy of the detection is the best in the "Heavy Rain" category, shown by a higher CSI and lower FAR. Furthermore, PDIR-Now slightly overestimates "Light Rain" but underestimates "Moderate Rain" and "Heavy Rain". The error in the different categories is more prone to occur in areas affected by orographic precipitation, specifically the southwest of the country. In this region, the ability of PDIR-Now to capture small-scale, orographically induced rainfall events diminished, highlighting the importance of spatial resolution and topographical considerations. Additionally, PDIR-Now similarly captured the RX1day and RX5day precipitation amounts throughout the gauge locations with only a slight underestimation. On the other hand, the CDD was underestimated compared to the rainfall gauges, and the CWD was overestimated. This reveals that PDIR-Now captures shorter dry spells and longer wet spells than the rain gauge network. Concerning different topographies, PDIR-Now shows greater accuracy at elevations below 1000m, indicated by higher CC and lower RMSE and MB. However, its performance declines above 1000m due to the orographic effect of the southwestern mountain range, leading to much lower CC values (0.18 for RX1day and 0.17 for RX5day) and increased RMSE and MB compared to stations at lower altitudes.

In conclusion, this manuscript provides insight on the ability of PDIR-Now to provide accurate near-real-time precipitation estimates. It underscores the complexities and challenges inherent to remote sensing-based precipitation products while highlighting their potential in diverse applications. The study's insights pave the way for ongoing research, seeking to harness the full potential of satellite-based precipitation estimation for improved hydrological and climatological studies, water resource management, and disaster preparedness over Saudi Arabia.

Author Contributions: R.S.A. conceptualized the study, performed the initial calculations, and wrote the first version of the manuscript. V.D. finalized the calculations, visualized all figures, and edited the manuscript. C.J.A. made revisions to the manuscript and edited the text. P.N. guided the collaboration and revised the finalized document.

Funding: This research was funded by Researchers Supporting Project number (RSP-2021/310) and (RSP2024R310), King Saud University, Riyadh, Saudi Arabia

Data Availability Statement: New data was not created through this work.

Acknowledgments: The authors would like to acknowledge the King Saud University, Riyadh, Saudi Arabia, for their support of this work. We extend our thanks to colleagues in the Department of Surface Water and Dams for providing essential daily precipitation data for our research. Our appreciation also goes to H.E. A. Alshaibani, the Deputy Minister for Water Affairs, for his support and encouragement.

Conflicts of Interest: The authors declare no conflicts of interest.

Abbreviations

The following abbreviations are used in this manuscript:

CC	Correlation coefficient
CDD	Consecutive dry days
CHIRPS	Climate Hazards Group InfraRed Precipitation with Station
CHRS	Center for Hydrometeorology and Remote Sensing
CMORPH	Climate Prediction Center Morphing Technique

CSI	Critical success index
CWD	Consecutive wet days
FAR	False alarm ratio
GSMap-MVK	Global Satellite Mapping of Precipitation Microwave-IR Combined Product
IMERG	Integrated Multi-satellitE Retrievals for Global Precipitation Measurement
MB	Mean bias
MEWA	Ministry of Environment, Water, and Agriculture
PERSIANN	Precipitation Estimation from Remotely Sensed Information using Artificial Neural Networks
PDIR-Now	PERSIANN–Dynamic Infrared near-real-time
PERSIANN-CCS	PERSIANN–Cloud Classification System
PERSIANN-CDR	PERSIANN–Climate Data Record
POD	Probability of detection
RMSE	Root-mean-squared error
SPP	Satellite precipitation product
Tb-R	Cloud-top brightness temperatures–rain rate
TRMM	Tropical Rainfall Measuring Mission
WRCP	World Climate Research Programme

References

1. Azeez, O.; Elfeki, A.; Kamis, A.S.; Chaabani, A. Dam Break Analysis and Flood Disaster Simulation in Arid Urban Environment: The Um Al-Khair Dam Case Study, Jeddah, Saudi Arabia. *Nat. Hazards* **2020**, *100*, 995–1011. [CrossRef]
2. Gao, Z.; Long, D.; Tang, G.; Zeng, C.; Huang, J.; Hong, Y. Assessing the Potential of Satellite-Based Precipitation Estimates for Flood Frequency Analysis in Ungauged or Poorly Gauged Tributaries of China's Yangtze River Basin. *J. Hydrol.* **2017**, *550*, 478–496. [CrossRef]
3. Katipoğlu, O.M.; Sarıgöl, M. Prediction of Flood Routing Results in the Central Anatolian Region of Türkiye with Various Machine Learning Models. *Stoch. Environ. Res. Risk Assess* **2023**, *37*, 2205–2224. [CrossRef]
4. Yeditha, P.K.; Kasi, V.; Rathinasamy, M.; Agarwal, A. Forecasting of Extreme Flood Events Using Different Satellite Precipitation Products and Wavelet-Based Machine Learning Methods. *Chaos Interdiscip. J. Nonlinear Sci.* **2020**, *30*, 063115. [CrossRef]
5. Zhao, Q.; Liu, K.; Sun, T.; Yao, Y.; Li, Z. A novel regional drought monitoring method using GNSS-derived ZTD and precipitation. *Remote Sens. Environ.* **2023**, *297*, 113778. [CrossRef]
6. Katiraie-Boroujerdy, P.-S.; Nasrollahi, N.; Hsu, K.; Sorooshian, S. Evaluation of Satellite-Based Precipitation Estimation over Iran. *J. Arid. Environ.* **2013**, *97*, 205–219. [CrossRef]
7. Nashwan, M.S.; Shahid, S.; Wang, X. Assessment of Satellite-Based Precipitation Measurement Products over the Hot Desert Climate of Egypt. *Remote Sens.* **2019**, *11*, 555. [CrossRef]
8. Qin, Y.; Chen, Z.; Shen, Y.; Zhang, S.; Shi, R. Evaluation of Satellite Rainfall Estimates over the Chinese Mainland. *Remote Sens.* **2014**, *6*, 11649–11672. [CrossRef]
9. Hong, Y.; Gochis, D.; Cheng, J.; Hsu, K.; Sorooshian, S. Evaluation of PERSIANN-CCS Rainfall Measurement Using the NAME Event Rain Gauge Network. *J. Hydrometeorol.* **2007**, *8*, 469–482. [CrossRef]
10. Miao, C.; Ashouri, H.; Hsu, K.-L.; Sorooshian, S.; Duan, Q. Evaluation of the PERSIANN-CDR Daily Rainfall Estimates in Capturing the Behavior of Extreme Precipitation Events over China. *J. Hydrometeorol.* **2015**, *16*, 1387–1396. [CrossRef]
11. Eini, M.R.; Rahmati, A.; Piniewski, M. Hydrological Application and Accuracy Evaluation of PERSIANN Satellite-Based Precipitation Estimates over a Humid Continental Climate Catchment. *J. Hydrol. Reg. Stud.* **2022**, *41*, 101109. [CrossRef]
12. Najmi, A.; Igmoullan, B.; Namous, M.; El Bouazzaoui, I.; Brahim, Y.A.; El Khalki, E.M.; Saidi, M.E.M. Evaluation of PERSIANN-CCS-CDR, ERA5, and SM2RAIN-ASCAT Rainfall Products for Rainfall and Drought Assessment in a Semi-Arid Watershed, Morocco. *J. Water Clim. Chang.* **2023**, *14*, 1569–1584. [CrossRef]
13. Rahmati Ziveh, A.; Bakhtar, A.; Shayeghi, A.; Kalantari, Z.; Bavani, A.M.; Ghajarnia, N. Spatio-Temporal Performance Evaluation of 14 Global Precipitation Estimation Products across River Basins in Southwest Iran. *J. Hydrol. Reg. Stud.* **2022**, *44*, 101269. [CrossRef]
14. Arsenault, R.; Brissette, F. Determining the Optimal Spatial Distribution of Weather Station Networks for Hydrological Modeling Purposes Using RCM Datasets: An Experimental Approach. *J. Hydrometeorol.* **2014**, *15*, 517–526. [CrossRef]
15. Guo, R.; Liu, Y. Evaluation of Satellite Precipitation Products with Rain Gauge Data at Different Scales: Implications for Hydrological Applications. *Water* **2016**, *8*, 281. [CrossRef]
16. Kummerow, C.; Barnes, W.; Kozu, T.; Shiue, J.; Simpson, J. The Tropical Rainfall Measuring Mission (TRMM) Sensor Package. *J. Atmos. Oceanic Technol.* **1998**, *15*, 809–817. [CrossRef]
17. Joyce, R.J.; Janowiak, J.E.; Arkin, P.A.; Xie, P. CMORPH: A Method That Produces Global Precipitation Estimates from Passive Microwave and Infrared Data at High Spatial and Temporal Resolution. *J. Hydrometeor* **2004**, *5*, 487–503. [CrossRef]
18. Funk, C.; Peterson, P.; Landsfeld, M.; Pedreros, D.; Verdin, J.; Rowland, J.; Romero, B.E.; Husak, G.; Michaelsen, J.; Verdin, A. A Quasi-Global Precipitation Time Series for Drought Monitoring. *USGS Sci. Chang. World* **2014**, *832*, 4. [CrossRef]

19. Hsu, K.; Gao, X.; Sorooshian, S.; Gupta, H.V. Precipitation Estimation from Remotely Sensed Information Using Artificial Neural Networks. *J. Appl. Meteor.* **1997**, *36*, 1176–1190. [CrossRef]
20. Hong, Y.; Hsu, K.-L.; Sorooshian, S.; Gao, X. Precipitation Estimation from Remotely Sensed Imagery Using an Artificial Neural Network Cloud Classification System. *J. Appl. Meteorol.* **2004**, *43*, 1834–1853. [CrossRef]
21. Ashouri, H.; Hsu, K.-L.; Sorooshian, S.; Braithwaite, D.K.; Knapp, K.R.; Cecil, L.D.; Nelson, B.R.; Prat, O.P. PERSIANN-CDR: Daily Precipitation Climate Data Record from Multisatellite Observations for Hydrological and Climate Studies. *Bull. Am. Meteorol. Soc.* **2015**, *96*, 69–83. [CrossRef]
22. Eini, M.R.; Olyaei, M.A.; Kamyab, T.; Teymoori, J.; Brocca, L.; Piniewski, M. Evaluating Three Non-Gauge-Corrected Satellite Precipitation Estimates by a Regional Gauge Interpolated Dataset over Iran. *J. Hydrol. Reg. Stud.* **2021**, *38*, 100942. [CrossRef]
23. AghaKouchak, A.; Mehran, A.; Norouzi, H.; Behrangi, A. Systematic and Random Error Components in Satellite Precipitation Data Sets. *Geophys. Res. Lett.* **2012**, *39*, 2012GL051592. [CrossRef]
24. Sorooshian, S.; Hsu, K.-L.; Gao, X.; Gupta, V.; Imam, B.; Braithwaite, D. Evaluation of PERSIANN system satellite-based estimates of tropical rainfall. *Bull. Amer. Meteor. Soc.* **2000**, *81*, 2035–2046. [CrossRef]
25. Sun, Q.; Miao, C.; Duan, Q.; Ashouri, H.; Sorooshian, S.; Hsu, K. A Review of Global Precipitation Data Sets: Data Sources, Estimation, and Intercomparisons. *Rev. Geophys.* **2018**, *56*, 79–107. [CrossRef]
26. Zhou, L.; Koike, T.; Takeuchi, K.; Rasmy, M.; Onuma, K.; Ito, H.; Selvarajah, H.; Liu, L.; Li, X.; Ao, T. A Study on Availability of Ground Observations and Its Impacts on Bias Correction of Satellite Precipitation Products and Hydrologic Simulation Efficiency. *J. Hydrol.* **2022**, *610*, 127595. [CrossRef]
27. Al-Ahmadi, K.; Al-Ahmadi, S. Spatiotemporal variations in rainfall–topographic relationships in southwestern Saudi Arabia. *Arab J. Geosci.* **2014**, *7*, 3309–3324. [CrossRef]
28. Nguyen, P.; Ombadi, M.; Gorooh, V.A.; Shearer, E.J.; Sadeghi, M.; Sorooshian, S.; Hsu, K.; Bolvin, D.; Ralph, M.F. PERSIANN Dynamic Infrared–Rain Rate (PDIR-Now): A Near-Real-Time, Quasi-Global Satellite Precipitation Dataset. *J. Hydrometeorol.* **2020**, *21*, 2893–2906. [CrossRef]
29. Salehi, H.; Sadeghi, M.; Golian, S.; Nguyen, P.; Murphy, C.; Sorooshian, S. The Application of PERSIANN Family Datasets for Hydrological Modeling. *Remote Sens.* **2022**, *14*, 3675. [CrossRef]
30. Yaswanth, P.; Kannan, B.A.M.; Bindhu, V.M.; Balaji, C.; Narasimhan, B. Evaluation of Remote Sensing Rainfall Products, Bias Correction and Temporal Disaggregation Approaches, for Improved Accuracy in Hydrologic Simulations. *Water Resour. Manag.* **2023**, *37*, 3069–3092. [CrossRef]
31. Kheimi, M.M.; Gutub, S. Assessment of Remotely-Sensed Precipitation Products Across the Saudi Arabia Region. *Int. J. Water Resour. Arid. Environ.* **2015**, *4*, 76–88.
32. Mahmoud, M.T.; Al-Zahrani, M.A.; Sharif, H.O. Assessment of Global Precipitation Measurement Satellite Products over Saudi Arabia. *J. Hydrol.* **2018**, *559*, 1–12. [CrossRef]
33. Sultana, R.; Nasrollahi, N. Evaluation of Remote Sensing Precipitation Estimates over Saudi Arabia. *J. Arid. Environ.* **2018**, *151*, 90–103. [CrossRef]
34. Helmi, A.M.; Abdelhamed, M.S. Evaluation of CMORPH, Persiann-CDR, CHIRPS v2.0, TMPA 3b42 V7, and GPM IMERG V6 satellite precipitation datasets in Arabian arid regions. *Water* **2022**, *15*, 92. [CrossRef]
35. Almazroui, M. Temperature Variability over Saudi Arabia and Its Association with Global Climate Indices. *JKAU Met. Environ. Arid. Land Agric. Sci.* **2011**, *23*, 85–108. [CrossRef]
36. Almazroui, M. Sensitivity of a Regional Climate Model on the Simulation of High Intensity Rainfall Events over the Arabian Peninsula and around Jeddah (Saudi Arabia). *Theor. Appl. Climatol.* **2011**, *104*, 261–276. [CrossRef]
37. Almazroui, M. Rainfall Trends and Extremes in Saudi Arabia in Recent Decades. *Atmosphere* **2020**, *11*, 964. [CrossRef]
38. Al-Taher, A.A. Drought and Human Adjustment in Saudi Arabia. *GeoJournal* **1994**, *33*, 411–422. [CrossRef]
39. Al-Ahmadi, K.; Al-Ahmadi, S. Rainfall-Altitude Relationship in Saudi Arabia. *Adv. Meteorol.* **2013**, *2013*, 363029. [CrossRef]
40. Hasanean, H.; Almazroui, M. Rainfall: Features and Variations over Saudi Arabia, A Review. *Climate* **2015**, *3*, 578–626. [CrossRef]
41. Chen, C.; Li, Z.; Song, Y.; Duan, Z.; Mo, K.; Wang, Z.; Chen, Q. Performance of Multiple Satellite Precipitation Estimates over a Typical Arid Mountainous Area of China: Spatiotemporal Patterns and Extremes. *J. Hydrometeorol.* **2020**, *21*, 533–550. [CrossRef]
42. Katiraie-Boroujerdy, P.-S.; Ashouri, H.; Hsu, K.; Sorooshian, S. Trends of Precipitation Extreme Indices over a Subtropical Semi-Arid Area Using PERSIANN-CDR. *Theor. Appl. Climatol.* **2017**, *130*, 249–260. [CrossRef]
43. Liu, J.; Xia, J.; She, D.; Li, L.; Wang, Q.; Zou, L. Evaluation of Six Satellite-Based Precipitation Products and Their Ability for Capturing Characteristics of Extreme Precipitation Events over a Climate Transition Area in China. *Remote Sens.* **2019**, *11*, 1477. [CrossRef]
44. Rojas, Y.; Minder, J.R.; Campbell, L.S.; Massmann, A.; Garreaud, R. Assessment of GPM IMERG satellite precipitation estimation and its dependence on microphysical rain regimes over the mountains of south-central Chile. *Atmos. Res.* **2021**, *253*, 105454. [CrossRef]
45. Barros, A.P.; Kuligowski, R.J. Orographic Effects during a Severe Wintertime Rainstorm in the Appalachian Mountains. *Mon. Wea. Rev.* **1998**, *126*, 2648–2672. [CrossRef]
46. Chen, H.; Sun, L.; Cifelli, R.; Xie, P. Deep Learning for Bias Correction of Satellite Retrievals of Orographic Precipitation. *IEEE Trans. Geosci. Remote Sens.* **2022**, *60*, 4104611. [CrossRef]

47. Zhao, Q.; Wang, W.; Li, Z.; Du, Z.; Yang, P.; Yao, W.; Yao, Y. A high-precision ZTD interpolation method considering large area and height differences. *GPS Solut.* **2024**, *4*, 28. [CrossRef]
48. Derin, Y.; Anagnostou, E.; Berne, A.; Borga, M.; Boudevillain, B.; Buytaert, W.; Chang, C.-H.; Chen, H.; Delrieu, G.; Hsu, Y.; et al. Evaluation of GPM-Era Global Satellite Precipitation Products over Multiple Complex Terrain Regions. *Remote Sens.* **2019**, *11*, 2936. [CrossRef]
49. Hobouchian, M.P.; Salio, P.; García Skabar, Y.; Vila, D.; Garreaud, R. Assessment of satellite precipitation estimates over the slopes of the subtropical Andes. *Atmos. Res.* **2017**, *190*, 43–54. [CrossRef]
50. Tao, Y.; Gao, X.; Hsu, K.; Sorooshian, S.; Ihler, A. A Deep Neural Network Modeling Framework to Reduce Bias in Satellite Precipitation Products. *J. Hydrometeorol.* **2016**, *17*, 931–945. [CrossRef]

remote sensing

Article

A Multi-Model Ensemble Pattern Method to Estimate the Refractive Index Structure Parameter Profile and Integrated Astronomical Parameters in the Atmosphere

Hanjiu Zhang [1,2,3], Liming Zhu [2,3,4], Gang Sun [2,3,4,*], Kun Zhang [2,3], Ying Liu [1,2,3], Xuebin Ma [2,3,4], Haojia Zhang [2,3,4], Qing Liu [2,3], Shengcheng Cui [2,3,4], Tao Luo [2,3,4], Xuebin Li [2,3,4] and Ningquan Weng [1,2,3,4]

1 School of Environmental Science and Optoelectronic Technology, University of Science and Technology of China, Hefei 230026, China; zhj999@mail.ustc.edu.cn (H.Z.)
2 Advanced Laser Technology Laboratory of Anhui Province, Hefei 230037, China
3 Key Laboratory of Atmospheric Optics, Anhui Institute of Optics and Fine Mechanics, Hefei Institutes of Physical Science (HFIPS), Chinese Academy of Sciences, Hefei 230031, China
4 Science Island Branch of Graduate School, University of Science and Technology of China, Hefei 230026, China
* Correspondence: gsun@aiofm.ac.cn

Abstract: In this study, we devised a constraint method, called multi-model ensemble pattern (MEP), to estimate the refractive index structure parameter (C_n^2) profiles based on observational data and multiple existing models. We verified this approach against radiosonde data from field campaigns in China's eastern and northern coastal areas. Multi-dimensional statistical evaluations for the C_n^2 profiles and integrated astronomical parameters have proved MEP's relatively reliable performance in estimating optical turbulence in the atmosphere. The correlation coefficients of MEP and measurement overall C_n^2 in two areas are up to 0.65 and 0.76. A much higher correlation can be found for a single radiosonde profile. Meanwhile, the difference evaluation of integrated astronomical parameters also shows its relatively robust performance compared to a single model. The prowess of this reliable approach allows us to carry out regional investigation on optical turbulence features with routine meteorological data soon.

Keywords: optical turbulence; refractive index structure parameter; vertical profile; radiosonde; routine meteorological parameters; multi-model ensemble pattern; integrated astronomical parameters

Citation: Zhang, H.; Zhu, L.; Sun, G.; Zhang, K.; Liu, Y.; Ma, X.; Zhang, H.; Liu, Q.; Cui, S.; Luo, T.; et al. A Multi-Model Ensemble Pattern Method to Estimate the Refractive Index Structure Parameter Profile and Integrated Astronomical Parameters in the Atmosphere. *Remote Sens.* **2023**, *15*, 1584. https://doi.org/10.3390/rs15061584

Academic Editors: Xiaoming Wang, Yuriy Kuleshov, Suelynn Choy, Mayra I. Oyola-Merced and Haobo Li

Received: 15 February 2023
Revised: 5 March 2023
Accepted: 9 March 2023
Published: 14 March 2023

1. Introduction

Optical turbulence (OT), caused by atmospheric inhomogeneities and fluctuations, is one of the most critical factors that limit the transmission and performance of imaging systems [1]. Researchers involved in light propagation in the atmosphere, especially laser physicists and astronomers, have been concerned with this issue for decades [2–11]. A turbulent atmosphere impacts light wave propagation in various aspects, such as phase changes and intensity fluctuations. These distortions lead to significant blurring, scintillations, broadening, arrival angle fluctuations, and laser beam wander [1,2,8]. Hence, parameterization and characterization of OT are essential for designing and operating photoelectric systems.

Among all the parameters assessing the influence on optoelectronic systems from the turbulent atmosphere, the refractive index structure parameter (C_n^2) is commonly used to characterize the optical turbulence in the atmosphere. The past decades have witnessed researchers' efforts to measure, parameterize, and estimate C_n^2. Up to now, different techniques (direct or indirect) using optical or non-optical principles have developed to obtain C_n^2 [12]. Among these techniques, a pair of micro-thermometers (MT) is the most common equipment used to obtain C_n^2 by invoking several hypotheses [1]. Utilizing a balloon-borne MT (in situ measurements), usually accompanied by measurements of routine meteorological parameters, is extensively employed for getting the C_n^2 profile in photoelectric applications, for example, site testing [6,13,14] and astronomical observatories routine scheduling [3,15]. Other remote sensing methods and instruments, for example, Multi-Aperture Scintillation Sensor (MASS), Slope Detection And Ranging (Slodar) and Solar Differential Image Motion Monitor+ (S-Dimm+), are also of vital importance for the development of modeling and the refinement of empirical dependencies for astronomy [16–20].

Meanwhile, methods parameterizing and estimating C_n^2 profiles are established to meet the need of engineering practices. Empirical, physically-based, statistical, and data-driven learning methods to estimate C_n^2 were subsequently developed. Simple empirical methods, such as the submarine laser communication (SLC) model [21], are only involved in a single elevation parameter. Physically-based models referring to thermodynamics or dynamics factors exist in lots of literature. Owing to their abundant physical connotations, these models are competitive in characterizing OT in terms of its physical mechanism. Hufnagel developed the Hufnagel model based on meteorology and stellar scintillation data [22]. The Hufnagel-Valley5/7 (HV5/7) model [22,23] is one of the most popular forms related to wind velocity in the free atmosphere. Ruggiero and DeBenedictis proposed the Hmnsp99 outer scale model, referring to gradients of temperature and wind shear [24]. Dewan developed a similar turbulence outer scale method utilizing wind shear [25]. Thorpe investigated the relationship between potential temperature inversion and the Thorpe scale; Basu proposed a simple approach to estimate C_n^2 profiles with the coarse-resolution potential temperature profiles [21,26]. The Ellison scale was developed to quantify the scales of water body overturns. This theory was also used to calculate C_n^2 [27,28]. Recently, several modified models, such as the wind shear and potential temperature (WSPT) model [29] and wind shear and temperature gradient (WSTG) model [30], were also applied to estimate C_n^2 profiles under different experimental environments. Other methods were developed in a statistical view, for example, statistical models devised by Vanzandt [31] and Trinquet [32]. Along with the development of computer science, deep learning tools have shown their advantage in handling high-dimensional and nonlinear issues. Researchers also applied this useful tool in estimating C_n^2 [33–35].

However, no one of the existing estimating approaches are superior to any of the others, to the best of our knowledge. Each existing approach has its own merits and limitations [21]. The universality and robustness of most existing approaches and models should be improved. However, the turbulent atmosphere with random, nonlinear, and infinite-element features makes it difficult to completely specify the precise mathematical expression of C_n^2 from the routine macroscopic meteorological parameters—for now, at least. The existing physical-based approaches were established on several hypotheses and statistical evidence, more or less. Here, we propose a multi-model ensemble pattern (MEP) method to estimate C_n^2 based on several existing physically-based methods. The purpose of this study is to take advantage of different existing approaches. The proposed model performance is not always the best. However, it can ensure that the C_n^2 and integrated astronomical parameters estimated by the MEP are competitive compared to the best of the existing models if it is not.

This paper is organized as follows: Section 2 describes the experimental site, instruments, and radiosonde data. Section 3 presents the theory of several existing approaches to estimate C_n^2 that we adopted and the proposed MEP method. Section 4.1 depicts the results of C_n^2 using different models. Section 4.2 exhibits the evaluation of different models in calculating integrated astronomical parameters. The summary and conclusions are given in Section 5.

2. Experimental Principles and Scientific Data

2.1. Experimental Principles

According to the Gladstone law [12,36] and neglecting the water vapor concentration contribution, the refractive index structure parameter C_n^2 ($m^{-2/3}$) can be computed via pressure P (hPa), absolute temperature T (K), and temperature structure parameter C_T^2 ($K^2m^{-2/3}$) as follows:

$$C_n^2 = \left(79 \times 10^{-6}\frac{P}{T^2}\right)^2 C_T^2. \quad (1)$$

C_T^2 can be calculated by the temperature structure function D_T^2 based on the Kolmogorov–Obukhov turbulence assumption [1]. D_T^2 is defined as:

$$D_T(r) = \left\langle [T(\vec{x}) - T(\vec{x}+\vec{r})]^2 \right\rangle = C_T^2 r^{2/3} (l_0 \ll r \ll L_0), \quad (2)$$

where triangle brackets denote an ensemble average; $\vec{x}$ and $\vec{x}+\vec{r}$ are the positions of the temperature probes; l_0 and L_0 represent the inner and outer scales, respectively; and r represents the distance of two probes that should be in the inertial sub-region. Radiosonde balloons equipped with micro-thermometers (MT) and routine meteorological sensors are used worldwide to obtain optical turbulence and meteorology parameters profiles.

In our case, the temperature probes (red rectangular boxes) used are shown in Figure 1b. The two platinum probes were isolated 1 m (r = 1 m) horizontally. T and P data necessary for calculating C_n^2 were measured by onboard temperature and pressure sensors. A Global Positioning System (GPS) was used to obtain the position information, and wind velocity was calculated from GPS data with a precision of 0.3 m/s. The Anhui Institute of Optics and Fine Mechanics (AIOFM) designed the whole system. The instruments' performance was summarized in Ref. [37]. The platinum wire probe resistance was 10 Ω with 10 μm diameter. The minimum detectable value of C_T^2 was 4.0×10^{-6} $K^2m^{-2/3}$. The sampling frequency of the processor was up to 100 Hz, and the data were averaged with a time interval of 1 s. The precision of the temperature and pressure sensors were 0.2 K and 1.5 hPa. The balloons ascend with a vertical velocity of approximately 5 m/s. The data were re-processed with a space interval of 10 m.

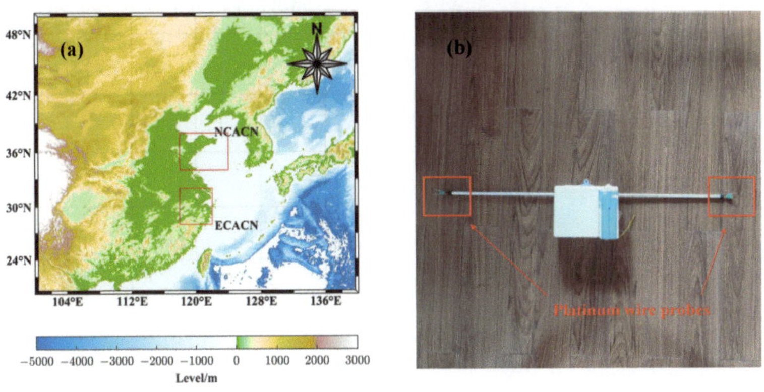

Figure 1. Field observation areas and instrument. (**a**) Sites locations: the eastern coastal area of China (ECACN) and the northern coastal area of China (NCACN). (**b**) Instrument: platinum wire probes (red rectangular boxes).

2.2. *Scientific Data*

The field observations were carried on two areas (Figure 1a) during April 2018. One observation was undertaken in the eastern coastal area of China (hereinafter ECACN), and the other in the northern coastal area of China (hereinafter NCACN). After removing several incomplete datasets with low termination altitude or missing data, we chose 16 and 20 profiles of ECACN and NCACN, respectively. The data collections of two areas are summarized in Table 1. More details are documented in Appendix B Tables A1 and A2.

Table 1. Radiosonde data collection of two areas.

Areas	Morning Launches	Evening Launches	Total Launches
ECACN	0	16	16
NCACN	9	11	20

3. Methodology of MEP

3.1. Theory of the Adopted Models

Seven different approaches (HV: Hufnagel-Valley 5/7; H9: Hmnsp99; DN: Dewan; TE: Thorpe; EN: Ellison; WT: WSPT; WG: WSTG) estimating C_n^2 with routine meteorological parameters were adopted in our study. We have summarized theories of these approaches in Appendix A to avoid interrupting the fluency of this article. More details can be found in the corresponding literature.

In data processing, all approaches except for HV involved gradient variables (the measured meteorological parameters or their derived parameters). Several approaches (TE, EN, and WT) calculated C_n^2 related to the sizes of localized overturns of the potential temperature. It was hard to distinguish these overturns for coarse resolution data because potential temperature profiles have an increasing tendency with height most of the time. Hence, we adopted the original resolution data in these approaches. Meanwhile, the other approaches (H9, DN, and WSTG) were calculated in the vertical resolution of 60 m. All seven approach estimations were re-processed on the scale of 60 m for consistency and convenience. Meanwhile, data exceeding 1 km above the ground level (AGL) were selected. Hence, the feature of C_n^2 and integrated astronomical parameters represent the free atmosphere results in our case.

3.2. MEP Method

Before introducing the principles of MEP, several theoretical basics should be elaborated first. For two variables, r and f, r_n is the reference variable (MT measured C_n^2 in this study), and f_n is the corresponding pattern result (estimated as C_n^2 in this study). The correlation coefficient (R) and their root-mean-square difference between two fields (E', also known as the centered root-mean-square difference) are defined as:

$$R = \frac{\frac{1}{N}\sum_{n=1}^{N}(f_n - \bar{f})(r_n - \bar{r})}{\sigma_f \sigma_r}, \tag{3}$$

$$E' = \left\{\frac{1}{N}\sum_{n=1}^{N}\left[(f_n - \bar{f}) - (r_n - \bar{r})\right]^2\right\}^{1/2}, \tag{4}$$

where σ_r ($\sigma_r = 1/N\sqrt{\sum_{n=1}^{N}(r_n - \bar{r})^2}$) and σ_f ($\sigma_f = 1/N\sqrt{\sum_{n=1}^{N}(f_n - \bar{f})^2}$) denote the reference variable standard deviation and pattern result standard deviation, respectively; $\bar{r}$ and $\bar{f}$ represent the average of two variables. Thus, we can deduce the relationship between the reference standard deviation σ_r, pattern standard deviation σ_f, and correlation coefficient R as:

$$E'^2 = \sigma_f{}^2 + \sigma_r{}^2 - 2\sigma_f\sigma_r R, \tag{5}$$

Taylor devised the Taylor diagram to provide a concise statistical summary of how well the patterns match each other in terms of the above four statistics (σ_r, σ_f, R and E') [38]. In our study, we have normalized the statistics ($\hat{\sigma}_f = \sigma_f/\sigma_r$, $\hat{\sigma}_r = 1$, $\hat{E}' = E'/\sigma_r$) referring to σ_r for convenience. According to Taylor's work, a skill function was also developed to assess the models' performance as follows:

$$S(\alpha,\beta) = \frac{2^{\alpha}(1+R)^{\beta}}{\left(\hat{\sigma}_f + 1/\hat{\sigma}_f\right)^{\alpha}(1+R_0)^{\beta}}, \quad (6)$$

where R_0 represents the maximum of R in a set of the same model and we set $R_0 = 1$; α and β are penalty coefficients that can adjust the proportion of skill function via model variance and correlation coefficient. A more considerable value of α or β means that the corresponding statistic ($\hat{\sigma}$ or R) has a more significant influence on the result of $S(\alpha,\beta)$.

Further, a weight function is defined as:

$$W_j(\gamma) = \frac{{S_j}^{\gamma}}{\sum_{i=1}^{N} {S_i}^{\gamma}}. \quad (7)$$

Note that all skill values are in the range of 0–1. We set a penalty parameter γ to distinguish the model's performance. Consequently, the multi-model ensemble pattern (MEP) method process is divided into three steps, as shown in Figure 2. We have summarized them as follows:

1. Using routine meteorological parameters estimating C_n^2 with multiple models;
2. Obtaining models skills $S(\alpha,\beta)$ against MT results in Equation (6);
3. Calculating weights $W_j(\gamma)$ of different models and MEP results.

Meanwhile, parameters (α,β) are used for modulating the weights of different statistics, and γ is used to distinguish the different models' performance. These penalty parameters can be changed as the research focus changes in practice. For example, we can increase the α value to increase the weight of data fluctuation in the evaluation system. It is the same for the β for correlations, and we chose the latter condition in our case. Moreover, a considerable γ means a more significant influence on the evaluation of skills. In our case, we set $\alpha = 2, \beta = 6, \gamma = 4$.

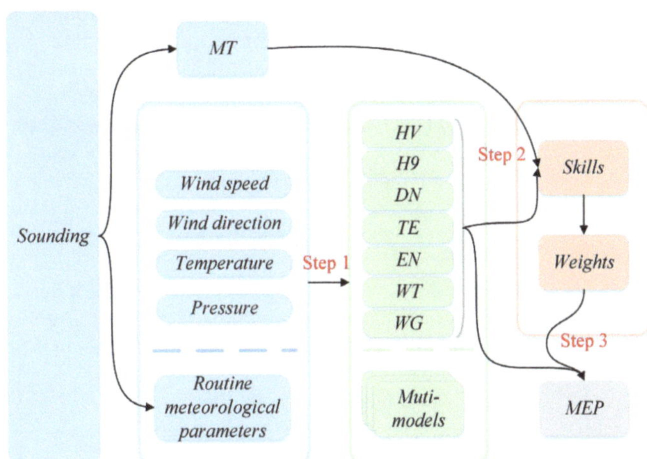

Figure 2. Process of MEP. Abbreviation meanings are as follows after this. MT: micro-thermal; HV: Hufnagel-Valley 5/7; H9: Hmnsp99; DN: Dewan; TE: Thorpe; EN: Ellison; WT: WSPT; WG: WSTG; MEP or ME: multi-model ensemble pattern method.

3.3. Statistical Analysis

In addition to the correlation coefficient (R), the root mean square error ($RMSE$), bias ($Bias$), and mean absolute error (MAE) were calculated to evaluate the performance of the different approaches. The definitions of these statistics are as follows:

$$RMSE = \sqrt{\frac{1}{N}\sum_{n=1}^{N}(r_n - f_n)^2}, \quad (8)$$

$$Bias = \frac{1}{N}\sum_{n=1}^{N}(r_n - f_n), \quad (9)$$

$$MAE = \frac{1}{N}\sum_{n=1}^{N}|r_n - f_n|. \quad (10)$$

4. Results and Discussion

4.1. Measured and Estimated C_n^2 Profiles

We employed 16 and 20 radiosonde datasets of two areas when evaluating the performance of the proposed and adopted methods. We used $log_{10}(C_n^2)$ instead of C_n^2 to generate readable data and curves. We provide two C_n^2 profiles of all approaches against the MT of each dataset in the primary text. See Appendix B for all days details in two sites (Figures A1–A3 in ECACN; Figures A4–A7 in NCACN).

Figure 3a,b displays MT and estimations C_n^2 profiles from the ECACN radiosonde campaign. The overall trends of estimations are consistent with MT. The C_n^2 magnitude of estimations and MT are mainly distributed in the range of $10^{-15} - 10^{-19}m^{-2/3}$. Distinct differences can also be seen between different estimations. HV has a very high correlation with MT within the troposphere. However, it underestimates C_n^2 significantly above approximately 20 km, which indicates that a more turbulent and complex atmospheric state might exist above the troposphere in this area. TE, EN, and WT have better performance in magnitude owing to the calibration of unknown proportionality constants according to MT measurements. H9, DN, and WG have a similar trend in the overall trend, while these estimations fluctuate a little bit more around the mean value against TE, EN, WT, and ME. By combining the corresponding Taylor diagrams in Figure 3c,d, we can also easily find that the values of normalized standard deviations of HV, H9, DE, and WG are much bigger than MT most of the time. Meanwhile, closer normalized standard deviation values to 1 (or MT) of TE, EN, and ME means that these approaches have similar behavior in C_n^2 fluctuation magnitude. In addition, ME also shows its advantage in correlation evaluation. Among all 16 launches, correlation coefficients between ME and MT are mainly distributed around 0.6–0.8 and the best one is up to approximately 0.9.

Figures A1–A3 in Appendix B display all the C_n^2 profiles in ECACN. Scatter figures of all approaches against MT for all launches were plotted to further study the overall statistical features. Figure 4 shows all launches estimated C_n^2 statistical feature in ECACN. The relevant statistics are summarized in Table 2. Although the overall $Bias$ of ME is slightly larger than WT, R, MAE and $RMSE$ of ME present the best performance of all approaches.

Table 2. ECACN 16 C_n^2 profiles statistics.

Statistics	HV	H9	DN	TE	EN	WT	WG	ME
R	0.64	0.61	0.53	0.60	0.63	0.53	0.60	0.65
$Bias$	0.60	0.57	0.08	−0.30	0.18	−0.005	−0.15	−0.11
MAE	0.81	0.73	0.76	0.56	0.53	0.54	0.64	0.51
$RMSE$	1.13	0.92	0.95	0.70	0.67	0.68	0.81	0.64

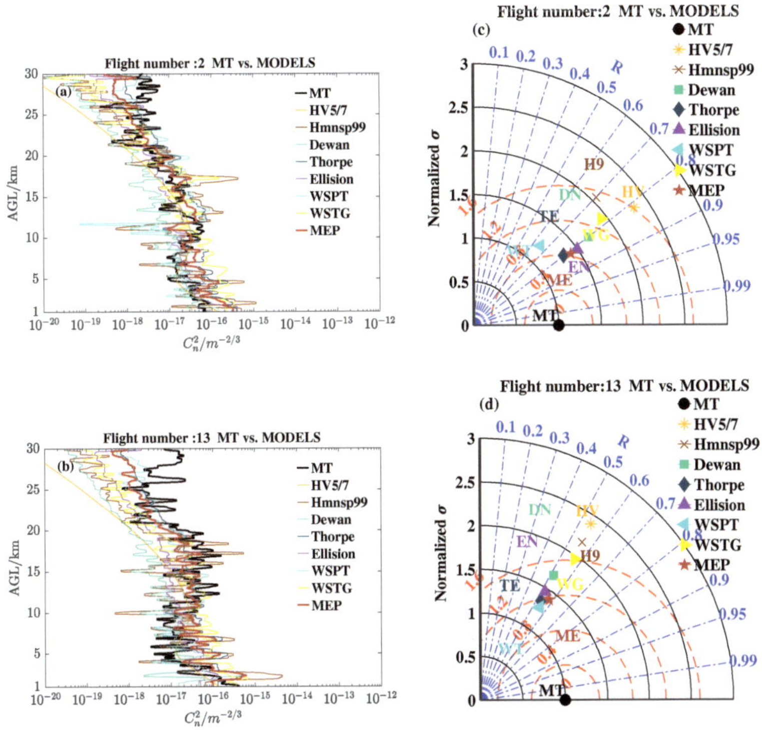

Figure 3. Measured and estimated C_n^2 profiles in ECACN. (**a**,**b**): single day C_n^2 profiles; (**c**,**d**): corresponding Taylor diagram of a single day C_n^2 statistics. The black solid curves in circles mark MT and models normalized standard deviation $\hat{\sigma}$; the blue dashed lines mark the correlation coefficient R; the red dashed lines mark the root-mean-square difference E' between the models and MT. Intuitively, the closer the model is to the reference (MT, the black circle solid point) in the diagram, the better the estimation is.

Validation was also done to the radiosonde data from the campaign carried out in NCACN. Twenty sounding datasets were selected in this area. Figure 5 exhibits two launches C_n^2 profiles and their corresponding Taylor diagrams. The characteristics of different approaches estimation profiles are similar to those in ECACN. MEP correlation coefficients of a single launch in NCACN are mainly distributed around 0.7–0.9, and the best one is more than 0.95. Figures A4–A7 in Appendix B display all the C_n^2 profiles in NCACN. Figure 6 shows all 20 launches estimated C_n^2 against the MT statistical feature in NCACN. The relevant statistics are summarized in Table 3. The overall correlation criteria R of MEP is the best of all approaches, up to 0.7632. Meanwhile, the deviation criteria *Bias*, *MAE*, and *RMSE* of MEP are the smallest. The above results of all the approaches in ECACN and NCACN have proved the potential of MEP in estimating C_n^2 utilizing radiosonde data.

Table 3. NCACN 20 C_n^2 profiles statistics.

Statistics	HV	H9	DN	TE	EN	WT	WG	ME
R	0.69	0.72	0.62	0.74	0.75	0.64	0.73	0.76
Bias	0.52	0.61	0.06	−0.22	0.26	0.02	−0.13	−0.09
MAE	0.74	0.71	0.73	0.49	0.49	0.52	0.56	0.45
RMSE	1.01	0.89	0.90	0.62	0.64	0.66	0.69	0.58

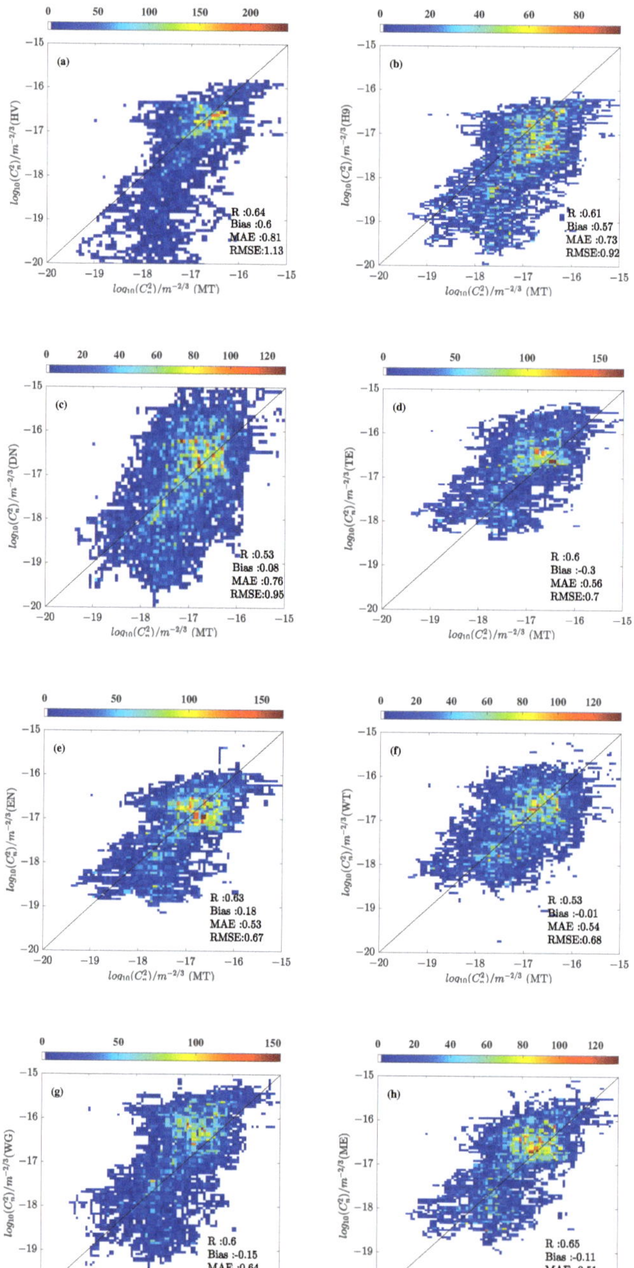

Figure 4. ECACN all launches MT vs. estimations C_n^2 scatter diagrams. (**a**): MT vs. Hafnagel-Valley 5/7; (**b**) MT vs. Hmnsp99; (**c**): MT vs. Dewan; (**d**): MT vs. Thorpe; (**e**): MT vs. Ellison; (**f**): MT vs. WSPT; (**g**): MT vs. WSTG; (**h**): MT vs. MEP. The color indicates the frequency distribution of C_n^2.

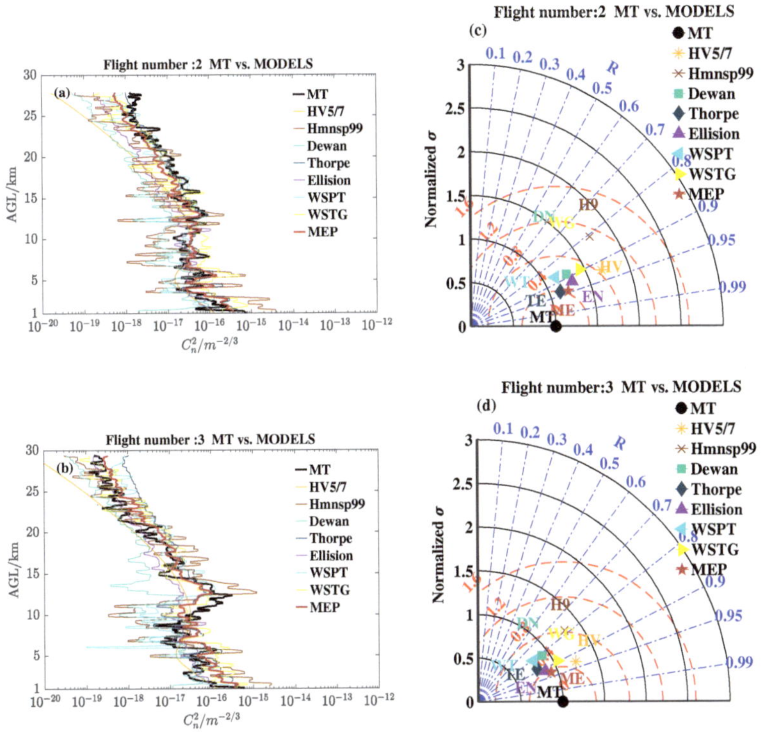

Figure 5. Measured and estimated C_n^2 profiles in NCACN. (**a**,**b**) A single day C_n^2 profiles; (**c**,**d**) corresponding Taylor diagram of a single day C_n^2 statistics.

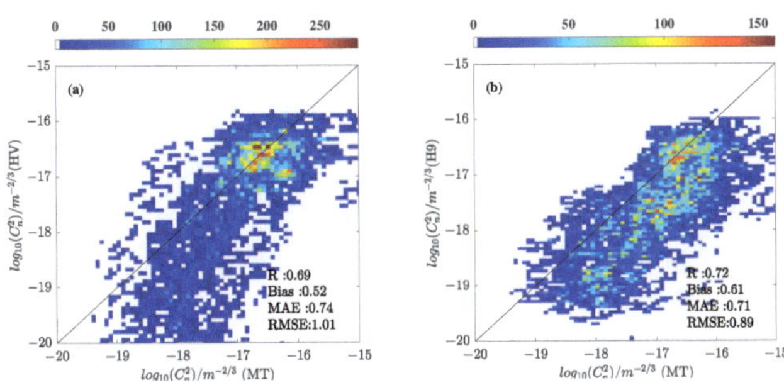

Figure 6. *Cont.*

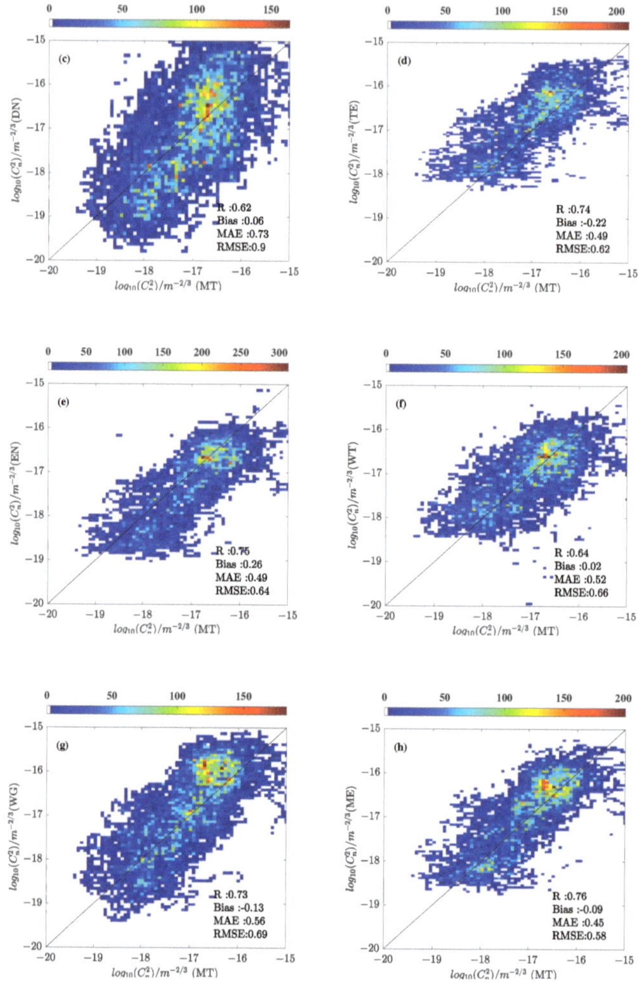

Figure 6. NCACN all launches MT vs. estimations C_n^2 scatter diagram. (**a–h**) The same as Figure 4 but for NCACN.

4.2. Integrated Astronomical Parameters from Measured and Estimated C_n^2 Profiles

Evaluating the optical turbulence influence on optoelectronic facilities (ground-based observatories, laser transmission, and free atmosphere optical communication systems) in the atmosphere is one of the primary aims of researchers. Hence, we also calculated the integrated astronomical parameters (Fried parameter r_0, seeing ϵ, isoplanatic angle θ_{AO} and scintillation rate σ_I^2) to evaluate the performance of the proposed method. These parameters are defined as [5,9,12]:

$$r_0 = \left[0.423\left(\frac{2\pi}{\lambda}\right)^2 \sec\varphi \int_{h_0}^{\infty} C_n^2(h)dh\right]^{-3/5}, \tag{11}$$

$$\varepsilon = 5.25\lambda^{-1/5}\left[\int_{h_0}^{\infty} C_n^2(h)dh\right]^{3/5}, \tag{12}$$

$$\theta_{AO} = 0.057\lambda^{6/5}\left[\int_{h_0}^{\infty} C_n^2(h)h^{5/3}dh\right]^{-3/5}, \tag{13}$$

$$\sigma_I^2 = 19.12\lambda^{-7/6}\int_{h_0}^{\infty} C_n^2(h)h^{5/6}dh. \tag{14}$$

φ is the solar zenith angle set as 0°; λ is a given wavelength (we set λ = 550 nm); h denotes the elevation above ground level (AGL) of the sites; h_0 represents the initial elevation (we set h_0 = 1000 m). Therefore, the conclusions of the integrated astronomical parameters included in this study can only represent the influence of the free atmosphere.

Details of these integrated astronomical parameters of all launches in ECACN are listed in the Appendix B, Tables A3–A6. The median values represent regional features and are of referential value for photoelectric applications. Median values of r_0, ϵ, θ_{AO}, and σ_I^2 calculated from MT are 10.10 cm, 1.10″, 0.67″, and 0.54″, respectively. These parameters calculated from ME are 8.93 cm, 1.25″, 0.73″, and 0.56″. The relative errors of median values are rather small. All the integrated astronomical parameters are depicted in Figure 7, and the relevant statistical feature of these parameters are summarized in Table 4. HV and DN overestimated r_0 and θ_{AO} and underestimated ϵ and σ_I^2 can be easily found both from the figures and their *Bias* from the table against MT. The ME correlation coefficients of R_0, ϵ and σ_I^2 are quite good. Meanwhile, the deviations are rather small compared to the other approaches.

Table 4. ECACN integrated astronomical parameters statistics (@λ = 550 nm).

	Statistics	HV	H9	DN	TE	EN	WT	WG	ME
r_0	*R*	0.46	0.32	0.70	0.61	0.46	0.62	0.46	0.69
	Bias	−4.07	6.50	−8.87	4.56	−1.23	0.005	5.17	3.13
	MAE	5.01	6.60	8.87	4.89	4.17	3.27	5.31	4.01
	RMSE	6.11	8.13	9.86	6.48	4.79	4.28	7.17	5.32
ϵ	*R*	0.14	0.34	0.53	0.53	0.44	0.65	0.40	0.51
	Bias	0.39	−1.14	0.57	−0.45	0.15	0.15	−0.56	−0.20
	MAE	0.42	1.17	0.57	0.54	0.39	0.29	0.60	0.38
	RMSE	0.58	1.37	0.70	0.58	0.52	0.39	0.69	0.42
θ_{AO}	*R*	0.45	−0.42	0.56	0.49	0.45	0.29	0.03	0.38
	Bias	−0.59	0.35	−0.95	0.21	−0.28	0.05	0.21	0.08
	MAE	0.66	0.43	0.97	0.30	0.41	0.28	0.31	0.28
	RMSE	0.74	0.59	1.03	0.45	0.47	0.40	0.47	0.40
σ_I^2	*R*	0.14	−0.08	0.51	0.72	0.62	0.40	0.31	0.50
	Bias	0.37	−0.87	0.47	−0.21	0.25	0.14	−0.28	−0.004
	MAE	0.38	1.01	0.47	0.33	0.29	0.27	0.40	0.28
	RMSE	0.53	1.18	0.60	0.37	0.39	0.38	0.46	0.34

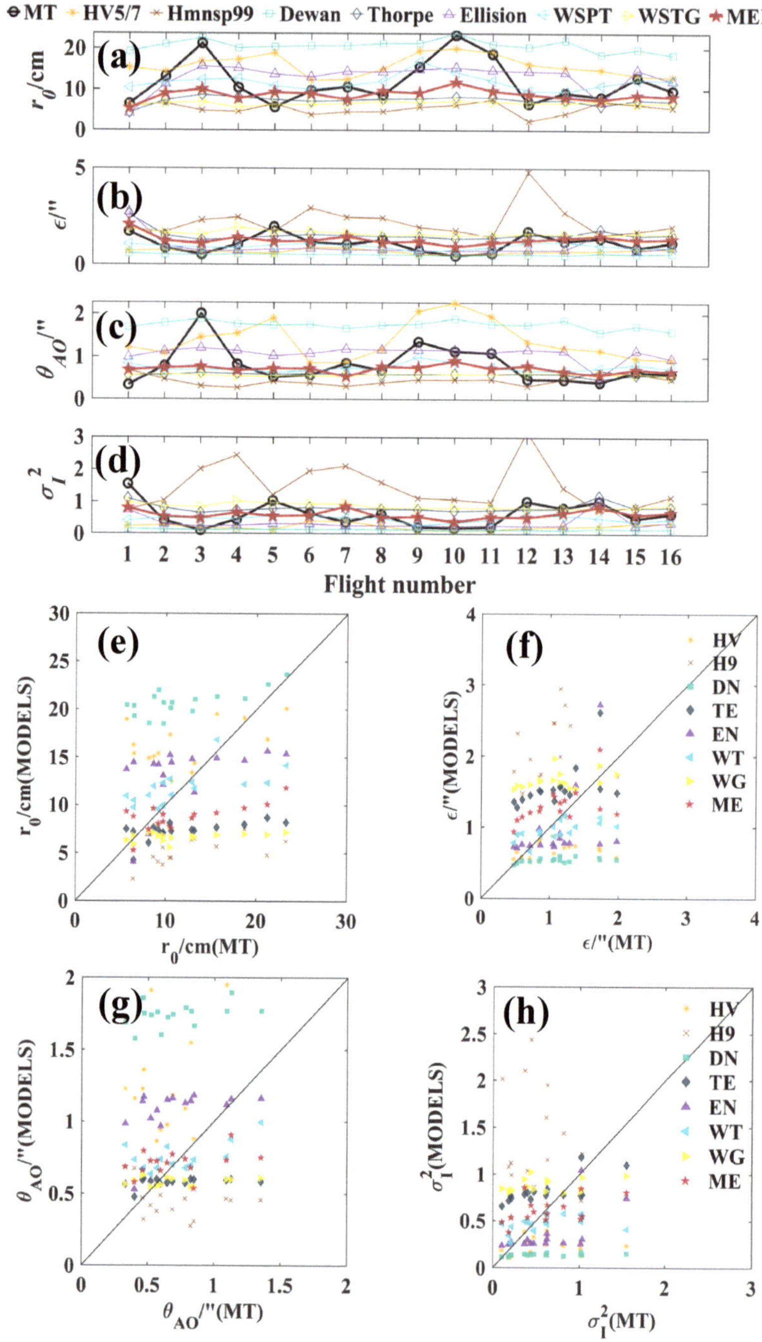

Figure 7. ECACN integrated astronomical parameters (@λ = 550 nm). (**a**) Fried parameter r_0 (in cm); (**b**) seeing ϵ (arcsec in $''$); (**c**) isoplanatic angle θ_{AO} (arcsec in $''$); (**d**) scintillation rate σ_I^2; (**e**–**h**): corresponding scatter diagram.

The same computation process was done for the 20 radiosonde data from NCACN. The parameters calculated in NCACN are listed in the Appendix B, Tables A7–A10. Median values of r_0, ϵ, θ_{AO}, and σ_I^2 calculated from MT are 10.31 cm. 1.08″, 0.63″, and 0.66″, respectively. The parameters calculated from ME are 8.83 cm, 1.26″, 0.72″, and 0.58″. All the integrated astronomical parameters are portrayed in Figure 8, and the relevant statistical feature of these parameters are summarized in Table 5. The features of these parameters' statistics are similar to the results in ECACN. The ME correlation coefficients of all parameters are even better in general compared to ECACN. Meanwhile, the deviation statistics are relatively small overall compared to the other approaches. A comprehensive comparison in two experimental areas between MT and the best estimations of all the parameter statistics were summarized in Table 6. Although MEP is not always the best estimation among these eight approaches, its gap with the optimal approach is minimal. All the above show MEP's considerable universality in studying optical turbulence characteristics.

Table 5. NCACN integrated astronomical parameters statistics (@λ = 550 nm).

	Statistics	HV	H9	DN	TE	EN	WT	WG	ME
r_0	R	−0.22	0.51	−0.006	0.64	0.59	0.40	0.42	0.60
	$Bias$	−4.68	5.90	−9.75	2.46	−3.90	−1.42	4.11	1.74
	MAE	6.28	5.98	9.75	4.04	4.19	3.69	4.86	3.63
	$RMSE$	7.61	7.17	10.89	4.87	5.37	4.44	6.03	4.27
ϵ	R	−0.16	0.62	0.12	0.58	0.64	0.24	0.53	0.58
	$Bias$	0.63	−1.32	0.84	−0.007	0.58	0.43	−0.37	0.08
	MAE	0.72	1.40	0.84	0.60	0.60	0.57	0.76	0.55
	$RMSE$	1.09	1.60	1.20	0.80	0.95	0.94	0.88	0.75
θ_{AO}	R	0.01	0.03	−0.03	0.38	0.47	0.32	0.20	0.45
	$Bias$	−0.56	0.36	−1.07	0.11	−0.41	0.01	0.16	0.03
	MAE	0.67	0.41	1.07	0.29	0.43	0.29	0.29	0.29
	$RMSE$	0.85	0.52	1.13	0.36	0.51	0.33	0.38	0.32
σ_I^2	R	−0.07	0.81	−0.05	0.31	0.39	0.11	0.24	0.59
	$Bias$	0.69	−1.24	0.84	0.26	0.69	0.48	0.02	0.34
	MAE	0.76	1.31	0.84	0.64	0.69	0.64	0.69	0.60
	$RMSE$	1.36	1.54	1.42	1.16	1.31	1.24	1.13	1.11

Table 6. Performance of MEP/ME against the best one (within parentheses) (the integrated astronomical parameters were calculated for the wavelength of light at λ = 550 nm. The values retain two decimal places).

Areas	Parameters	R	$Bias$	MAE	$RMSE$
ECACN	C_n^2	0.65(0.65:ME)	−0.11(−0.005:WT)	0.51(0.51:ME)	0.64(0.64:ME)
	r_0	0.69(0.70:DN)	3.13(0.005:WT)	4.01(3.27:WT)	5.32(4.28:WT)
	ϵ	0.51(0.65:WT)	−0.20(0.15:EN)	0.38(0.29:WT)	0.42(0.39:WT)
	θ_{AO}	0.38(0.56:DN)	0.08(0.05:WT)	0.28(0.28:WT)	0.40(0.40:ME)
	σ_I^2	0.50(0.72:TE)	−0.004(−0.004:ME)	0.28(0.27:WT)	0.34(0.34:ME)
NCACN	C_n^2	0.76(0.76:ME)	−0.09(0.02:WT)	0.45(0.45:ME)	0.58(0.58:ME)
	r_0	0.60(0.64:TE)	1.74(−1.42:WT)	3.63(3.63:ME)	4.27(4.27:ME)
	ϵ	0.58(0.64:EN)	0.08(−0.007:TE)	0.55(0.55:ME)	0.75(0.75:ME)
	θ_{AO}	0.45(0.47:EN)	0.03(0.01:WT)	0.29(0.29:ME)	0.32(0.32:ME)
	σ_I^2	0.59(0.59:ME)	0.34(0.02:WG)	0.60(0.60:ME)	1.11(1.11:ME)

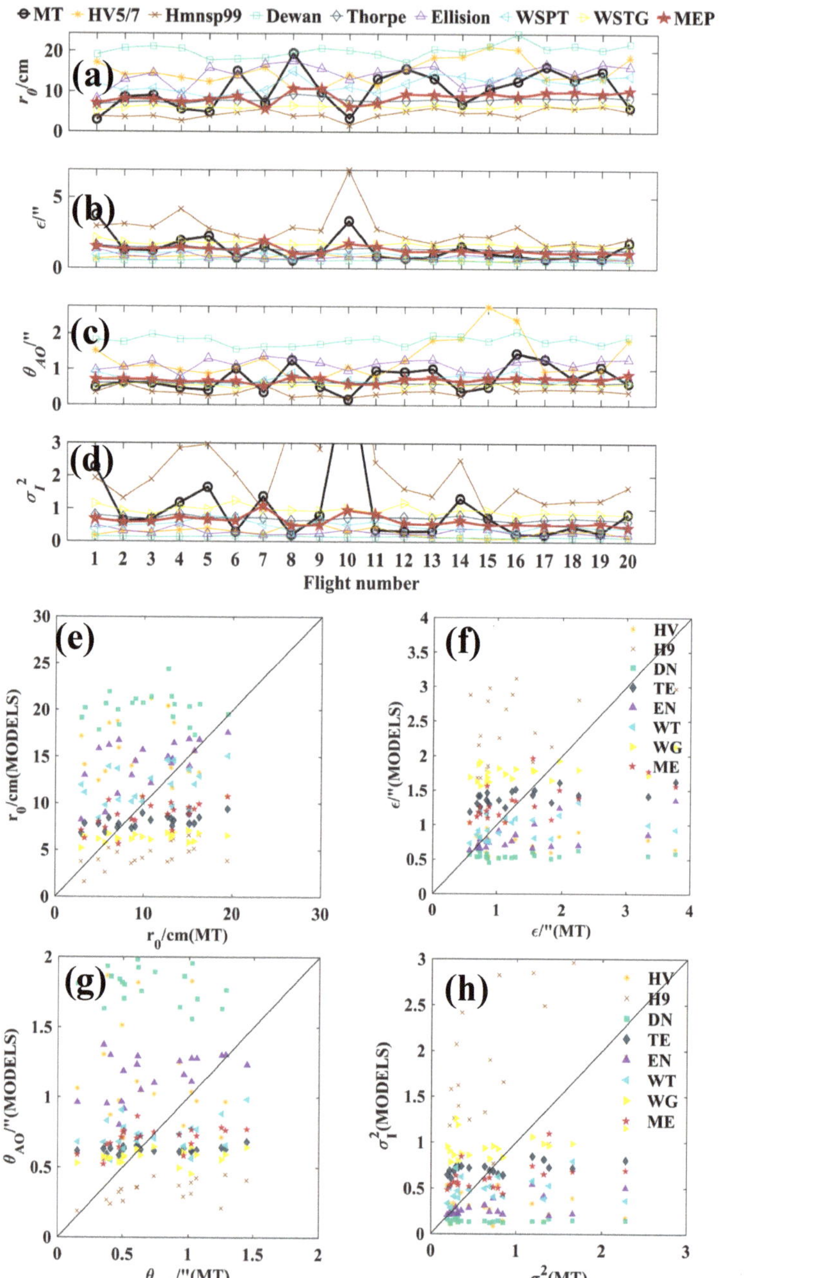

Figure 8. NCACN integrated astronomical parameters (@λ = 550 nm). (**a**) Fried parameter r_0; (**b**) seeing ϵ; (**c**) isoplanatic angle θ_{AO}; (**d**) scintillation rate σ_I^2; (**e**–**h**): corresponding scatter diagram.

5. Conclusions

In this study we propose a multi-model ensemble pattern method to estimate the C_n^2 based on several existing physical-based approaches. Balloon radiosonde data were collected in two areas of China to validate this method. Multiple dimensions evaluation including C_n^2 and the integrated astronomical parameters (r_0, ϵ, θ_{AO} and σ_I^2) of all approaches were done against the MT measured results. Statistical analysis of these methods' performance mainly focuses on the overall trend (R) and deviation (Bias, MAE, and RMSE). The best performance of all approaches against the MEP is summarized in Table 6. The C_n^2 correlation coefficients of MEP are up to 0.65 and 0.76. The overall agreements of the C_n^2 profiles in two areas are quite good. A single profile has an even higher correlation coefficient of more than 0.95. Several statistical assessments of deviations of C_n^2 are relatively small. These indicate that MEP has the capacity to estimate C_n^2 well. Meanwhile, the evaluations of integrated astronomical parameters also show its promising potential for calculating these parameters as C_n^2 does. Although MEP was not always the best method in all parameters statistical evaluations, it showed competitive performance in these evaluations. Hence, the MEP method presented good stability and universality, and even the validation radiosonde data were collected in different areas, which meant significantly different atmospheric conditions. The MEP appreciably contains more information than a single method, including thermodynamic and dynamic factors of the optical turbulence. Moreover, the MEP method could be less sensitive to different parametric settings caused by each method, producing a more robust C_n^2 estimate.

It should be noted that a single approach performed relatively well after well-designing the relevant parameters according to field radiosonde measurement in previous practice [21,28–30]. However, the designed parameters might be less effective for other sites. This weakness makes a single model challenging to extend without sufficient prior data. Reliable and universal methods estimating C_n^2 from routine meteorological parameters are critical to evaluate the optical turbulence influence on adaptive optics systems. The most obvious example is a forecasting study in which the astronomer can not obtain optical turbulence in advance directly. Generally, researchers can forecast C_n^2 via weather forecasting models, combining different estimating approaches [8,12,39,40]. In addition, it also provides us with an applicable method to study regional optical turbulence characteristics from historical meteorological data. To be certain, more validation work should be done up until that point.

Author Contributions: H.Z. ((Hanjiu Zhang): methodology, software, data analysis and writing–draft preparation; L.Z.: software, data analysis, and editing; G.S.: funding acquisition, guidance, data collection, data curation and writing—review; K.Z., Y.L., X.M. and H.Z. (Haojia Zhang): investigation and editing; Q.L.: instrument design and manufacture; S.C., T.L. and X.L.: funding acquisition, experiment guarantee, and data collection; N.W.: guidance and writing—review. All authors have read and agreed to the published version of the manuscript.

Funding: This research was funded by the National High-tech Research and Development Program (No. E23D0HA65S2).

Data Availability Statement: Data underlying the results presented in this paper are not publicly available at this time but may be obtained from the authors upon reasonable request.

Acknowledgments: The authors express their thanks to all AIOFM colleagues carrying out the experiments. The authors would like to thank the reviewers for their remarks and suggestions on the manuscript.

Conflicts of Interest: The authors declare no conflict of interest.

Abbreviations

The following abbreviations are used in this manuscript:

MEP or ME	Multi-model Ensemble Pattern
OT	Optical Turbulence
MT	Micro-thermometers
SLC	Submarine Laser Communication
ECACN	Eastern coastal area of China
NCACN	Northern coastal area of China
MASS	Multi-Aperture Scintillation Sensor
Slodar	Slope Detection And Ranging
S-Dimm+	Solar Differential Image Motion Monitor+
GPS	Global Positioning System
AGL	Above the ground level
HV5/7 or HV	Hufnagel-Valley5/7 Model
H9	Hmnsp99 Model
DN	Dewan Model
TE	Thorpe Model
EN	Ellison Model
WT	WSPT Model
WG	WSTG Model
AIOFM	Anhui Institute of Optics and Fine Mechanics

Appendix A. Theories of Adopted Approaches

Appendix A provides theories of adopted approaches estimating C_n^2. We place the contents of Section 3.1 (primary manuscript) in this appendix document to avoid interrupting the fluency of the original article. In Section 3 (primary manuscript), we introduce several adopted physical-based approaches to estimate C_n^2. Theories of these approaches are briefly summarized below in the appendix document. For more details, one may refer to the corresponding literature. We roughly classify these approaches into three types. The first type is the semi-physical and semi-statistical approach. The typical approach is the Hufnagel model. The second type focuses on the calculation of the turbulence outer scale. We adopted the Hmnsp99, Dewan, and wind shear and temperature gradient (WSTG) models. The third type estimates C_n^2 via the temperature structure parameter. The Thorpe, Ellison, and wind shear and potential temperature (WSPT) models are examples.

Appendix A.1. Hufnagel-Valley 5/7 Model

The Hufnagel-Valley model was developed based on the statics of radio sounding and stellar scintillation [22,23]. The most commonly used form is the Hufnagel-Valley 5/7 model related to wind speed. The calculated formula is expressed as:

$$C_n^2(h) = 8.2 \times 10^{-26} W h^{10} e^{-h} + 2.7 \times 10^{-16} e^{-h/1.5} + A e^{-h/0.1}, \tag{A1}$$

where e is the Euler number; h (unit: m) is the height above the ground. In our case, the parameters $A = 1.7 \times 10^{-4}$ and $W = (1/15) \int_5^{20} V^2(h) dh$, where V is the wind speed (unit: m/s) between 5 and 20 km.

Appendix A.2. The Outer-Scale Method

Other approaches (Hmnsp99, Dewan, WSTG) estimating C_n^2 are based on the outer scale according to the work by Tatarskii [1]:

$$C_n^2 = 2.8 L_0^{4/3} M^2, \tag{A2}$$

where L_0 is the turbulence outer scale and M is the vertical gradient of the potential refractive index. The value of M^2 can be calculated from the temperature (T in K) and pressure (P in hPa) profiles as below:

$$M^2 = \left[\frac{-79 \times 10^{-6} P}{T^2}\left(\frac{d\theta}{dh}\right)\right]^2, \tag{A3}$$

where θ (unit: K) is the potential temperature defined as $\theta = T(1000/P)^{0.286}$. Hence, the key to the C_n^2 estimation becomes the calculation of the outer scale. Most methods that parameterize the outer scale or turbulence with macroscopic quantities rely not only on the existing theoretical basis of turbulence but also on the statistics of large amounts of experiment and numerical simulation data and physical intuition and perspicacity of the founders of these models. The adopted approaches in our study calculating L_0 are listed below.

Appendix A.2.1. Hmnsp99 Model

The Hmnsp99 model defines L_0 with wind shear (S) and temperature gradients (dT/dh) [24]. The expressions are different in the troposphere and stratosphere as

$$L_0^{4/3} = \begin{cases} 0.1^{4/3} \times 10^{0.362+16.728S-192.347\frac{dT}{dh}}, Troposphere \\ 0.1^{4/3} \times 10^{0.757+13.819S-57.784\frac{dT}{dh}}, Stratosphere \end{cases} \tag{A4}$$

where $S = \sqrt{(du/dh)^2 + (dv/dh)^2}$ (hereinafter), u and v are the north and east horizontal wind components, respectively.

Appendix A.2.2. Dewan Model

Dewan model deduces L_0 from only one parameter (wind shear) [25]. Meanwhile, it is similar to the Hmnsp99 model, which has a different form in the troposphere and stratosphere:

$$L_0^{4/3} = \begin{cases} 0.1^{4/3} \times 10^{1.64+42S}, Troposphere \\ 0.1^{4/3} \times 10^{0.506+50S}.Stratosphere \end{cases} \tag{A5}$$

Appendix A.2.3. WSTG Model

The WSTG model is a modified model that comes from Hmnsp99. The calculation of the outer scale is related to the dynamic and thermodynamic state of the atmosphere [30]. The expression is as follows:

$$L_0^{4/3} = \begin{cases} 0.1^{4/3} \times 10^{0.835-37.464S-306.034\frac{dT}{dh}}, S < 0.016 \cap dT/dh < 0 \\ 0.1^{4/3} \times 10^{0.825+66.9S-52.783\frac{dT}{dh}}, S < 0.016 \cap dT/dh > 0 \\ 0.1^{4/3} \times 10^{0.715+52.907S-102.515\frac{dT}{dh}}, S > 0.016 \cap dT/dh < 0 \\ 0.1^{4/3} \times 10^{2.215-9.882S-101.666\frac{dT}{dh}}, S > 0.016 \cap dT/dh > 0 \end{cases} \tag{A6}$$

Appendix A.3. The Temperature Structure Parameter Method

The remaining approaches (Thorpe, Ellison, WSPT) deduce C_n^2 from the Gladstone relationship [12] as follows:

$$C_n^2 = \left(79 \times 10^{-6}\frac{P}{T^2}\right)^2 C_T^2, \tag{A7}$$

and the temperature structure parameter (C_T^2) expressed as:

$$C_T^2 = c_0 L_0^{4/3}\left(\frac{\partial\bar{\theta}}{\partial h}\right)^2. \tag{A8}$$

c_0 is a constant that should be determined by experiment. $c1$, $c2$, and $c3$ are also unknown proportionality constants determined by experiment data in the following Equations (A9), (A11), and (A13).

Appendix A.3.1. Thorpe Model

The Thorpe model quantifies C_T^2 with the Thorpe scale (L_T) and sorted potential temperature gradients ($\partial\theta_s/\partial h$) [26] as follows:

$$C_T^2 = c_1 L_T^{4/3}\left(\frac{\partial\theta_s}{\partial h}\right)^2, \tag{A9}$$

$$L_T = \begin{cases} \sqrt{\left|h_{original} - h_{sorted}\right|} \\ or \\ \left|h_{original} - h_{sorted}\right| \end{cases}. \tag{A10}$$

θ_s(hereinafter) is the sorted potential temperature rearranged in ascending order; $h_{original}$ and h_{sorted} are the corresponding heights of the original potential temperature and sorted potential temperature, respectively. In our case, we chose the latter formula of Equation (A10) to calculate L_T.

Appendix A.3.2. Ellison Model

Ellison proposed the Ellison scale, which refers to density or potential temperature, to study the overturning of fluid caused by turbulences [27,28]. The calculation formula is as follows:

$$C_T^2 = c_2 L_E^{4/3}\left(\frac{\partial\theta_s}{\partial h}\right)^2, \tag{A11}$$

$$L_E = \left|\frac{\Delta\theta}{\partial\theta_s/\partial h}\right|. \tag{A12}$$

L_E is the Ellison scale; $\Delta\theta$ (hereinafter) represents the difference value of the original and sorted (ascending) potential temperature.

Appendix A.3.3. WSPT Model

The WSPT model involves both the wind speed and potential temperature information, calculating C_T^2 [29] as follows:

$$C_T^2 = c_3 L_W^{4/3}\left(\frac{\partial\theta_s}{\partial h}\right)^2, \tag{A13}$$

$$L_W = \sqrt{\frac{|\Delta\theta|}{\partial\theta_s/\partial h}\cdot\left(\frac{uv}{S^2}\right)^{1/2}}. \tag{A14}$$

Appendix B. Details of Two Areas Radiosonde, Models Estimations, and Integrated Astronomical Parameters Results

This Appendix B provides the details of two areas radiosonde, models estimations, and integrated astronomical parameters results. Variables, symbols, and abbreviations used in this document have the same meanings as the primary manuscript.

Appendix B.1. Radiosonde Details

Appendix B.1.1. ECACN

Radiosonde details of the eastern coastal area of China (ECACN) are included in Table A1.

Table A1. ECACN radiosonde details (BJT: Beijing time).

Site	Flight Number	Date	Release Time (BJT)	Flight Duration /s	Termination Altitude (AGL)/m
ECACN	1	5 Apirl 2018	19:30	5010	29,020
	2	9 Apirl 2018	19:30	5027	31,210
	3	10 Apirl 2018	19:30	4767	30,410
	4	11 Apirl 2018	19:30	4788	27,880
	5	12 Apirl 2018	19:30	5014	29,810
	6	15 Apirl 2018	19:30	4088	25,740
	7	17 Apirl 2018	19:30	4355	26,410
	8	18 Apirl 2018	19:30	4686	28,150
	9	19 Apirl 2018	19:30	5037	30,120
	10	20 Apirl 2018	19:30	5371	31,620
	11	22 Apirl 2018	19:30	4820	28,220
	12	24 Apirl 2018	19:30	5176	30,710
	13	25 Apirl 2018	19:30	5051	30,910
	14	26 Apirl 2018	19:30	5088	29,410
	15	27 Apirl 2018	19:30	5443	31,750
	16	28 Apirl 2018	19:30	5144	31,170

Appendix B.1.2. NCACN

Radiosonde details of the northern coastal area of China (NCACN) are included in Table A2.

Table A2. NCACN radiosonde details.

Site	Flight Number	Date	Release Time (BJT)	Flight Duration /s	Termination Altitude (AGL)/m
NCACN	1	3 Apirl 2018	19:30	4464	28,660
	2	4 Apirl 2018	7:30	4210	27,970
	3	4 Apirl 2018	19:30	4747	29,460
	4	5 Apirl 2018	7:30	4808	29,320
	5	8 Apirl 2018	19:30	4271	27,370
	6	9 Apirl 2018	7:30	4855	28,880
	7	9 Apirl 2018	19:30	4780	29,660
	8	10 Apirl 2018	19:30	5275	29,780
	9	12 Apirl 2018	7:30	4591	27,810
	10	13 Apirl 2018	7:30	4633	28,710
	11	14 Apirl 2018	19:30	5069	29,680
	12	16 Apirl 2018	7:30	5360	29,380
	13	16 Apirl 2018	19:30	5292	30,050
	14	17 Apirl 2018	7:30	5176	28,850
	15	20 Apirl 2018	7:30	5155	29,660
	16	21 Apirl 2018	19:30	4853	29,400
	17	25 Apirl 2018	19:30	5012	30,750
	18	26 Apirl 2018	7:30	4714	28,790
	19	26 Apirl 2018	19:30	4901	30,660
	20	27 Apirl 2018	19:30	4798	28,530

Appendix B.2. The Refractive Index Structure Parameter of MT and Estimations

Appendix B.2.1. ECACN MT and Models Estimations

The refractive structure index parameter profiles of MT and estimations in ECACN are exhibited in Figures A1–A3.

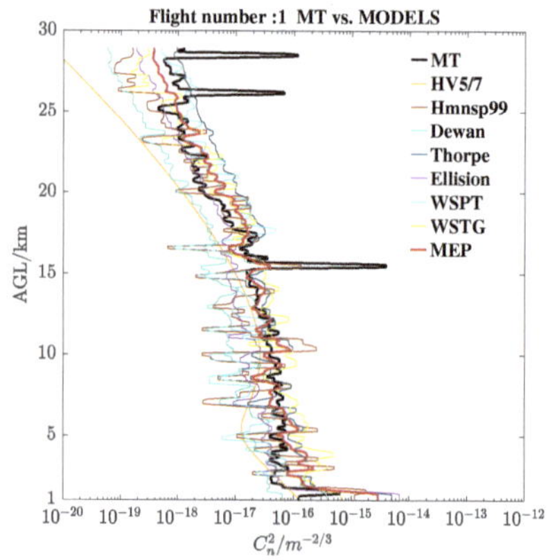

(**a**) Flight number: 1.

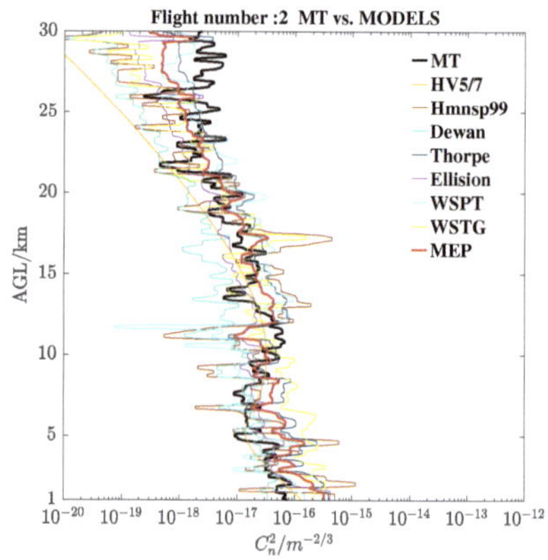

(**b**) Flight number: 2.

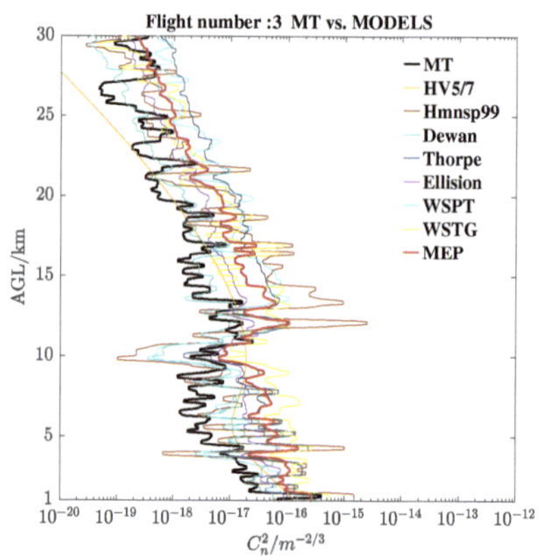

(**c**) Flight number: 3.

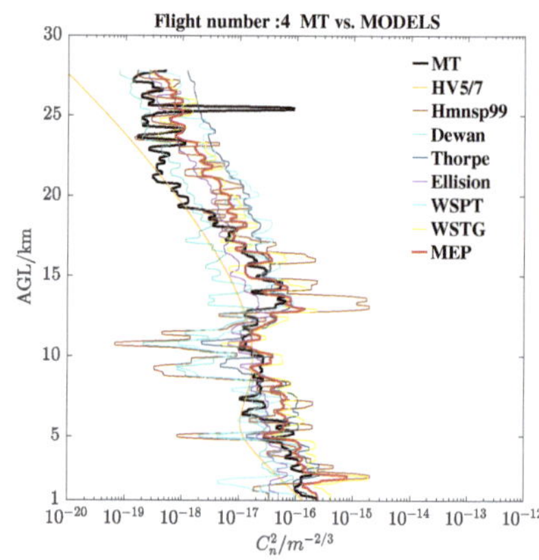

(**d**) Flight number: 4.

Figure A1. The refractive index structure parameter profiles of MT and estimations in ECACN: Figure A1 sub-figures (**a**–**d**) are flight numbers 1–4 in Table A1.

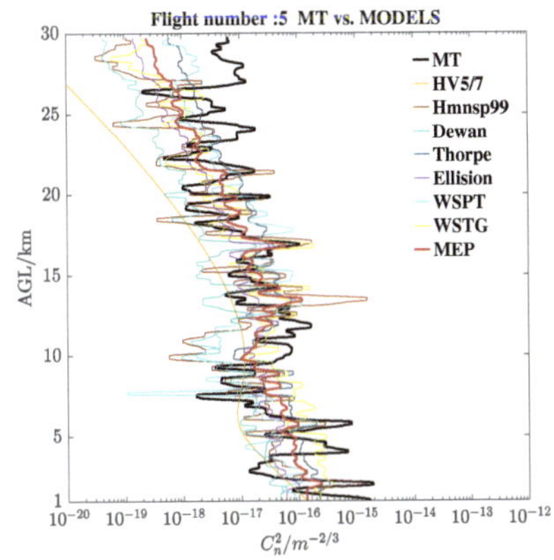

(**a**) Flight number: 5.

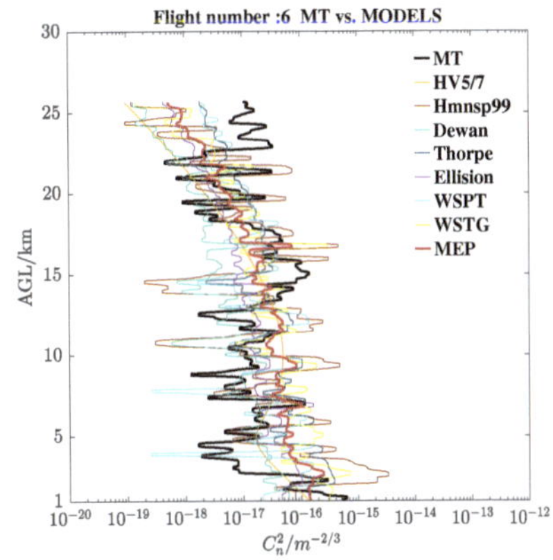

(**b**) Flight number: 6.

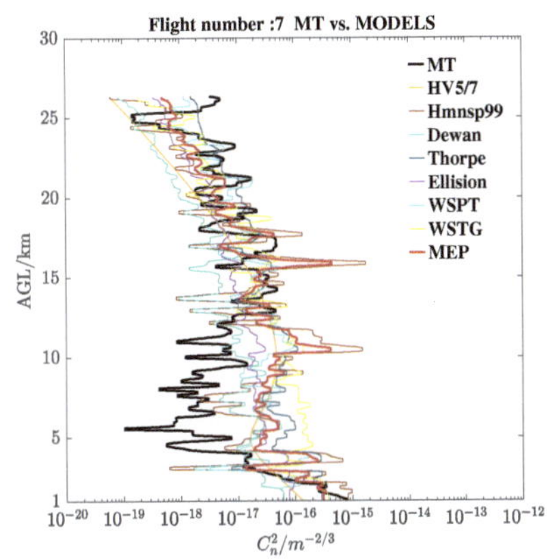

(**c**) Flight number: 7.

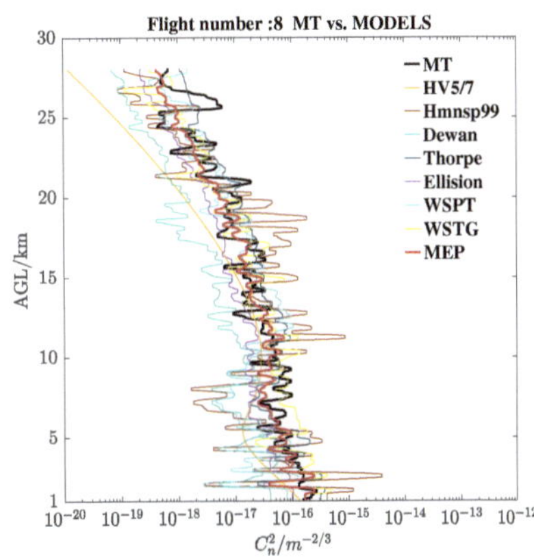

(**d**) Flight number: 8.

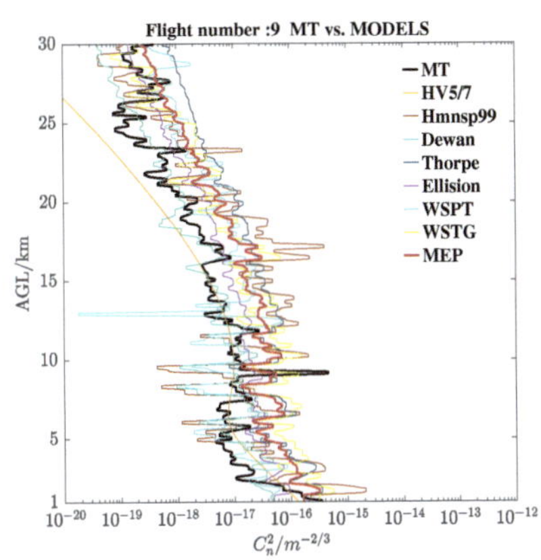

(**e**) Flight number: 9.

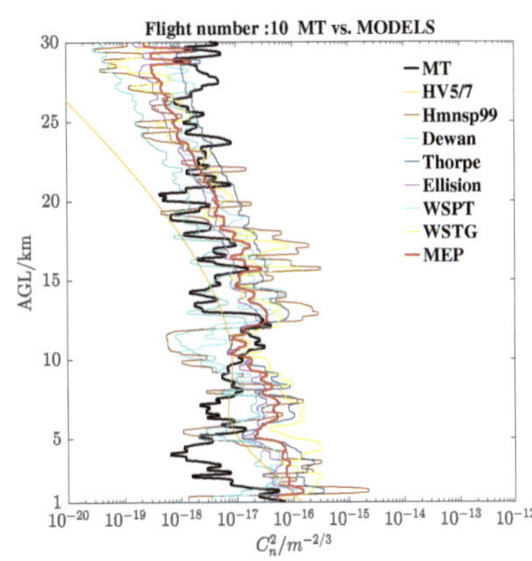

(**f**) Flight number: 10.

Figure A2. The refractive index structure parameter profiles of MT and estimations in ECACN: Figure A2 sub-figures (**a**–**f**) are flight numbers 5–10 in Table A1.

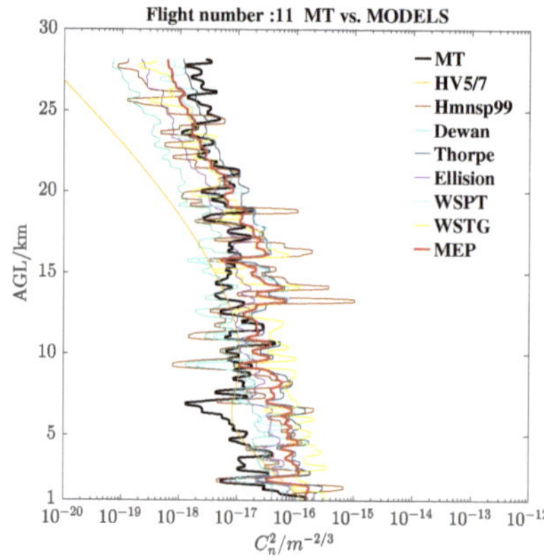

(**a**) Flight number: 11.

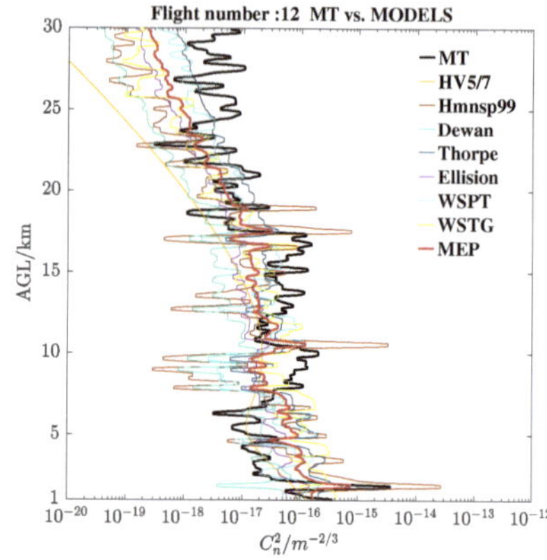

(**b**) Flight number: 12.

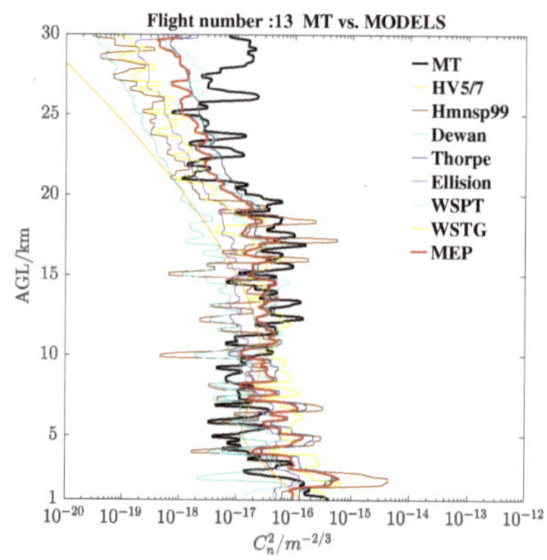

(**c**) Flight number: 13.

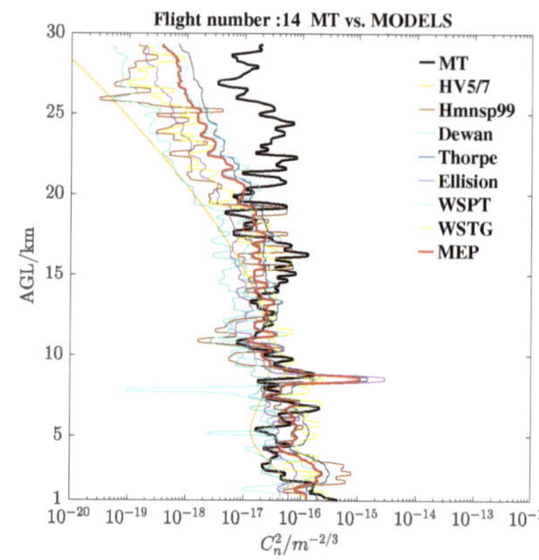

(**d**) Flight number: 14.

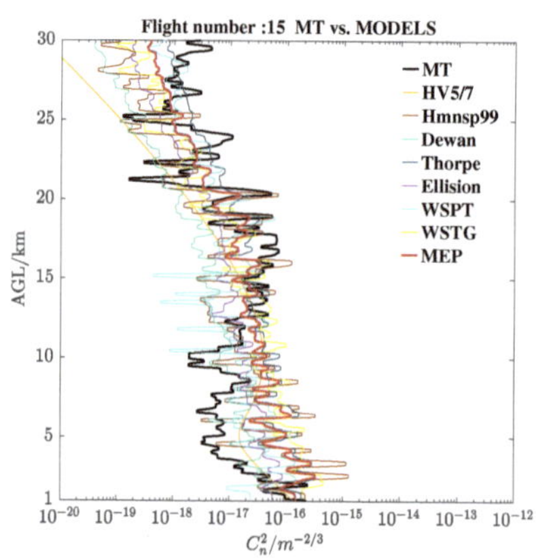

(**e**) Flight number: 15.

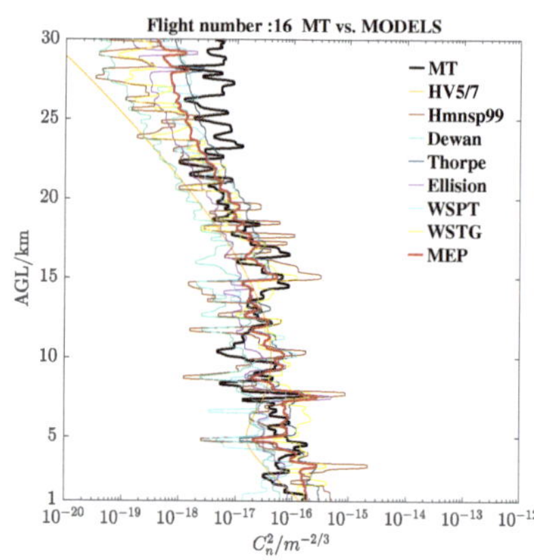

(**f**) Flight number: 16.

Figure A3. The refractive index structure parameter profiles of MT and estimations in ECACN: Figure A3 sub-figures (**a**–**f**) are flight numbers 11–16 in Table A1.

Appendix B.2.2. NCACN MT and Models Estimations

The refractive structure index parameter profiles of MT and estimations in NCACN are exhibited in Figures A4–A7.

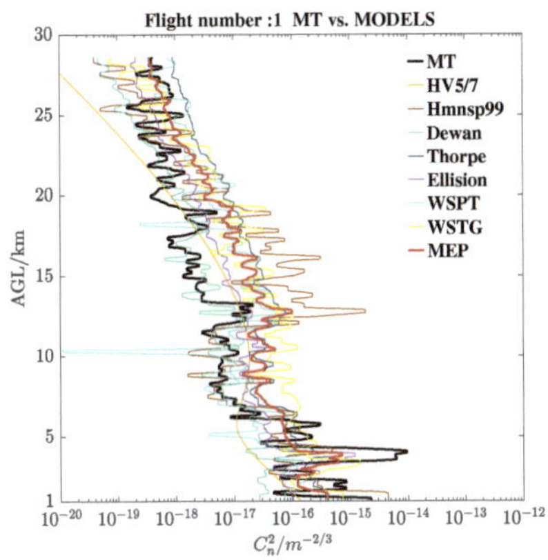

(**a**) Flight number: 1.

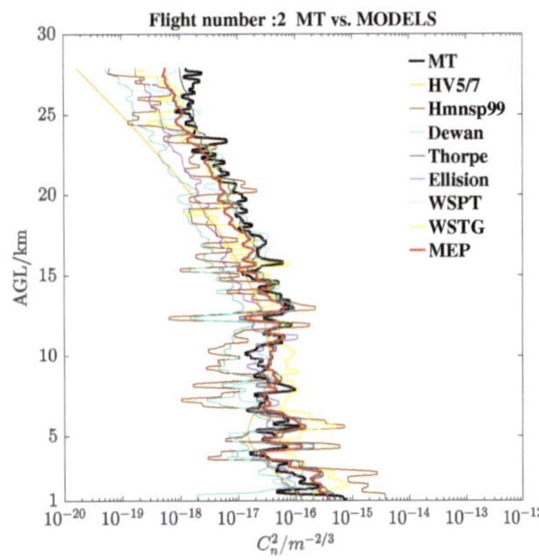

(**b**) Flight number: 2.

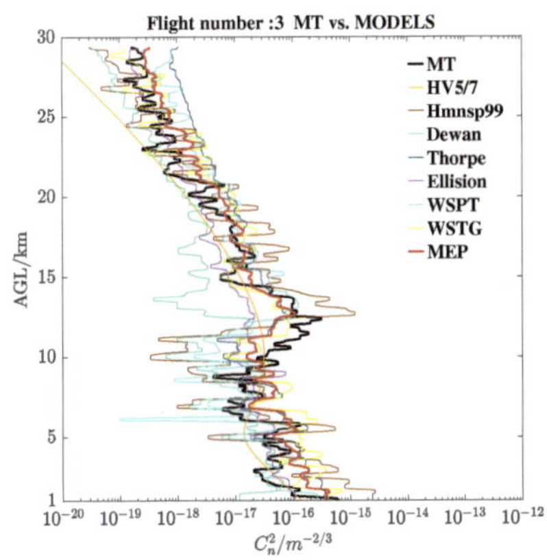

(**c**) Flight number: 3.

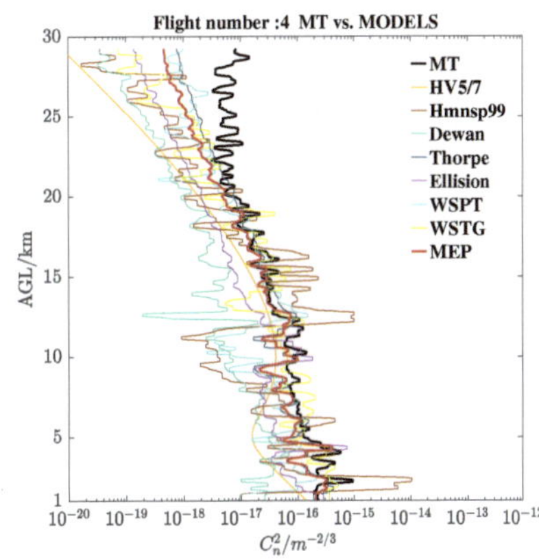

(**d**) Flight number: 4.

Figure A4. The refractive index structure parameter profiles of MT and estimations in NCACN: Figure A4 sub-figures (**a–d**) are flight numbers 1–4 in Table A2.

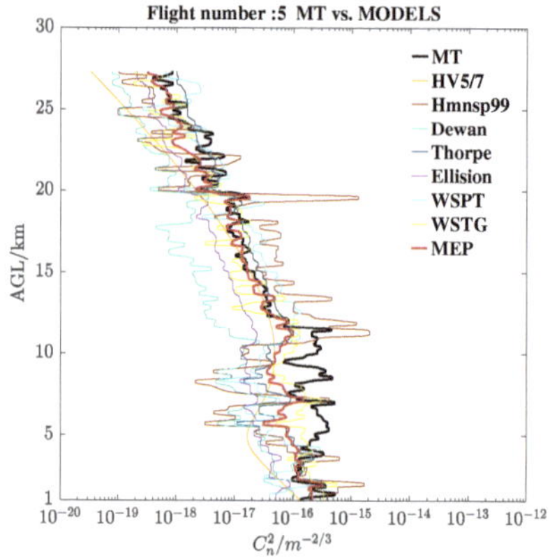

(**a**) Flight number: 5.

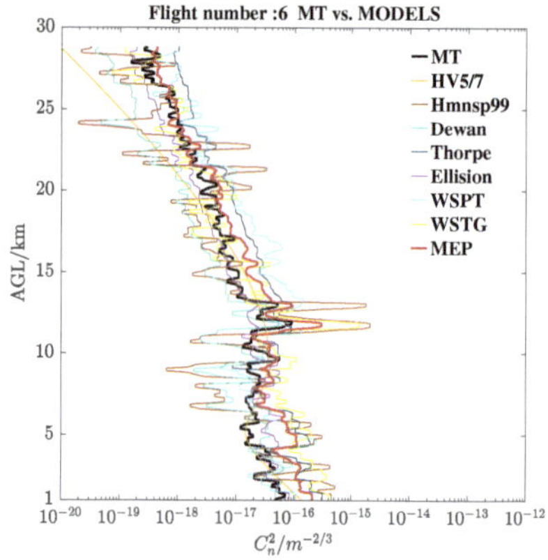

(**b**) Flight number: 6.

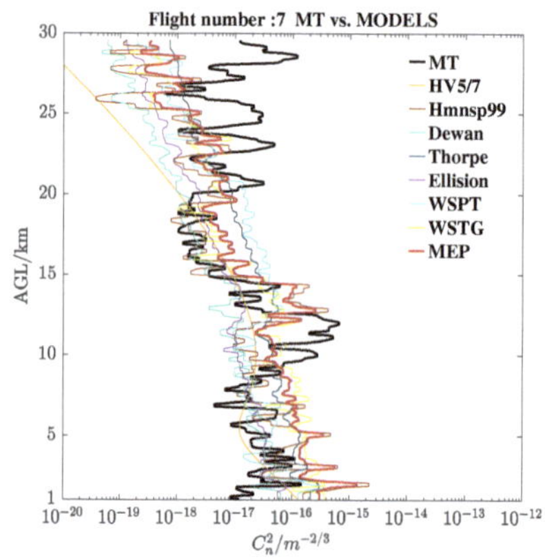

(**c**) Flight number: 7.

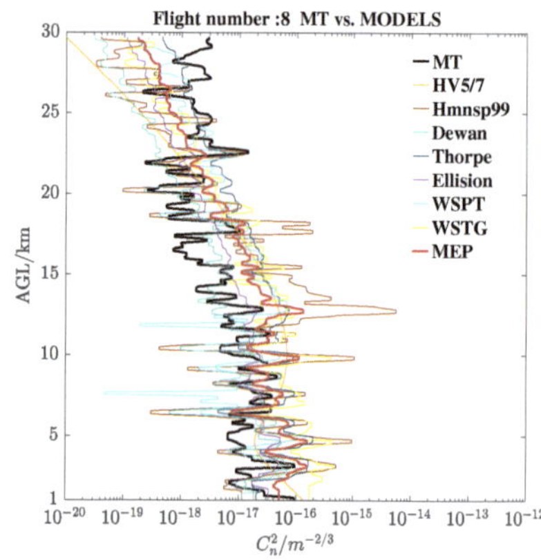

(**d**) Flight number: 8.

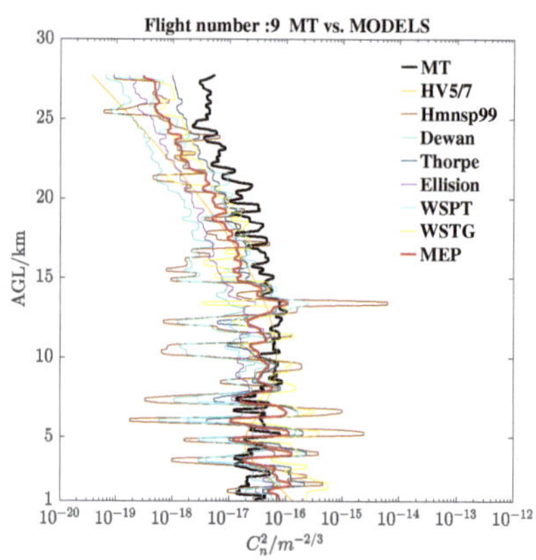

(**e**) Flight number: 9.

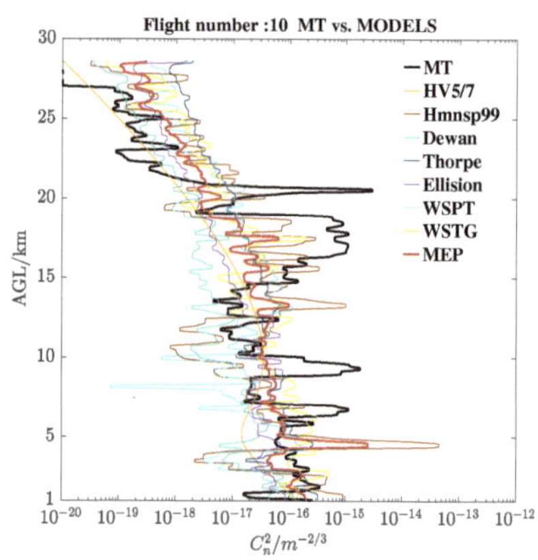

(**f**) Flight number: 10.

Figure A5. The refractive index structure parameter profiles of MT and estimations in NCACN: Figure A5 sub-figures (**a**–**f**) are flight numbers 5–10 in Table A2.

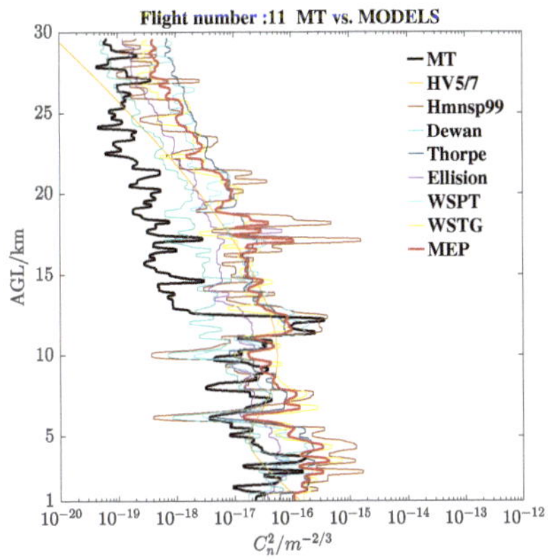

(**a**) Flight number: 11.

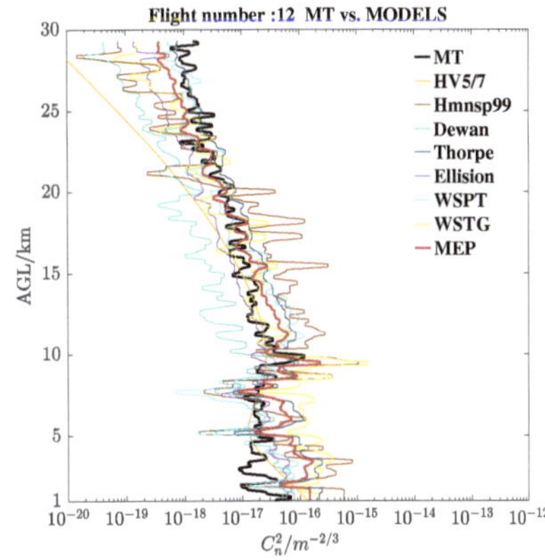

(**b**) Flight number: 12.

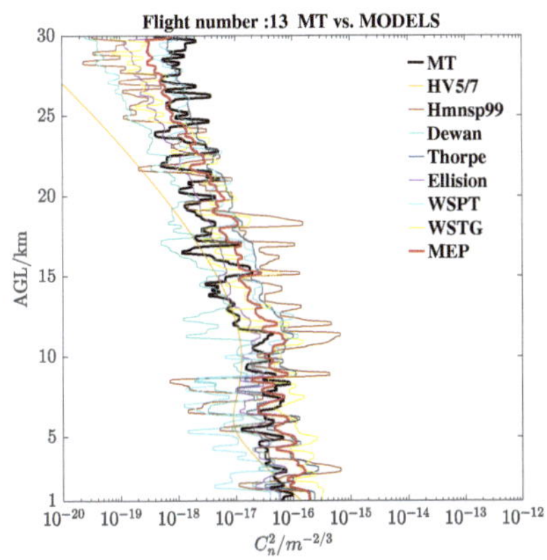

(**c**) Flight number: 13.

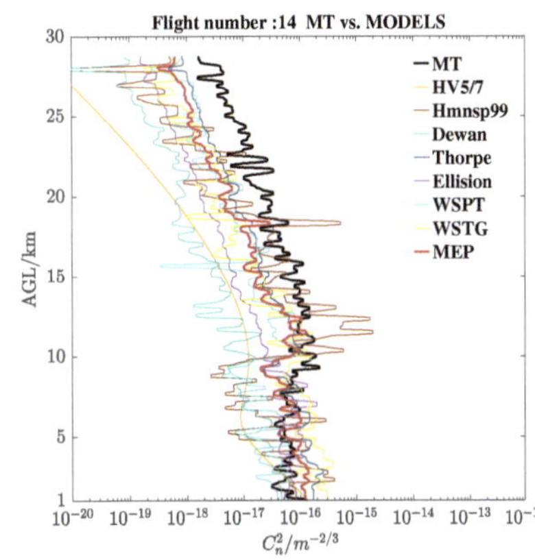

(**d**) Flight number: 14.

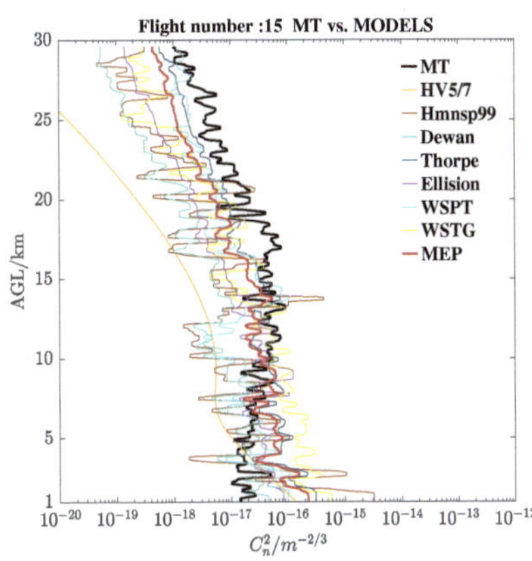

(**e**) Flight number: 15.

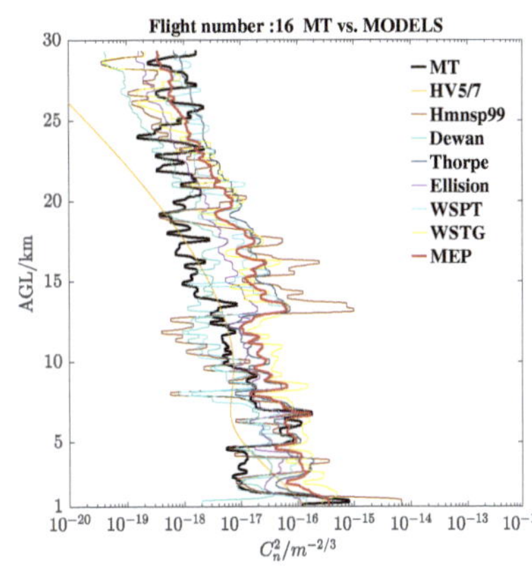

(**f**) Flight number: 16.

Figure A6. The refractive index structure parameter profiles of MT and estimations in NCACN: Figure A6 sub-figures (**a–f**) are flight numbers 11–16 in Table A1.

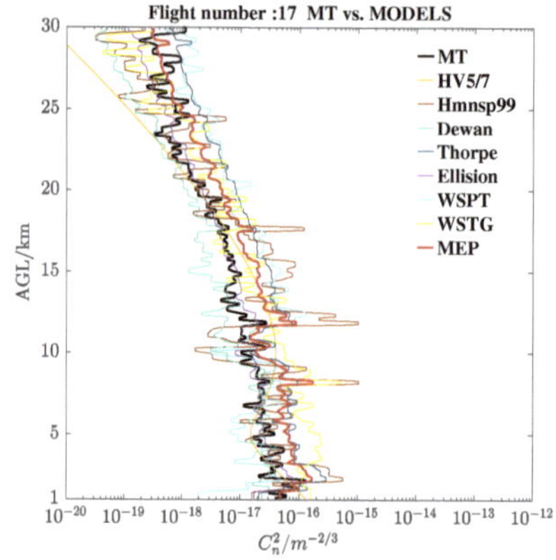

(a) Flight number: 17.

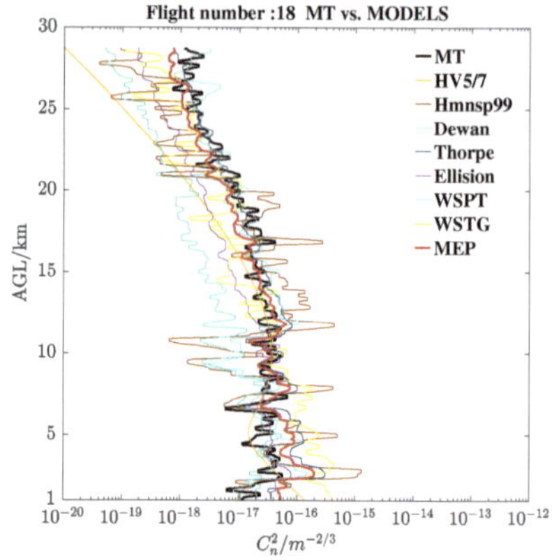

(b) Flight number: 18.

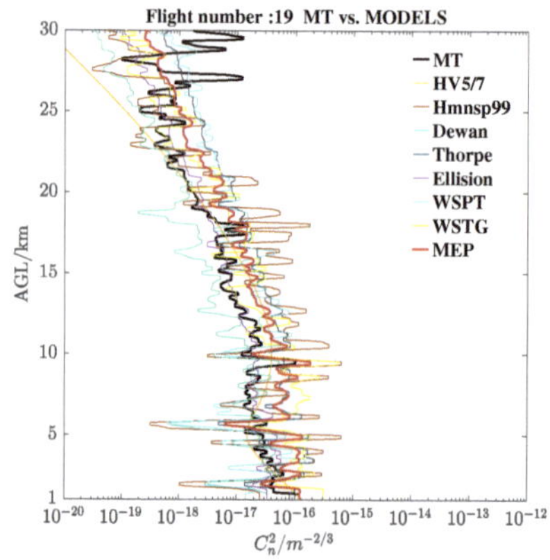

(c) Flight number: 19.

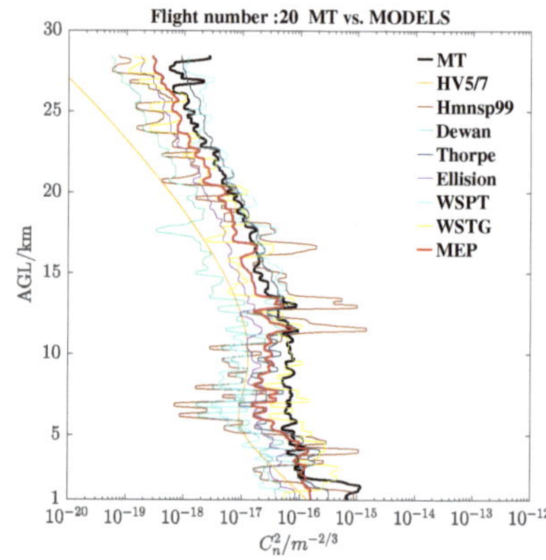

(d) Flight number: 20.

Figure A7. The refractive index structure parameter profiles of MT and estimations in NCACN: Figure A7 sub-figures (**a–d**) are flight numbers 17–20 in Table A2.

Appendix B.3. The Integrated Astronomical Parameters

Appendix B.3.1. ECACN Integrated Astronomical Parameters Details

The integrated astronomical parameters details calculated from radiosonde and model results in ECACN are included in Tables A3–A6.

Table A3. ECACN integrated astronomical parameters details (r_0 @λ = 550 nm).

Parameter	Flight Number	Method								
		MT	HV	H9	DN	TE	EN	WT	WG	ME
r_0/cm	1	6.45	15.41	5.95	19.31	4.25	4.08	10.48	5.92	5.30
	2	13.15	14.38	6.52	21.05	7.33	11.32	11.79	6.67	9.05
	3	21.22	16.91	4.80	22.64	8.72	15.68	12.40	6.98	10.10
	4	10.47	17.37	4.51	20.13	8.13	15.24	12.73	5.64	7.78
	5	5.63	18.97	6.43	20.46	7.50	13.76	10.94	6.34	9.34
	6	9.72	12.37	3.77	20.69	7.05	13.10	9.98	6.33	9.03
	7	10.63	12.53	4.51	20.72	7.28	14.47	11.04	6.64	7.56
	8	8.66	15.11	4.57	21.32	7.64	14.30	12.02	7.14	9.65
	9	15.61	19.53	5.68	21.35	7.69	14.93	16.86	6.92	9.21
	10	23.28	20.11	6.23	23.69	8.21	15.39	14.21	7.22	11.84

Table A3. *Cont.*

Parameter	Flight Number	MT	HV	H9	DN	Method TE	EN	WT	WG	ME
r_0/cm	11	18.69	19.12	7.52	21.14	8.01	14.69	12.23	7.03	9.71
	12	6.43	16.28	2.31	20.37	7.21	14.47	9.76	6.78	8.82
	13	9.19	15.40	4.08	22.02	7.38	14.27	9.58	6.82	8.06
	14	8.12	14.94	7.06	18.54	6.05	6.97	11.01	7.19	7.47
	15	12.90	13.44	6.36	19.82	7.41	14.82	12.43	6.70	8.66
	16	9.72	13.12	5.57	18.49	7.11	12.14	12.34	6.78	8.27
	Median	10.10	15.41	5.63	20.70	7.39	14.38	11.91	6.78	8.93
	Mean	11.87	15.94	5.37	20.73	7.31	13.10	11.86	6.69	8.74

Table A4. ECACN integrated astronomical parameters details (ϵ @λ = 550 nm).

Parameter	Flight Number	MT	HV	H9	DN	Method TE	EN	WT	WG	ME
$\epsilon/''$	1	1.72	0.72	1.87	0.58	2.61	2.73	1.06	1.88	2.10
	2	0.85	0.77	1.71	0.53	1.52	0.98	0.94	1.67	1.23
	3	0.52	0.66	2.32	0.49	1.28	0.71	0.90	1.59	1.10
	4	1.06	0.64	2.47	0.55	1.37	0.73	0.87	1.97	1.43
	5	1.97	0.59	1.73	0.54	1.48	0.81	1.02	1.75	1.19
	6	1.14	0.90	2.95	0.54	1.58	0.85	1.11	1.76	1.23
	7	1.05	0.89	2.47	0.54	1.53	0.77	1.01	1.67	1.47
	8	1.28	0.74	2.43	0.52	1.46	0.78	0.92	1.56	1.15
	9	0.71	0.57	1.96	0.52	1.45	0.74	0.66	1.61	1.21
	10	0.48	0.55	1.79	0.47	1.35	0.72	0.78	1.54	0.94
	11	0.59	0.58	1.48	0.53	1.39	0.76	0.91	1.58	1.15
	12	1.73	0.68	4.81	0.55	1.54	0.77	1.14	1.64	1.26
	13	1.21	0.72	2.72	0.50	1.51	0.78	1.16	1.63	1.38
	14	1.37	0.74	1.57	0.60	1.84	1.59	1.01	1.55	1.49
	15	0.86	0.83	1.75	0.56	1.50	0.75	0.89	1.66	1.28
	16	1.14	0.85	1.99	0.60	1.56	0.92	0.90	1.64	1.34
	Median	1.10	0.72	1.98	0.54	1.50	0.77	0.93	1.64	1.25
	Mean	1.11	0.71	2.25	0.54	1.56	0.96	0.96	1.67	1.31

Table A5. ECACN integrated astronomical parameters details (θ_{AO} @λ = 550 nm).

Parameter	Flight Number	MT	HV	H9	DN	Method TE	EN	WT	WG	ME
$\theta_{AO}/''$	1	0.33	1.23	0.69	1.69	0.57	0.99	0.84	0.57	0.69
	2	0.78	1.09	0.47	1.79	0.57	1.13	0.68	0.60	0.74
	3	2.01	1.47	0.31	1.89	0.62	1.20	0.70	0.58	0.77
	4	0.83	1.55	0.28	1.77	0.60	1.14	0.70	0.55	0.68
	5	0.52	1.91	0.42	1.74	0.59	1.02	0.64	0.55	0.73
	6	0.57	0.87	0.38	1.76	0.60	1.08	0.68	0.56	0.73
	7	0.85	0.88	0.31	1.67	0.60	1.18	0.74	0.55	0.54
	8	0.69	1.18	0.39	1.75	0.58	1.16	0.71	0.61	0.77
	9	1.36	2.07	0.46	1.77	0.58	1.16	1.00	0.61	0.75
	10	1.13	2.27	0.46	1.90	0.60	1.16	0.88	0.60	0.91
	11	1.10	1.95	0.47	1.77	0.59	1.12	0.76	0.60	0.73
	12	0.47	1.36	0.32	1.75	0.61	1.17	0.68	0.63	0.80
	13	0.46	1.23	0.47	1.86	0.59	1.14	0.67	0.63	0.66
	14	0.40	1.16	0.68	1.58	0.48	0.53	0.74	0.59	0.58
	15	0.65	0.98	0.61	1.73	0.58	1.15	0.83	0.62	0.71
	16	0.60	0.94	0.49	1.60	0.57	0.97	0.74	0.56	0.66
	Median	0.67	1.23	0.46	1.76	0.59	1.14	0.72	0.60	0.73
	Mean	0.80	1.38	0.45	1.75	0.58	1.08	0.75	0.59	0.72

Table A6. ECACN integrated astronomical parameters details (σ_I^2 @λ = 550 nm).

Parameter	Flight Number	MT	HV	H9	DN	TE	EN	WT	WG	ME
		Method								
	1	1.54	0.24	0.76	0.15	1.10	0.75	0.41	0.99	0.81
	2	0.40	0.28	1.04	0.13	0.80	0.30	0.49	0.87	0.54
	3	0.10	0.19	2.02	0.12	0.66	0.24	0.47	0.85	0.49
	4	0.44	0.17	2.44	0.14	0.73	0.26	0.46	1.03	0.67
	5	1.03	0.13	1.22	0.14	0.78	0.31	0.56	0.97	0.55
	6	0.63	0.40	1.95	0.14	0.82	0.31	0.57	0.95	0.58
	7	0.37	0.39	2.10	0.14	0.79	0.26	0.49	0.95	0.86
	8	0.61	0.25	1.61	0.14	0.78	0.26	0.49	0.82	0.51
	9	0.21	0.12	1.12	0.13	0.77	0.26	0.27	0.82	0.54
σ_I^2	10	0.18	0.11	1.08	0.12	0.72	0.25	0.34	0.82	0.38
	11	0.22	0.13	0.99	0.14	0.75	0.27	0.44	0.84	0.54
	12	1.01	0.21	3.18	0.14	0.78	0.26	0.57	0.82	0.52
	13	0.82	0.24	1.44	0.12	0.78	0.27	0.58	0.81	0.66
	14	1.02	0.26	0.72	0.17	1.18	1.05	0.49	0.84	0.85
	15	0.47	0.33	0.87	0.14	0.82	0.26	0.40	0.84	0.60
	16	0.63	0.35	1.16	0.16	0.84	0.37	0.46	0.92	0.67
	Median	0.54	0.24	1.19	0.14	0.78	0.26	0.48	0.85	0.56
	Mean	0.61	0.24	1.48	0.14	0.82	0.35	0.47	0.88	0.61

Appendix B.3.2. NCACN Integrated Astronomical Parameters Details

The integrated astronomical parameters details calculated from radiosonde and model results in NCACN are included in Tables A7–A10.

Table A7. NCACN integrated astronomical parameters details (r_0 @λ = 550 nm).

Parameter	Flight Number	MT	HV	H9	DN	TE	EN	WT	WG	ME
		Method								
	1	2.95	17.18	3.74	19.16	6.84	8.21	12.00	5.21	7.12
	2	8.63	14.05	3.57	20.75	7.36	13.01	10.20	6.15	8.27
	3	9.04	14.56	3.85	21.19	7.50	14.53	10.75	6.64	8.16
	4	5.67	13.40	2.67	20.68	6.90	8.99	9.81	5.77	7.40
	5	4.93	12.47	3.96	17.86	7.79	15.88	8.42	6.19	8.03
	6	15.11	13.97	4.86	18.12	7.82	14.03	9.47	5.81	8.84
	7	7.19	15.95	5.82	18.41	7.76	16.78	10.38	6.20	5.66
	8	19.46	10.79	3.86	19.55	9.41	17.60	15.12	6.57	10.77
	9	9.86	10.72	4.16	20.76	8.94	15.73	10.18	6.40	10.73
r_0	10	3.32	14.17	1.61	20.21	7.85	13.02	11.18	6.46	6.27
	11	13.17	11.67	3.99	19.27	7.60	14.78	9.66	6.57	7.11
	12	15.67	15.55	5.16	17.36	7.84	15.65	12.09	5.93	9.40
	13	13.35	18.65	6.17	20.62	8.16	16.50	14.61	6.91	9.29
	14	7.06	18.80	4.77	20.02	7.42	11.02	13.74	6.24	8.79
	15	10.77	21.23	4.92	21.43	8.19	12.19	12.68	6.11	9.73
	16	12.67	20.45	3.74	24.38	8.56	14.97	14.64	6.84	8.83
	17	16.23	13.34	6.61	20.65	8.50	16.84	12.93	6.82	9.93
	18	13.03	13.83	6.00	21.46	8.35	14.32	12.10	6.31	10.11
	19	15.06	13.40	6.60	20.38	8.86	16.90	13.65	7.10	9.62
	20	6.06	18.60	5.22	21.95	8.43	16.24	13.98	6.75	10.36
	Median	10.31	14.11	4.46	20.50	7.84	14.88	12.04	6.36	8.83
	Mean	10.46	15.14	4.56	20.21	8.00	14.36	11.88	6.35	8.72

Table A8. NCACN integrated astronomical parameters details (ϵ @λ = 550 nm).

Parameter	Flight Number	MT	HV	H9	DN	Method TE	EN	WT	WG	ME
$\epsilon/''$	1	3.77	0.65	2.97	0.58	1.63	1.35	0.93	2.13	1.56
	2	1.29	0.79	3.12	0.54	1.51	0.85	1.09	1.81	1.34
	3	1.23	0.76	2.88	0.52	1.48	0.76	1.03	1.67	1.36
	4	1.96	0.83	4.17	0.54	1.61	1.24	1.13	1.93	1.50
	5	2.26	0.89	2.81	0.62	1.43	0.70	1.32	1.80	1.39
	6	0.74	0.80	2.29	0.61	1.42	0.79	1.17	1.91	1.26
	7	1.55	0.70	1.91	0.60	1.43	0.66	1.07	1.79	1.97
	8	0.57	1.03	2.88	0.57	1.18	0.63	0.74	1.69	1.03
	9	1.13	1.04	2.67	0.54	1.24	0.71	1.09	1.74	1.04
	10	3.35	0.78	6.92	0.55	1.42	0.85	0.99	1.72	1.77
	11	0.84	0.95	2.79	0.58	1.46	0.75	1.15	1.69	1.56
	12	0.71	0.71	2.16	0.64	1.42	0.71	0.92	1.87	1.18
	13	0.83	0.60	1.80	0.54	1.36	0.67	0.76	1.61	1.20
	14	1.57	0.59	2.33	0.56	1.50	1.01	0.81	1.78	1.26
	15	1.03	0.52	2.26	0.52	1.36	0.91	0.88	1.82	1.14
	16	0.88	0.54	2.97	0.46	1.30	0.74	0.76	1.62	1.26
	17	0.69	0.83	1.68	0.54	1.31	0.66	0.86	1.63	1.12
	18	0.85	0.80	1.85	0.52	1.33	0.78	0.92	1.76	1.10
	19	0.74	0.83	1.68	0.55	1.25	0.66	0.81	1.57	1.16
	20	1.83	0.60	2.13	0.51	1.32	0.68	0.80	1.65	1.07
	Median	1.08	0.79	2.50	0.54	1.42	0.75	0.92	1.75	1.26
	Mean	1.39	0.76	2.71	0.55	1.40	0.81	0.96	1.76	1.31

Table A9. NCACN integrated astronomical parameters details (θ_{AO} @λ = 550 nm).

Parameter	Flight Number	MT	HV	H9	DN	Method TE	EN	WT	WG	ME
$\theta_{AO}/''$	1	0.49	1.52	0.35	1.82	0.62	0.97	0.91	0.54	0.72
	2	0.64	1.05	0.62	1.76	0.63	1.05	0.73	0.59	0.72
	3	0.61	1.11	0.36	1.98	0.66	1.23	0.69	0.63	0.71
	4	0.47	0.97	0.33	1.84	0.59	0.80	0.68	0.54	0.65
	5	0.41	0.88	0.24	1.86	0.63	1.30	0.55	0.57	0.67
	6	1.02	1.04	0.31	1.56	0.62	1.11	0.56	0.46	0.67
	7	0.35	1.31	0.56	1.63	0.63	1.38	0.68	0.58	0.53
	8	1.26	0.72	0.21	1.63	0.64	1.30	0.88	0.56	0.78
	9	0.51	0.71	0.26	1.71	0.65	1.19	0.63	0.57	0.75
	10	0.15	1.06	0.19	1.81	0.62	0.96	0.68	0.53	0.59
	11	0.97	0.80	0.29	1.86	0.64	1.16	0.65	0.61	0.59
	12	0.93	1.25	0.37	1.65	0.61	1.26	0.74	0.50	0.73
	13	1.02	1.83	0.39	1.96	0.61	1.28	0.78	0.65	0.77
	14	0.38	1.87	0.27	1.93	0.59	0.95	0.83	0.57	0.66
	15	0.51	2.75	0.75	1.80	0.63	0.92	0.79	0.58	0.76
	16	1.45	2.40	0.41	2.06	0.68	1.23	0.99	0.65	0.77
	17	1.29	0.97	0.45	1.77	0.63	1.31	0.66	0.59	0.77
	18	0.74	1.02	0.44	1.89	0.62	1.10	0.71	0.65	0.75
	19	1.06	0.97	0.43	1.70	0.62	1.28	0.77	0.59	0.73
	20	0.61	1.82	0.36	1.93	0.66	1.29	0.74	0.63	0.87
	Median	0.63	1.06	0.36	1.82	0.63	1.21	0.72	0.58	0.72
	Mean	0.74	1.30	0.38	1.81	0.63	1.15	0.73	0.58	0.71

Table A10. NCACN integrated astronomical parameters details (σ_I^2 @λ = 550 nm).

		Method								
Parameter	**Flight Number**	**MT**	**HV**	**H9**	**DN**	**TE**	**EN**	**WT**	**WG**	**ME**
	1	2.28	0.18	1.94	0.14	0.80	0.50	0.36	1.16	0.69
	2	0.63	0.30	1.33	0.14	0.74	0.32	0.50	0.94	0.60
	3	0.69	0.27	1.90	0.12	0.69	0.25	0.52	0.82	0.62
	4	1.19	0.33	2.85	0.14	0.85	0.54	0.58	1.06	0.74
	5	1.66	0.39	2.96	0.15	0.72	0.22	0.79	0.99	0.68
	6	0.29	0.30	2.07	0.17	0.73	0.29	0.72	1.26	0.64
	7	1.38	0.22	0.99	0.16	0.73	0.20	0.53	0.97	1.10
	8	0.19	0.53	3.69	0.16	0.64	0.21	0.33	0.96	0.49
	9	0.79	0.54	2.83	0.14	0.65	0.25	0.63	0.93	0.50
σ_I^2	10	5.37	0.29	8.17	0.14	0.73	0.35	0.52	1.01	0.96
	11	0.36	0.45	2.42	0.14	0.74	0.26	0.62	0.88	0.85
	12	0.31	0.23	1.62	0.17	0.75	0.23	0.47	1.19	0.56
	13	0.32	0.14	1.40	0.13	0.72	0.22	0.40	0.80	0.54
	14	1.32	0.14	2.49	0.13	0.81	0.41	0.38	0.99	0.65
	15	0.74	0.09	0.77	0.13	0.68	0.39	0.41	0.96	0.51
	16	0.23	0.10	1.58	0.11	0.61	0.24	0.29	0.79	0.54
	17	0.22	0.34	1.18	0.14	0.69	0.21	0.51	0.91	0.52
	18	0.45	0.31	1.25	0.13	0.72	0.29	0.48	0.86	0.52
	19	0.27	0.33	1.26	0.15	0.69	0.22	0.41	0.86	0.57
	20	0.85	0.14	1.66	0.12	0.64	0.22	0.41	0.84	0.44
	Median	0.66	0.30	1.78	0.14	0.72	0.25	0.49	0.95	0.58
	Mean	0.98	0.28	2.22	0.14	0.72	0.29	0.49	0.96	0.64

References

1. Tatarskii, V.I. *Wave Propagation in a Turbulent Medium*; McGraw-Hill: New York, NY, USA, 1961.
2. Beckers, J.M. Adaptive optics for astronomy: Principles, performance, and applications. *Annu. Rev. Astron. Astrophys.* **1993**, *31*, 13–62. [CrossRef]
3. Coulman, C.E. Fundmental and applied aspects of astronomical seeing. *Annu. Rev. Astron. Astrophys.* **1985**, *23*, 19–57. [CrossRef]
4. Davies, R.; Kasper, M. Adaptive Optics for Astronomy. *Annu. Rev. Astron. Astrophys.* **2012**, *50*, 305–351. [CrossRef]
5. Fried, D.L. Limiting resolution looking down through the atmosphere. *J. Opt. Soc. Am.* **1966**, *56*, 1380–1385. [CrossRef]
6. Deng, L.; Yang, F.; Chen, X.; He, F.; Liu, Q.; Zhang, B.; Zhang, C.; Wang, K.; Liu, N.; Ren, A.; et al. Lenghu on the Tibetan Plateau as an astronomical observing site. *Nature* **2021**, *596*, 353. [CrossRef] [PubMed]
7. Fugate, R.Q.; Fried, D.L.; Ameer, G.A.; Boeke, B.R.; Browne, S.L.; Roberts, P.H.; Ruane, R.E.; Tyler, G.A.; Wopat, L.M. Measurement of atmospheric wave-front distortion using scatterd-light from a laser guide-star. *Nature* **1991**, *353*, 144–146. [CrossRef]
8. Mahalov, A.; Moustaoui, M. Characterization of atmospheric optical turbulence for laser propagation. *Laser Photonics Rev.* **2010**, *4*, 144–159. [CrossRef]
9. Roddier, F. V the Effects of Atmospheric Turbulence in Optical Astronomy. In *Progress in Optics*; Wolf, E., Ed.; Elsevier: Amsterdam, The Netherlands, 1981; Volume 19, pp. 281–376. [CrossRef]
10. Zhang, R.Z.; Hu, N.Z.; Zhou, H.B.; Zou, K.H.; Su, X.Z.; Zhou, Y.Y.; Song, H.Q.; Pang, K.; Song, H.; Minoofar, A.; et al. Turbulence-resilient pilot-assisted self-coherent free-space optical communications using automatic optoelectronic mixing of many modes. *Nat. Photonics* **2021**, *15*, 743–750. [CrossRef]
11. Zhu, Z.; Janasik, M.; Fyffe, A.; Hay, D.; Zhou, Y.; Kantor, B.; Winder, T.; Boyd, R.W.; Leuchs, G.; Shi, Z. Compensation-free high-dimensional free-space optical communication using turbulence-resilient vector beams. *Nat. Commun.* **2021**, *12*, 1666. [CrossRef]
12. Masciadri, E.; Vernin, J.; Bougeault, P. 3D mapping of optical turbulence using an atmospheric numerical model I. A useful tool for the ground-based astronomy. *Astron. Astrophys. Suppl. Ser.* **1999**, *137*, 185–202. [CrossRef]
13. Marks, R.D.; Vernin, J.; Azouit, M.; Briggs, J.W.; Burton, M.G.; Ashley, M.C.B.; Manigault, J.F. Antarctic site testing—Microthermal measurements of surface-layer seeing at the South Pole. *Astron. Astrophys. Suppl. Ser.* **1996**, *118*, 385–390. [CrossRef]
14. Qing, C.; Wu, X.; Li, X.; Luo, T.; Su, C.; Zhu, W. Mesoscale optical turbulence simulations above Tibetan Plateau: First attempt. *Opt. Express* **2020**, *28*, 4571–4586. [CrossRef] [PubMed]
15. Guryanov, A.E.; Kallistratova, M.A.; Kutyrev, A.S.; Petenko, I.V.; Shcheglov, P.V.; Tokovinin, A.A. The contribution of the lower atmospheric layers to the seeing at some mountain observatories. *Astron. Astrophys.* **1992**, *262*, 373–381.

16. Tokovinin, A.; Kornilov, V. Measuring turbulence profile from scintillations of single stars. In *Astronomical Site Evaluation in the Visible and Radio Range Workshop, IAU Technical Workshop, Proceedings of a Workshop Held at Cadi Ayyad University, Marrakech, Morocco, 13–17 November 2000*; Astronomical Society of the Pacific Conference Series; Astronomical Soc Pacific: San Francisco, CA, USA, 2002; Volume 266, pp. 104–112.
17. Osborn, J.; Wilson, R.; Butterley, T.; Shepherd, H.; Sarazin, M. Profiling the surface layer of optical turbulence with SLODAR. *Mon. Not. R. Astron. Soc.* **2010**, *406*, 1405–1408. [CrossRef]
18. Wang, Z.Y.; Zhang, L.Q.; Kong, L.; Bao, H.; Guo, Y.M.; Rao, X.J.; Zhong, L.B.; Zhu, L.; Rao, C.H. A modified S-DIMM plus: Applying additional height grids for characterizing daytime seeing profiles. *Mon. Not. R. Astron. Soc.* **2018**, *478*, 1459–1467. [CrossRef]
19. Kovadlo, P.G.; Shikhovtsev, A.Y.; Kopylov, E.A.; Kiselev, A.V.; Russkikh, I.V. Study of the Optical Atmospheric Distortions using Wavefront Sensor Data. *Russ. Phys. J.* **2021**, *63*, 1952–1958. [CrossRef]
20. Deng, J.; Song, T.; Liu, Y. Development Overview of Daytime Atmospheric Optical Turbulence Profile Detection Technology. *Prog. Astron.* **2022**, *40*, 345–363.
21. Basu, S. A simple approach for estimating the refractive index structure parameter (C_n^2) profile in the atmosphere. *Opt. Lett.* **2015**, *40*, 4130–4133. [CrossRef]
22. Hufnagel, R.E.; Stanley, N.R. Modulation Transfer Function Associated with Image Transmission through Turbulent Media. *J. Opt. Soc. Am.* **1964**, *54*, 52–61. [CrossRef]
23. Barletti, R.; Ceppatelli, G.; Paternò, L.; Righini, A.; Speroni, N. Mean vertical profile of atmospheric turbulence relevant for astronomical seeing. *J. Opt. Soc. Am.* **1976**, *66*, 1380–1383. [CrossRef]
24. Ruggiero, H.; DeBenedictis, D.A. Forecasting optical turbulence from mesoscale numerical weather prediction models. In Proceedings of the DoD High Performance Modernization Program Users Group Conference, Vicksburg, MS, USA, June 2002.
25. Dewan, E.M.; Grossbard, N. The inertial range "outer scale" and optical turbulence. *Environ. Fluid Mech.* **2007**, *7*, 383–396. [CrossRef]
26. Thorpe, S.A. *The Turbulent Ocean*; Cambridge University Press: Cambridge, UK, 2005; Volume 9780521835435. [CrossRef]
27. Ellison, T.H. Turbulent transport of heat and momentum from an infinite rough plane. *J. Fluid Mech.* **1957**, *2*, 456–466. [CrossRef]
28. Wu, S.; Su, C.D.; Wu, X.Q.; Luo, T.; Li, X.B. A Simple Method to Estimate the Refractive Index Structure Parameter (C_n^2) in the Atmosphere. *Publ. Astron. Soc. Pac.* **2020**, *132*, 11. [CrossRef]
29. Wu, S.; Wu, X.Q.; Su, C.D.; Yang, Q.K.; Xu, J.Y.; Luo, T.; Huang, C.; Qing, C. Reliable model to estimate the profile of the refractive index structure parameter (C_n^2) and integrated astroclimatic parameters in the atmosphere. *Opt. Express* **2021**, *29*, 12454–12470. [CrossRef] [PubMed]
30. Xu, M.M.; Shao, S.Y.; Weng, N.Q.; Liu, Q. Analysis of the Optical Turbulence Model Using Meteorological Data. *Remote Sens.* **2022**, *14*, 3085. [CrossRef]
31. Vanzandt, T.E.; Gage, K.S.; Warnock, J.M. Statistical-model for probability of turbulence and calculation of vertical profiles of turbulence parameters. *Bull. Am. Meteorol. Soc.* **1978**, *59*, 1230.
32. Trinquet, H.; Vernin, J. A statistical model to forecast the profile of the index structure constant C_N^2. *Environ. Fluid Mech.* **2007**, *7*, 397–407. [CrossRef]
33. Bi, C.C.; Qing, C.; Wu, P.F.; Jin, X.M.; Liu, Q.; Qian, X.M.; Zhu, W.Y.; Weng, N.Q. Optical Turbulence Profile in Marine Environment with Artificial Neural Network Model. *Remote Sens.* **2022**, *14*, 2267. [CrossRef]
34. Bolbasova, L.A.; Andrakhanov, A.A.; Shikhovtsev, A.Y. The application of machine learning to predictions of optical turbulence in the surface layer at Baikal Astrophysical Observatory. *Mon. Not. R. Astron. Soc.* **2021**, *504*, 6008–6017. [CrossRef]
35. Wang, Y.; Basu, S. Using an artificial neural network approach to estimate surface-layer optical turbulence at Mauna Loa, Hawaii. *Opt. Lett.* **2016**, *41*, 2334–2337. [CrossRef]
36. Gladstone, J.H.; Dale, T.P. Researches on the Refraction, Dispersion, and Sensitiveness of Liquids. *Philos. Trans. R. Soc. Lond.* **1863**, *153*, 317–343.
37. Chen, X.W.; Li, X.B.; Sun, G.; Liu, Q.; Zhu, W.Y.; Weng, N.Q. Effects of intermittency and stratification on the evaluation of optical propagation. *Chin. Opt. Lett.* **2017**, *15*, 5. [CrossRef]
38. Taylor, K.E. Summarizing multiple aspects of model performance in a single diagram. *J. Geophys. Res.-Atmos.* **2001**, *106*, 7183–7192. [CrossRef]
39. Haslebacher, C.; Demory, M.E.; Demory, B.O.; Sarazin, M.; Vidale, P.L. Impact of climate change on site characteristics of eight major astronomical observatories using high-resolution global climate projections until 2050 Projected increase in temperature and humidity leads to poorer astronomical observing conditions. *Astron. Astrophys.* **2022**, *665*, 53. [CrossRef]
40. Yang, Q.K.; Wu, X.Q.; Wang, Z.Y.; Hu, X.D.; Guo, Y.M.; Qing, C. Simulating the night-time astronomical seeing at Dome A using Polar WRF. *Mon. Not. R. Astron. Soc.* **2022**, *515*, 1788–1794. [CrossRef]

remote sensing

MDPI

Article

Multiscale Spatiotemporal Variations of GNSS-Derived Precipitable Water Vapor over Yunnan

Minghua Wang [1,2,3], Zhuochen Lv [1], Weiwei Wu [4,*], Du Li [5], Rui Zhang [6] and Chengzhi Sun [1,3]

1 School of Remote Sensing and Geomatics Engineering, Nanjing University of Information Science and Technology, Nanjing 210044, China; mhwang@nuist.edu.cn (M.W.); zclyu@nuist.edu.cn (Z.L.); 003453@nuist.edu.cn (C.S.)
2 Shanghai Key Laboratory of Space Navigation and Positioning Techniques, Shanghai Astronomical Observatory, Chinese Academy of Sciences, Shanghai 200030, China
3 Technology Innovation Center for Integration Applications in Remote Sensing and Navigation, Ministry of Natural Resources, Nanjing 210044, China
4 Institute of Earthquake Forecasting, China Earthquake Administration, Beijing 100036, China
5 Shanghai Astronomical Observatory, Chinese Academy of Sciences, Shanghai 200030, China; dli@nuist.edu.cn
6 College of Nature Resource and Environment, South China Agricultural University, Guangzhou 510642, China; rzhang@scau.edu.cn
* Correspondence: wwei@ief.ac.cn

Abstract: The geographical location of Yunnan province is at the upstream area of water vapor transportation from the Bay of Bengal and the South China Sea to inland China. Understanding the spatiotemporal variations of water vapor over this region holds significant importance. We utilized the Global Navigation Satellite System (GNSS) data collected from 12 stations situated in Yunnan, which are part of the Crustal Movement Observation Network of China, to retrieve hourly precipitable water vapor (PWV) data from 2011 to 2022. The retrieved PWV data at Station KMIN were evaluated by the nearby radiosonde data, and the results show that the mean bias and RMS of the differences between the two datasets are 0.08 and 1.78 mm, respectively. Average PWV values at these stations are in the range of 11.77 to 33.53 mm, which decrease from the southwest to the north of Yunnan and are negatively correlated with the stations' heights and latitudes. Differences between average PWV in the wet season and dry season range from 12 to 27 mm. These differences tend to increase as the average PWV increases. The yearly rates of PWV variations, averaging 0.18 mm/year, are all positive for the stations, indicating a year-by-year increase in water vapor. The amplitudes of the PWV annual cycles are 9.75–20.94 mm. The spatial variation of these amplitudes is similar to that of the average PWV over the region. Generally, monthly average PWV values increase from January to July and decrease from July to December, and the growth rate is less than the decline rate. Average diurnal PWV variations show unimodal PWV distributions over the course of the day at the stations except Station YNRL, where bimodal PWV distribution was observed.

Keywords: global navigation satellite system; precipitable water vapor; secular trend; annual cycle; diurnal variation; radiosonde

Citation: Wang, M.; Lv, Z.; Wu, W.; Li, D.; Zhang, R.; Sun, C. Multiscale Spatiotemporal Variations of GNSS-Derived Precipitable Water Vapor over Yunnan. *Remote Sens.* **2024**, *16*, 412. https://doi.org/10.3390/rs16020412

Academic Editors: Haobo Li, Suelynn Choy, Yuriy Kuleshov, Mayra I. Oyola-Merced and Xiaoming Wang

Received: 17 December 2023
Revised: 14 January 2024
Accepted: 17 January 2024
Published: 20 January 2024

1. Introduction

Water vapor in the atmosphere is a significant greenhouse gas [1,2] and it plays a crucial role in various atmospheric physical and chemical processes. Water vapor is the most active ingredient of the atmosphere. Its variations are closely associated with most weather phenomena. Additionally, water vapor exerts influence on the global water cycle and the heat balance between Earth and the atmosphere, as well as between the Earth–atmosphere system and outer space. It also facilitates the transport of heat from tropical regions to middle and high latitudes [3]. Thus, the observation of atmospheric water vapor content holds immense importance for weather, climate, and environment studies.

One term used to quantify the amount of water vapor is precipitable water vapor (PWV), which represents the depth of water in a column of the atmosphere if all the water in that column were precipitated as rain [4]. Atmospheric scientists have developed a variety of ways to measure PWV, but each has its limitations. Accurate PWV can be obtained by using radiosonde data, which include relative humidity, pressure, and temperature observations at different altitudes [5]. However, radiosondes are typically released only twice a day per site, and the distribution of these sites is relatively sparse. As a result, they cannot provide PWV data with high spatiotemporal resolution [6]. Ground-based water vapor radiometers are instruments that scan the sky and measure the microwave radiation emitted by atmospheric water vapor [7]. While they can provide PWV data with high temporal resolution, their spatial resolution is limited due to the limited number of these instruments in use. Satellite-based water vapor microwave radiometers can provide high-quality PWV data over oceans but face limitations over land. Although these microwave radiometers onboard low-earth-orbit satellites offer high spatial resolution, their temporal resolution is often compromised due to the long revisiting periods of the satellites. Infrared water vapor radiometers installed on geostationary satellites, e.g., Fengyun-4 satellites, retrieve PWV with high spatiotemporal resolutions (temporal resolution of 4–15 min and horizontal resolution of several kilometers) [8], but they are unable to accurately measure PWV on rainy or cloudy days.

Global Navigation Satellite System (GNSS) provides a cost-effective means of retrieving PWV with high temporal resolution, regardless of weather conditions [9]. The GNSS signals experience delays due to water vapor in the atmosphere. The relation between the zenith wet delay (ZWD) of the microwave signals and PWV was modeled by Askne and Nordius [10], establishing the basis for GNSS meteorology [11,12]. Since the usefulness of GNSS for water vapor retrieval was demonstrated [6,9], it has been widely used in meteorological and environmental studies. These include investigations into the relation between PWV and precipitation [13–15], deep convections [16–18], the effects of incorporating the GNSS-derived PWV into numerical weather prediction systems [19,20], summer monsoon and atmospheric rivers [21–23], and drought monitoring [24]. Moreover, the long-time and high temporal resolution GNSS-derived PWV data have been utilized to analyze secular trends and diurnal variations of PWV [25–29].

Yunnan province is situated in the southwestern region of China, bordering the southeastern side of the Tibetan Plateau. It occupies the headwater area of several major rivers, e.g., the Jinsha River and Nanpan River in Yunnan serve as the upper reaches of the Yangtze River and the Pearl River, respectively. Additionally, Yunnan is located in the upstream region of the water vapor transportation from the Bay of Bengal and the South China Sea to inland China. The variations in PWV across this area reflect changes in local weather patterns, climate conditions, and hydrological environment. Understanding these PWV variations is crucial for effective water resource and disaster management in Yunnan and downstream regions. Thus, many studies have focused on the PWV variations in this area. Fu et al. [30] and Hai et al. [31] utilized GPS data from 5–7 stations to analyze the PWV variations in Yunnan, but the PWV time series data used in their studies spanned only 1–3 years, which are insufficient for conducting secular PWV variation analyses. Shen and Duan [32] used the monthly NCEP/NCAR reanalysis data to examine the spatiotemporal variation of PWV in Yunnan, but the monthly data may not adequately capture short-term fluctuations, such as diurnal PWV variations. Li et al. [33] used GNSS data from 2010 to 2013 to analyze the multiscale temporal variations of PWV. Since their GNSS data were from a single station at Dali, their results and conclusions are confined to this small area rather than the entirety of Yunnan province. Hu et al. [34] investigated the variations of GNSS-derived PWV over the Yunnan-Guizhou Plateau. Their study focused on the relation between PWV and precipitation during convective weather in the summer season.

With the continuous advancements in both hardware and software of GNSS, the quality of GNSS observations has evidently improved. As the high-quality GNSS observations accumulate in Yunnan, it is of great importance to investigate PWV variations over this area with the recent GNSS observables and the latest data processing strategies. In this study, we adopted recent GNSS data over 11 years (2011–2022) from 12 stations located at Yunnan to derive hourly PWV data. First, we evaluated the GNSS-derived PWV data with radiosonde data. Then, we analyzed the geographical distributions of multiple-year-averaged PWV values across the region. Next, we determined and analyzed the secular trends and amplitudes of annual and semiannual cycles of PWV variations. At last, we investigated the monthly and diurnal variations of PWV at the individual stations. These analyses aim to provide a thorough understanding of the multiscale spatiotemporal variations of PWV and enhance our knowledge of the dynamic changes in PWV over Yunnan region.

2. Data and Methods

2.1. Data Description

The Crustal Movement Observation Network of China (CMONOC) has consistently conducted long-term observations of GNSS data, ensuring a reliable and high-quality dataset. In this study, we utilized GNSS data (sampling rate of 30 s) from CMONOC to retrieve hourly PWV data. The period of the used data extends from 1 July 2011 to 30 June 2022 (11 years). To ensure comprehensive coverage, we selected 12 stations located within Yunnan, spanning approximately 21°N to 29°N and 97°E to 107°E, to analyze PWV variations. The geographical distribution of the stations is shown in Figure 1 (red triangles). These stations are distributed with approximately even spacing. The longitudes, latitudes, and heights of the GNSS stations are shown in Table 1.

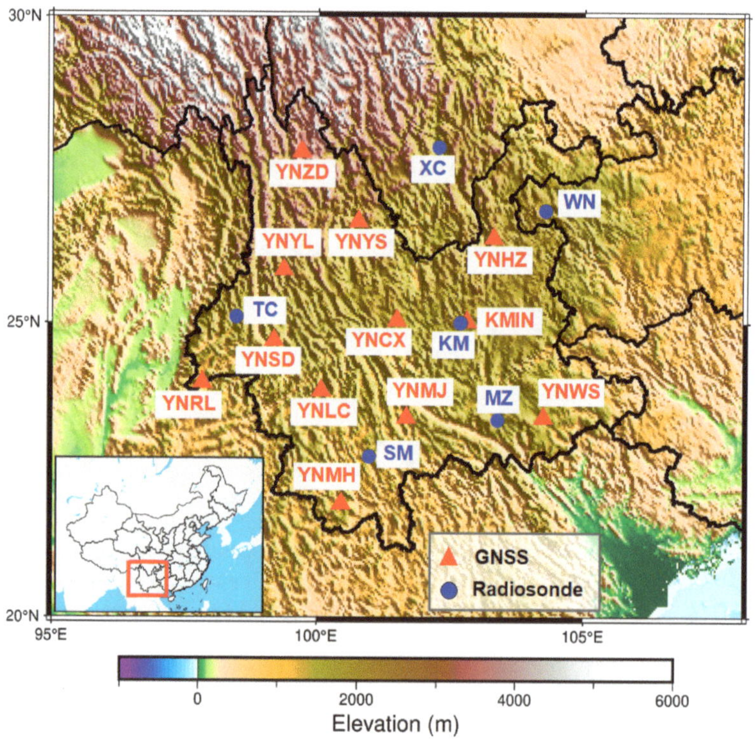

Figure 1. Geographical distributions of GNSS and radiosonde stations. The red triangles mark the locations of the GNSS stations, and the blue dots denote the sites of radiosondes. The inset presents a zoomed-out map highlighting the province of Yunnan, enclosed by a distinct red rectangle.

Table 1. The constructed Ts-Tm linear models and the coordinates of the related radiosonde sites and GNSS stations. The first column shows the coordinates and names of the radiosonde sites. Atmospheric profiles from these sites were utilized to construct Ts-Tm linear models shown in the second column. Each Ts-Tm linear model displayed on a row was used to calculate Tm for the GNSS stations presented on the same row, third column of the table.

Radiosonde Lon (°E), Lat (°N), Hgt (m)	Ts-Tm Model (K)	GNSS Lon(°E), Lat(°N), Hgt (m)
XC (Xichang) 102.26, 27.90, 1599	Tm = 0.58 Ts + 110.66	YNZD 99.70, 27.82, 3296 YNYS 100.75, 26.68, 2144
TC (Tengchong) 98.48, 25.11, 1649	Tm = 0.52 Ts + 130.75	YNRL 97.85, 24.00, 723 YNSD 99.19, 24.71, 1479 YNYL 99.37, 25.88, 1696 YNLC 100.08, 23.87, 1559
SM (Simao) 100.98, 22.76, 1303	Tm = 0.35 Ts + 181.68	YNMH 100.45, 21.95, 1166 YNMJ 101.67, 23.42, 1282
MZ (Mengzi) 103.38, 23.38, 1302	Tm = 0.49 Ts + 138.31	YNWS 104.25, 23.41, 1452
WN (Weining) 104.28, 26.86, 2236	Tm = 0.62 Ts + 102.15	YNHZ 103.29, 26.41, 2264
KM (Kunming) 102.68, 25.01, 1892	Tm = 0.45 Ts + 148.44	KMIN 102.80, 25.03, 1986 YNCX 101.49, 25.05, 1785

Each GNSS station is equipped with a TRIMBLE NETR9 receiver that is connected to a TRM59800.00 or TRM59900.00 antenna. Additionally, there is a collocated meteorological sensor at each station to record air pressures and temperatures. The air pressures and temperatures are measured with accuracies of 0.3 mbar and 0.1 K, respectively. These measurements are essential for converting zenith tropospheric delay (ZTD) to PWV. During the conversion process, a key parameter is water vapor weighted mean temperature (Tm), which directly affects the accuracy of the converted PWV. Tm is usually estimated from the surface temperature (Ts) by using a simple linear model [9]. The linear relations between Ts and Tm are highly location dependent in the region of Yunnan [8,35,36], indicating that different Ts-Tm linear models may be adopted to calculate accurate Tm values at different GNSS stations. We used the atmospheric profile data (from 2005 to 2018) observed from 6 radiosonde stations to construct 6 site-specific Ts-Tm linear models, respectively. These radiosonde sites are within or near the region of Yunnan (blue dots in Figure 1). Table 1 shows the coordinates of the radiosonde stations, as well as their corresponding Ts-Tm linear models. Each constructed Ts-Tm linear model was utilized to calculate Tm for accurate PWV conversion at the nearby GNSS stations. The coordinates of these GNSS stations, along with their corresponding Ts-Tm linear models, are also shown in Table 1.

2.2. Retrieval of PWV

GNSS signals experience delays as they pass through the neutral atmosphere, resulting in the measured distances between satellites and receiving antennas to be longer than the actual distances. The slant path tropospheric delay (in length) can be calculated by

$$SPD = \int_s N(s)ds \tag{1}$$

where SPD is the slant path delay in length, s denotes the slant path passed by the GNSS signal, and $N(s)$ is the refractive index of the atmosphere. The $N(s)$ is not a constant and it varies based on several factors, including air pressures and temperature. Due to the challenges in obtaining accurate vertical profiles of $N(s)$ in practice, Equation (1) is not commonly used to derive the SPD. In GNSS data processing, the tropospheric delay is estimated as an unknown quantity, and the ZTD instead of SPD is estimated to reduce the number of the unknowns. The relation between SPD and ZTD is

$$SPD = MF \cdot ZTD \tag{2}$$

where MF is an elevation-angle-dependent mapping function. It can be written in continued fraction form as [37,38]

$$MF = \frac{1 + \frac{a}{1+\frac{b}{1+c}}}{\sin e + \frac{a}{\sin e + \frac{b}{\sin e + c}}} \tag{3}$$

where e is the elevation angle of site-to-satellite direction, and coefficients a,b, and c are derived from radiosonde data [39] or numerical weather models [38,40]. Equation (2) is suitable for the stations with azimuthal-symmetry local atmosphere. Under the circumstance of an unsymmetrical atmosphere, two gradient parameters are added into Equation (2) to compensate for the asymmetry [41].

The estimated ZTD can be partitioned into zenith hydrostatic delay (ZHD) and ZWD. By using the real observed surface air pressure, the ZHD can be modeled with an accuracy of several millimeters. The commonly used ZHD mode is [42]

$$ZHD = 0.0022768 \frac{P_0}{1 - 0.00266\cos(2\varphi) - 0.28 \cdot 10^{-6} \cdot h} \tag{4}$$

where P_0 is the surface pressure (in mbar), φ is the latitude of the GNSS station, and h is the height of the station (in meters). The ZWD can be acquired by ZTD minus ZHD. The conversion of ZWD to PWV is

$$PWV = \Pi \cdot ZWD \tag{5}$$

where Π is a dimensionless coefficient, which is given by [9,43]

$$\Pi = \frac{10^6}{\rho_w R_v[-k_1 \frac{M_w}{M_d} + k_2 + \frac{k_3}{T_m}]} \tag{6}$$

where ρ_w is the density of liquid water (in Kg/m^3); R_v is the specific gas constant of water vapor (in J/(Kg·K)); k_1,k_2, and k_3 are constants (in K/mbar) [9]; M_w and M_d are molar masses of water vapor and dry air, respectively (in g/mol); and Tm is the weighted mean temperature of atmosphere (in K). The definition of Tm is [44]

$$T_m = \frac{\int \left(\frac{P_v}{T}\right) dz}{\int \left(\frac{P_v}{T^2}\right) dz} \tag{7}$$

where P_v is the partial pressure of water vapor (in mbar), and T is the temperature (in K). The practical application of Equation (7) is limited because it relies on having accurate profiles of P_v and T, which are not readily available in many cases. Bevis et al. [9] used the radiosonde data to find the relation between Tm and surface temperature (Ts), and they fitted a linear Ts-Tm model. Thus, with this model, one can calculate the Tm from the observed surface temperature. We used a similar method to generate the Ts-Tm models specific for the study area (see Table 1 for the Ts-Tm models).

We used the Bernese GNSS software version 5.2 [45] to estimate ZTD. The data processing basically followed the default strategy of the Center for Orbit Determination in Europe (CODE) (Table 2).

Table 2. Strategy of GNSS data processing.

Ephemeris	**Orbit**	CODE GPS satellite orbit final products
	Clock	CODE GPS satellite clock offset final products
Measurement models	**Basic observables**	GPS L1 + L2
	Modeled observables	Double differences Ionosphere-free linear combination
	Satellite antenna center of mass offsets	igs14.atx
	GPS attitude model	Nominal (yaw-steering) attitude implemented
	Troposphere a priori model	ECMWF-based hydrostatic delay mapped with hydrostatic VMF1
	Ionosphere	Second-order effect applied
Estimated parameters	**Adjustment method**	Weighted least-squares algorithms
	Station coordinates	Adjusted with minimum constraints
	Troposphere	Zenith tropospheric delay and two gradient parameters estimated every hour Loose relative constraints of 5 m are applied
	Gradient model	Chen–Herring model [46]
	Ambiguity	Partly fixed

The Bernese software always attempts to fix ambiguities to the maximum extent, but those ambiguities which do not satisfy the statistical threshold during ambiguity resolution are kept as floating-point numbers. In our data processing, on average, 70% of ambiguities were fixed as integers, the remaining were retained as floating-point values.

In this study, all the meteorological data used for converting GNSS ZTD to PWV (pressure for calculating ZHD and temperature for converting ZWD to PWV) are observed with the collocated meteorological equipment. We did not use other data (e.g., reanalysis data) or apply any interpolation to fill the gaps of the meteorological observations. Thus, the retrieved PWV time series are free of the potential biases caused by different meteorological data sources or by the interpolations.

3. Results

3.1. Continuity of GNSS ZTD and PWV Time Series

Using the 30 s-interval GNSS observations, we estimated ZTD on an hour-by-hour basis. Throughout the long-term observation period, the GNSS stations occasionally encounter some interruptions due to instrument and electrical failures. This causes the gaps in the observations. In addition, some noisy GNSS observations were eliminated in the phase of quality checking during data processing, which could further increase the gaps in the observations. As a result, ZTD on these corresponding epochs could not be estimated. For assessing the continuity of the derived ZTD and PWV time series, we set an evaluation index named Data Available Rate (DAR). The definition of DAR is the number of real retrieved data over the number of ideal continuous data. The DAR of ZTD for each station is shown in Figure 2 (red dots). The smallest DAR of ZTD is observed at Station YNMH, which is 88%. At the other 11 stations, DARs of ZTD are larger than 90%, with nine of them having a DAR greater than 98%. Regarding PWV, DAR values are comparable to those of ZTD at Stations YNCX, YNLC, YNMH, YNSD, YNWS, YNYL, YNYS, and YNZD. However, at the remaining stations, particularly at Station YNMJ, DARs of PWV are significantly lower than those of ZTD. These discrepancies arise due to a large number of missing meteorological observations at those stations and no interpolation being applied to fill the gaps of meteorological data for converting the corresponding ZTD into PWV.

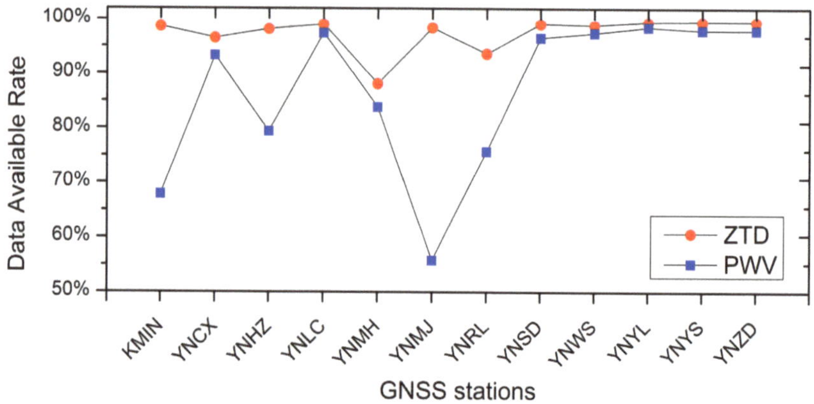

Figure 2. Data available rates (DARs) of ZTD and PWV at each GNSS station. DAR = the number of real retrieved data/the number of ideal continuous data.

3.2. Evaluation of the GNSS PWV with Radiosonde Data

We evaluated GNSS-derived PWV with the radiosonde data. The GNSS station KMIN and radiosonde station Kunming are located in the same city, and the distance between them is less than 15 km. Thus, the two stations are well spatially matched. We retrieved the PWV from the radiosonde profiles (hereafter referred to as RDS-derived PWV). The PWV is calculated as

$$PWV = \frac{1}{\rho_w} \int \frac{q}{g} dP \tag{8}$$

where q is the specific humidity (in g/g), P is the pressure of the atmosphere (in Pa), and g is the gravitational acceleration (in m/s^2). In the computation, g is not regarded as a constant since its value is dependent on the latitude and height.

The temporal resolution of RDS-derived PWV is 12 h, while that of GNSS-derived PWV is 1 h. To ensure fair comparisons, we only chose data with the same epochs from the two datasets. This selection process resulted in a total of 4040 paired data points for the comparisons. Figure 3a shows both the time series of GNSS-derived PWV (red) and RDS-derived PWV (blue). The two datasets match each other very well. The average bias between them is 0.08 mm, and the RMS of the differences between them is 1.78 mm. There is a data gap of GNSS PWV with the period spanning from 2013 to 2015 at Station KMIN. An instrument failure at this station caused the missing of pressure and temperature observations. Without these meteorological observables, we were not able to convert the ZTD to PWV, which caused the data gap. Figure 3b shows the scatter points of the two datasets and also the linear fitting result. The slope of the linear fitted model is close to 1 (1.024) and the goodness of fit is 0.97, which all indicate that the two datasets are highly consistent. These comparisons demonstrate that the GNSS-derived PWV has similar accuracy to the RDS-derived one, and hence can be used for analyzing water vapor variations.

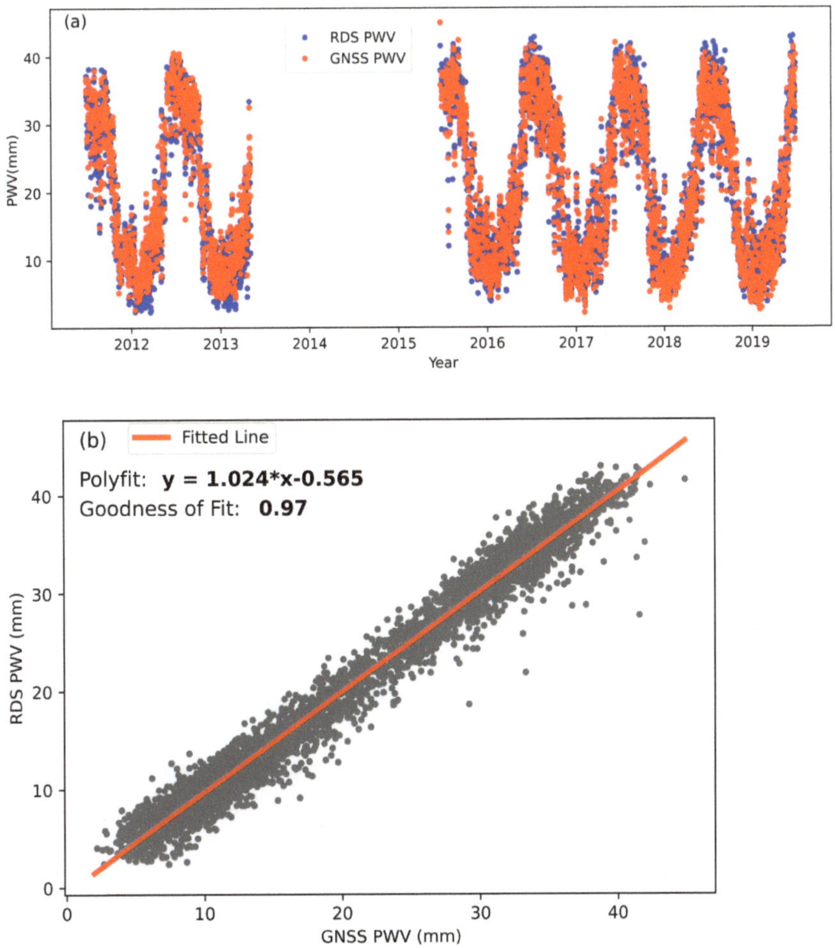

Figure 3. Comparison of GNSS and radiosonde PWV at Station KMIN. (**a**) The time series of GNSS-PWV and RDS-PWV. (**b**) the goodness of fit between GNSS-PWV and RDS-PWV.

3.3. Average PWV and Spatial Distribution

We averaged PWV values between 2011 and 2022 for each GNSS station. Figure 4 shows the geographical distribution of the average PWV. The largest average PWV is observed at Station YNRL located at the west boundary of Yunnan province, which is above 30 mm, while the smallest one is at Station YNZD located at the northwest of the area, which is only 11.8 mm (about 1/3 of the largest one). Average PWV values at the other stations are in the range of 18 to 29 mm. In general, the average PWV tends to decrease from southwest to the north of the area (Figure 4 contour lines). In addition to the all-season averaged PWV, we also calculated the average PWV in the wet season (June to October) and dry season (November to May) separately for the GNSS stations. The geographical distributions of wet-season and dry-season averaged PWV are similar to those of the all-season averaged PWV.

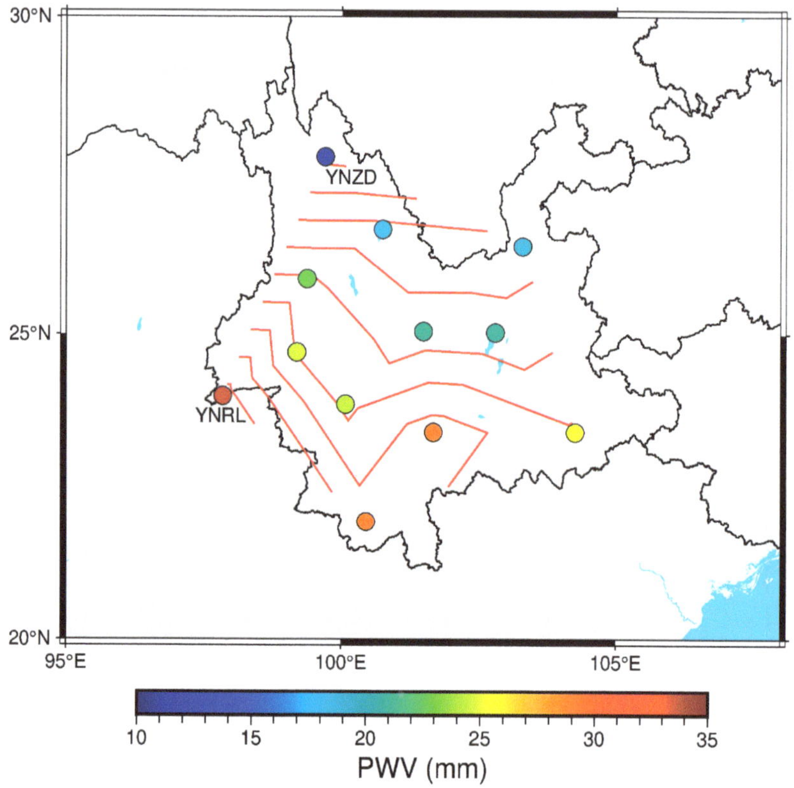

Figure 4. Geographical distribution of 11-year-averaged PWV observed at the GNSS stations. The color on a circle indicates the value of average PWV at that station. The red lines are the contour lines of average PWV.

Figure 5 shows the relationship between the variation in average PWV and both the station height and latitude. The largest average PWV is observed at the station with the lowest height (YNRL: 723 m), while the smallest average PWV is at the highest station (YNZD: 3297). It clearly shows that the average PWV decreases as the height of the GNSS station increases. As for the latitudes, it shows that, in general, the average PWV decreases as the latitude of the station increases, except for Station YNRL. The latitude of Station YNRL is not the lowest one among the 12 GNSS stations; however, the average PWV observed at this station is the largest. This is probably subject to the special local climate of Station YNRL.

Figure 6 shows the all-season, wet-season, and dry-season averaged PWV for each station. At most stations, the average PWV values in the wet season are about 10 mm larger than those of the all-season averaged PWV, while the average PWV values in the dry season are smaller than those of the all-season averaged PWV by about 10 mm. The larger all-season averaged PWV values generally correspond to the larger average PWV values in both the wet season and dry season. The differences between average PWV in the wet season and dry season at these stations are in the range of 12 to 27 mm (Figure 6 blue bars). Overall, these differences tend to decease as the average PWV decreases.

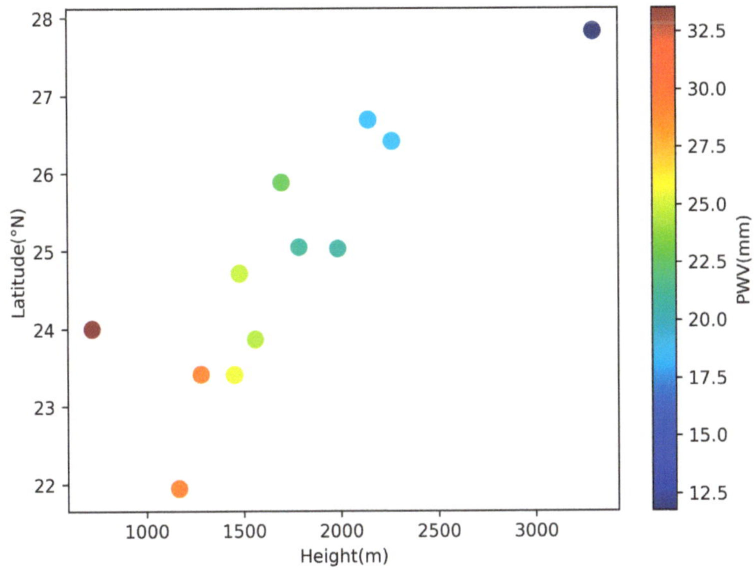

Figure 5. Relation of the variation in average PWV with the station height and latitude. The color represents the value of average PWV.

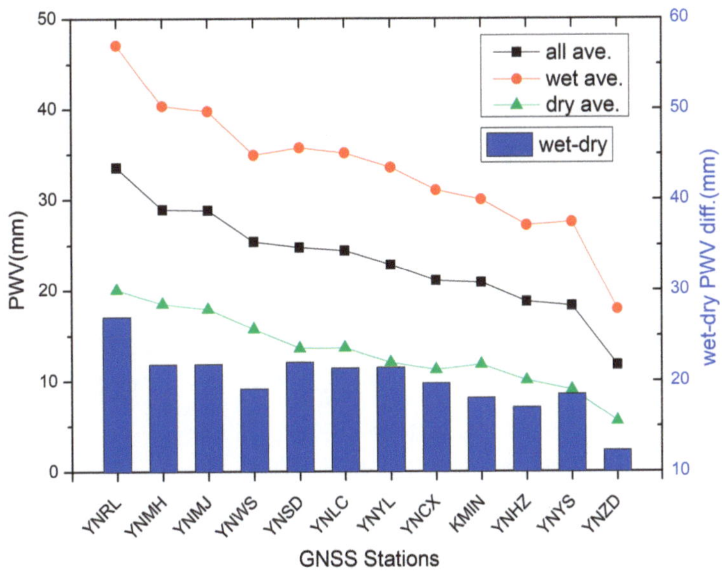

Figure 6. All-season (black), wet-season (red), and dry–season (green) averaged PWV for each GNSS station. A blue bar is the difference between the average PWV in wet season and dry season at the corresponding station. The stations are arranged in order of decreasing PWV.

3.4. Secular, Annual and Semiannual Variations of PWV

The time series of GNSS-derived PWV at these stations show significant annual cycles. Figure 7 shows the PWV variation at Station YNSD (other stations show similar variations). To quantitatively analyze the PWV time series at the stations, we modeled the PWV variations with a mathematical model that contains a secular trend, an annual cycle, and a semiannual cycle. The model is written as

$$PWV(t) = a_0 + a_1(t - 2011.0) + A_1 \sin(2\pi(t - 2011.0) + \varphi_1) + A_2 \sin(4\pi(t - 2011.0) + \varphi_2) \quad (9)$$

where t is the time in unit of year, a_0 and a_1 are the coefficients that describe the secular trend of PWV variation, A_1 and A_2 are the amplitudes of annual and semiannual PWV variations, and φ_1 and φ_2 are the initial phases.

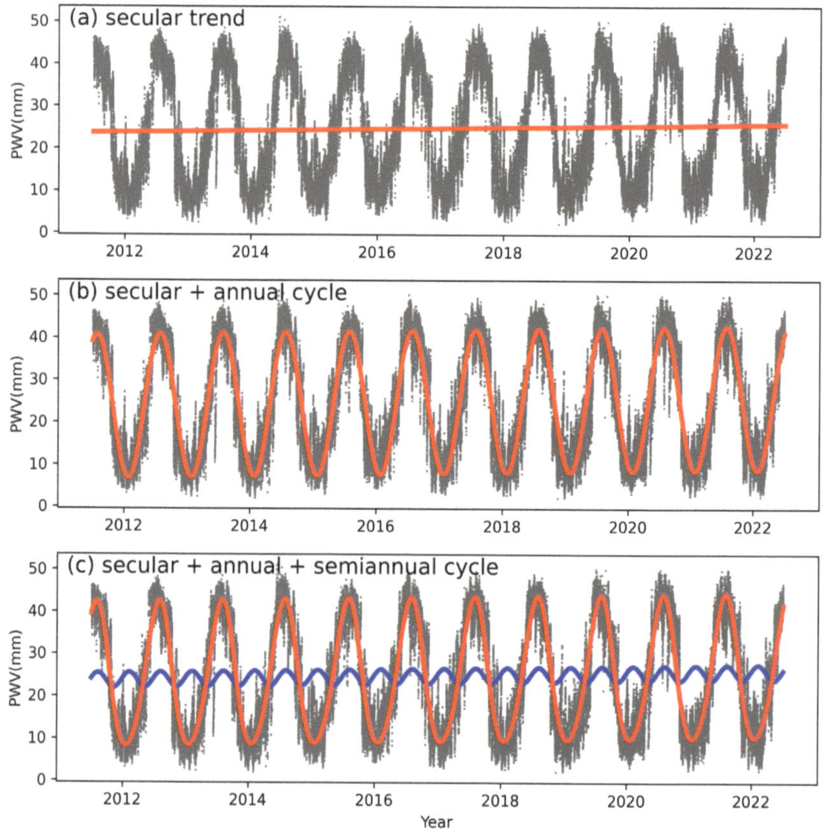

Figure 7. GNSS-derived PWV time series at Station YNSD. The red solid line in (**a**) indicates the secular trend of PWV variations. The red sinusoid in (**b**) consists of the secular and the annual variation. In (**c**), the blue sinusoid is the semiannual variation superimposed on the secular trend, and the red sinusoid consists of the secular, annual, and semiannual variations.

We estimated the coefficients of Equation (9) by the least-squared method for each station. Figure 7a shows the secular trend of PWV variations at Station YNSD (red line). The secular trend is indistinctive, which indicates that the year-to-year change of PWV quantity is very slow. The superposition of the annual and secular variation for Station YNSD is shown in Figure 7b (red sinusoid), which well describes the magnitude of the fluctuation of PWV. When the term of semiannual variation (blue sinusoid in Figure 7c) is added, the model (red sinusoid in Figure 7c) fits the GNSS-derived PWV even better.

From the secular trends of PWV variations, we derived the rate of interannual PWV variations for each station. The rates are positive at all stations (Figure 8), averaging 0.18 mm/year, which indicates that the average PWV observed at each station increases yearly from 2011 to 2022. The minimal rate is observed at Station KMIN, which is 0.11 mm/year, while at Stations YNYC, YNYL, and YNHZ, the rates are up to or above 0.22 mm/year (twice the minimal rate). The derived rate of interannual PWV variations at

Station YNMJ is 0.29 mm/year (not shown in Figure 8), which is significantly larger than the rates at the other stations. Due to the lack of collocated meteorological data before 2017 at Station YNMJ, the time span of the available GNSS-derived PWV data at this station (2017–2022) is much shorter than for the data from the other stations (2011–2022). Thus, we believe that the derived rate of interannual PWV variations at Station YNMJ is not as reliable as the derived rates at the other stations.

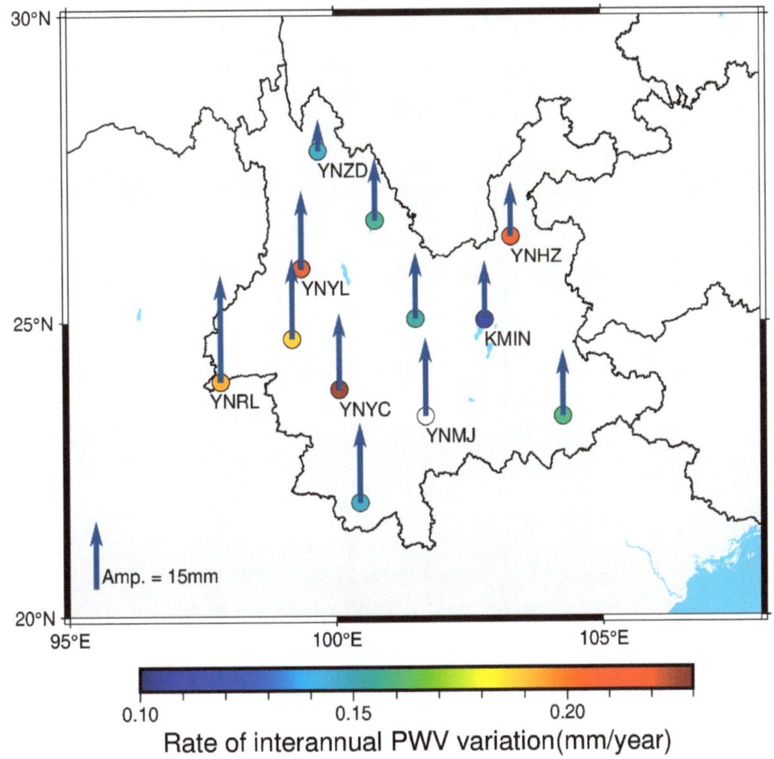

Figure 8. Rate of interannual PWV variations and amplitude of annual PWV variations at each station. The colors filled in the circles indicate the rates of interannual PWV variations and the arrows denote the amplitudes of PWV annual variations.

Figure 8 also shows the amplitude of the annual cycle of PWV variations at each station. The average of the annual PWV variation amplitudes at these stations is 15 mm. The maximal amplitude is observed at Station YNRL, which reaches 20.9 mm, while the minimum is at Station YNZD, which is 9.7 mm. The distribution of the annual cycle amplitudes shows that the magnitude of the amplitude decreases from the southwest to the north of Yunnan region. This phenomenon is highly similar to the distribution of average PWV (refer to Figure 4 for the average PWV distribution), indicating that the larger average PWV values correspond to the greater amplitude of the annual cycle of PWV variations. The amplitudes of semiannual PWV variations, in the range of 0.3 to 2.5 mm, are much smaller than the annual amplitudes (compare the blue sinusoid in Figure 7c with red sinusoid in Figure 7b). On average, the semiannual amplitudes are about one tenth of the annual amplitudes.

3.5. Monthly and Diurnal Variations of PWV

We averaged PWV values over the period from 2011 to 2022 for each individual month (January to December). Figure 9 shows the monthly variations in PWV at the

12 GNSS stations. The curves representing the month-to-month changes in PWV exhibit similar patterns across all stations. At most stations, the maximal monthly average PWV values are in July (Stations KMIN and YNWS in June), and the minimums are in January (Stations YNMH and YNMJ in February). The curves are not symmetric about the peaks the increasing rates of PWV from January to July are less than the decreasing rates from July to December. The distribution of PWV is uneven among the seasons, with the wet season typically accounting for approximately 70% of the total PWV over the entire year on average.

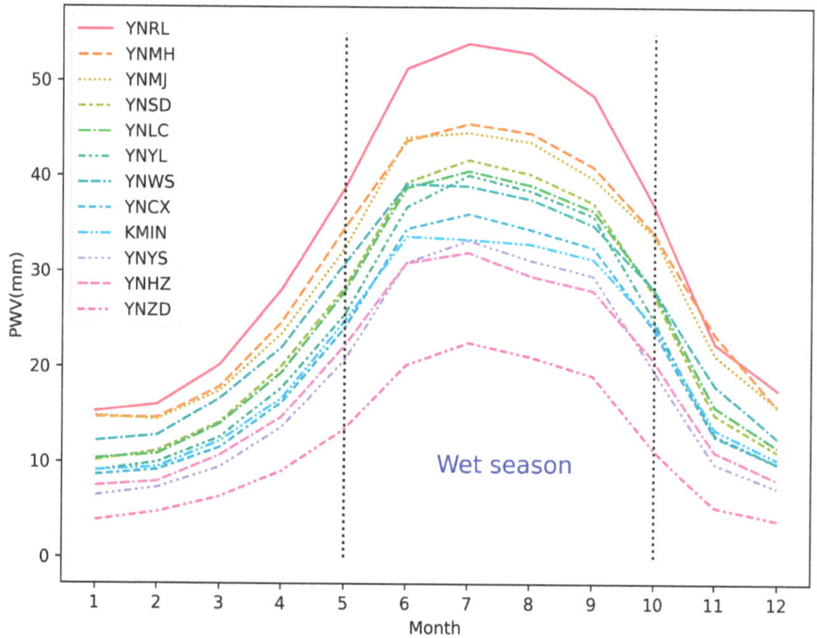

Figure 9. Monthly PWV variation at each station.

Using the hourly GNSS-derived PWV from 2011 to 2022, we calculated the all-season, wet-season, and dry-season average PWV values for each individual hour (0:00, 1:00, …, 23:00). Figure 10 shows the diurnal PWV variations at Stations KMIN and YNRL. The curves of all-season, wet-season, and dry-season averaged PWV diurnal variations are similar at the same station. Diurnal PWV variations at Station KMIN show the unimodal distribution pattern among the hours (Figure 10a–c). Data of the other stations show a similar pattern of PWV distribution as Station KMIN except for Station YNRL, where bimodal distribution of PWV is observed. At Station YNRL, the all-season, wet-season, and dry-season averaged diurnal variations of PWV all show that the high PWV values occur both in the afternoon and at night.

Though diurnal PWV variations observed at 11 out of 12 GNSS stations show a similar unimodal distribution, the diurnal peaks (or valleys) of PWV values are asynchronous among different stations. Table 3 shows the time of diurnal peak and valley for the all-season, wet-season, and dry-season averaged PWV at each station (except Station YNRL). For these stations, the diurnal maximums (peaks) of the all-season averaged hourly PWV appear at 17:00–23:00 local time, most at 17:00–19:00 (late afternoon), while the diurnal minimums (valleys) occur at 8:00–10:00 (morning). The differences between the times of diurnal PWV peaks in the wet season and dry season are 0 to 5 h (most no more than 2 h). The times of diurnal PWV minimum in the dry season are 0 to 4 h later than those in the wet season (most no more than 1 h).

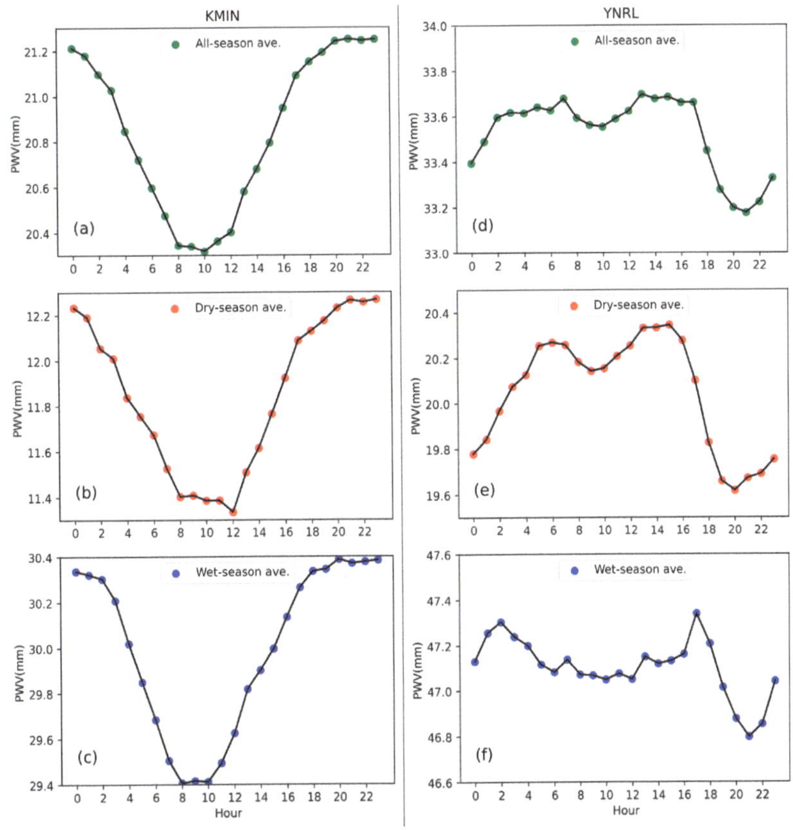

Figure 10. Diurnal PWV variations at Stations KMIN (left) and YNRL (right): (**a–c**) are all-season, dry-season, and wet-season averaged diurnal PWV variations at Station KMIN, respectively, while (**d–f**) are for Station YNRL. The hours are local time.

Table 3. Hours (local time) of the diurnal PWV maximum (peak) and minimum (valley) at each station (except Station YNRL).

Site	All-Season Ave.		Dry-Season Ave.		Wet-Season Ave.	
	Max	Min	Max	Min	Max	Min
KMIN	21	10	23	12	20	8
YNCX	17	8	18	8	17	8
YNHZ	18	9	21	10	17	9
YNLC	17	8	17	9	17	8
YNMH	18	9	18	9	20	8
YNMJ	23	8	21	8	23	8
YNSD	17	8	16	8	17	8
YNWS	19	8	20	9	18	8
YNYL	19	9	19	9	24	9
YNYS	18	9	18	9	18	9
YNZD	19	10	20	10	18	7

We calculated the magnitude of the diurnal PWV variation for each station. The magnitude, describing diurnal PWV fluctuation, is defined as the diurnal PWV peak minus valley. Figure 11 shows the geographical distribution of the magnitudes of diurnal PWV variations. The magnitudes of the diurnal PWV fluctuations, averaging 1.1 mm, are station dependent. The smallest magnitude (0.5 mm) is observed at Station YNRL, while at this station, both the average PWV and the amplitude of annual PWV variations are the largest among the other stations (refer to Figure 4 for the average PWV and Figure 8 for the amplitudes). The largest magnitude is at Station YNHZ (1.6 mm), which is about three times the magnitude at Station YNRL. However, Both the average PWV and the amplitude of annual PWV variations at Station YNHZ are much smaller than those at Station YNRL. The distribution of the magnitudes of diurnal PWV variations show no significant geographical pattern.

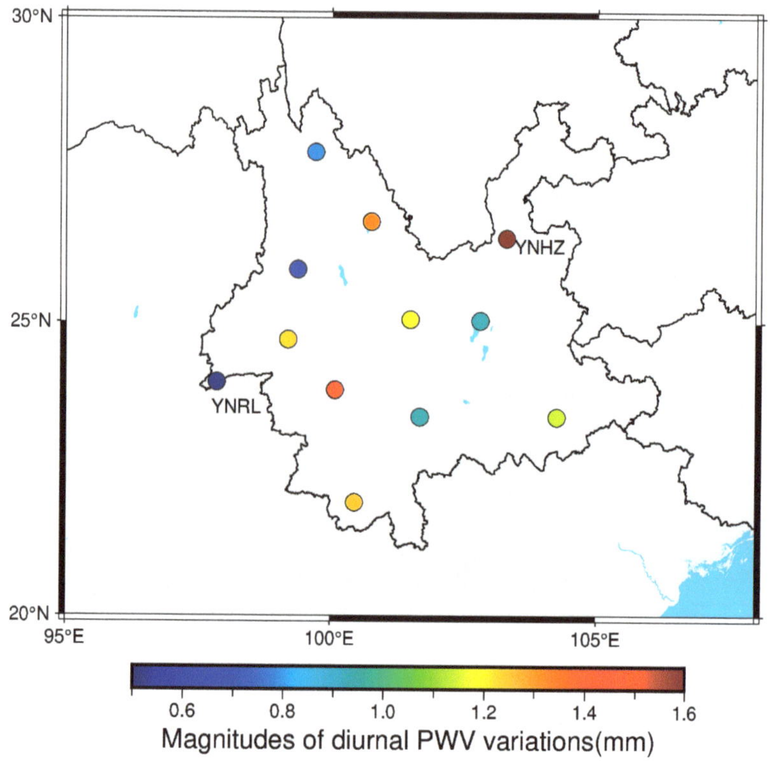

Figure 11. Magnitude of diurnal PWV variation at each station. The magnitude is defined as the difference between the diurnal PWV maximum and minimum.

4. Discussion

We evaluated the GNSS-derived PWV with radiosonde data at Station KMIN, and the mean bias and RMS of the differences between the two datasets are 0.08 mm and 1.78 mm, respectively. Some previous studies also made similar comparisons in the region of Yunnan. Fu et al. [30] and Hai et al. [31] assessed the GNSS-derived PWV with RDS-derived PWV at 3–5 stations (including station KMIN) located at Yunnan. Their results show high correlations between the time series of GNSS-derived PWV and RDS-derived PWV (correlation coefficients larger than 0.9), which are consistent with our results. Nonetheless, their comparisons show 5–7 mm RMS of differences between the two datasets, which are much larger than the 1.78 mm RMS in this study. The results of Hai et al. [31] shows significant biases between the GNSS-derived PWV and RDS-derived PWV, while there are no evident biases observed in the current study and in the study of Fu et al. [30]. Hu et al. [34] used the ERA5 reanalysis dataset of the European Center for Medium-Range Weather Forecasts (ECMWF) to evaluate the GNSS-derived PWV, and their results show 2–6 mm biases and 4.5–7 mm RMS between GNSS-derived PWV and ERA5-derived PWV. There are many factors for the different evaluation results in these studies. Normally, the consistency of different-sourced PWV data in the wet season is poorer than that in the dry season. Hai et al. [31] and Hu et al. [34] exclusively used the summer data (in the wet season) for the evaluation, which is partly responsible for the large biases and RMS in their PWV assessments. We used the all-season data for the evaluation, and applied the latest mapping function to estimate the ZTD. Moreover, we generated the site-specific weighted mean temperature models for the region and used these customized models to convert the ZWD to PWV. All these contribute to the high consistency between GNSS-derived PWV and RDS-derived PWV in current study.

The distribution of all-season averaged PWV values show a clear southwest-to-north decreasing pattern in the region of Yunnan, and both the average PWV in wet season and dry season show a similar decreasing pattern. These results are consistent with the study of Shen and Duan [32]. In their study, the used PWV data were from a different source (NCEP/NCAR monthly reanalysis data), which further demonstrates the reliability of our derived PWV variation pattern. Their data are from 1981 to 2011, while ours are from 2011 to 2022, indicating that the PWV spatial variation pattern in Yunnan has not changed during the last four decades. Our results show that the content of PWV is highly dependent on the station height and latitude: a higher station height (or latitude) generally corresponds to less PWV. This negative correlation between PWV and station heights (or latitudes) was also found in the studies of Jin et al. [25] and Shi et al. [28]. These results are reasonable. The observed PWV is a quantity that integrated the water vapor from the station height to the top of the troposphere. It is inherently negatively correlated with the station height. Globally, high latitudes are colder than low latitudes. The ability of the atmosphere to hold water vapor decreases are the temperature decreases. These explain the negative correlation between the PWV value and the latitude.

The diurnal variations in PWV at the stations in Yunnan exhibit a predominantly unimodal distribution over the course of the day. However, it is worth noting that Station YNRL, located at the west boundary of Yunnan, displays a distinctive bimodal distribution of PWV. Hai et al. [31] found this bimodal diurnal PWV distribution at a different station in Mengla county, which is at the south boundary of Yunnan (see Figure A1 in Appendix A). These bimodal diurnal PWV distributions were observed at both the west (Station YNRL at Ruili county) and south (Mengla county) boundary of Yunnan, indicating the special local climate at these border areas.

5. Conclusions

Using GNSS data observed at 12 CMONOC stations located at Yunnan, China, we retrieved the hourly PWV from 2011 to 2022 and analyzed multiscale spatiotemporal PWV variations over the region. Evaluating the GNSS-derived PWV with radiosonde data at Station KMIN shows good consistency between the two datasets, indicating that the GNSS-derived PWV is as accurate as RDS-derived PWV and hence it can be reliably used in meteorological studies. In the study area, the average PWV values observed at different stations can be quite different: the maximum is three times as large as the minimum. Generally, the average PWV increases with the decrease in station height, and also with the decrease in station latitude (excluding Station YNRL). For these stations, the higher average PWV in the wet season corresponding to higher average PWV in the dry season, and the mean of the differences between average PWV in the wet season and dry season is 20 mm. We analyzed the secular trends and cycles of the PWV time series. The yearly rates of PWV variations are all positive at the 12 stations. This phenomenon of increasing PWV year by year is in line with the context of climate warming. The average amplitude of PWV annual cycles is 15 mm, which is about 10 times as large as the average amplitude of PWV semiannual cycles. Monthly PWV variations show that the maximal monthly average PWV occurs in July or June, and the minimum appears in January or February. The content of average PWV in the wet season accounts for 70% of the sum of PWV over the entire year. Average diurnal PWV variations show unimodal distributions over the course of the day at the stations, while Station YNRL is an exception, where two diurnal PWV peaks were observed. At most stations, the average diurnal PWV maximums occur in the late afternoon (17:00–19:00), and the minimums appear in the morning (8:00–10:00).

In this study, both the largest average PWV and the greatest amplitude of the annual PWV cycle were observed at Station YNRL. The diurnal PWV distribution of this station (bimodal) is different from that of the other stations (unimodal). Furthermore, the average PWV at this station did not follow the rule of negative correlation between PWV and latitude as the other stations do. All these indications suggest that the local climate at Station YNRL differs from that of the other stations, which deserves further investigation.

Author Contributions: Conceptualization, M.W.; methodology, M.W., Z.L., D.L., and W.W.; software, W.W. and Z.L.; Validation R.Z. and C.S.; formal analysis, M.W. and C.S.; data curation, W.W. and R.Z.; writing—original draft preparation, M.W.; writing—review and editing, W.W. All authors have read and agreed to the published version of the manuscript.

Funding: This research was funded by the National Natural Science Foundation of China (NO. 42374022; 42304008), the Opening Project of Shanghai Key Laboratory of Space Navigation and Positioning Techniques (NO. 202103), the Research Program of China Seismic Experimental Site (NO. CEAIEF20220403), the Open Fund of Key Laboratory of Marine Environmental Survey Technology and Application, Ministry of Natural Resources (NO. MESTA-2020-B011), the Guangdong Basic and Applied Basic Research Foundation (NO. 2019A1515011268), and the Guangzhou Science and Technology Plan Project (NO. 202102020380).

Data Availability Statement: The radiosonde data used in this study are available from http://weather.uwyo.edu/upperair/sounding.html (accessed on 1 January 2023). The generated ZTD and PWV data are available from the corresponding author on reasonable request.

Acknowledgments: The authors are grateful to the Crustal Movement Observation Network of China (CMONC) for providing the GNSS data and the collocated meteorological observables, and they would like to thank three anonymous reviewers for their constructive comments and useful suggestions to improve the manuscript.

Conflicts of Interest: The authors declare no conflict of interest.

Appendix A

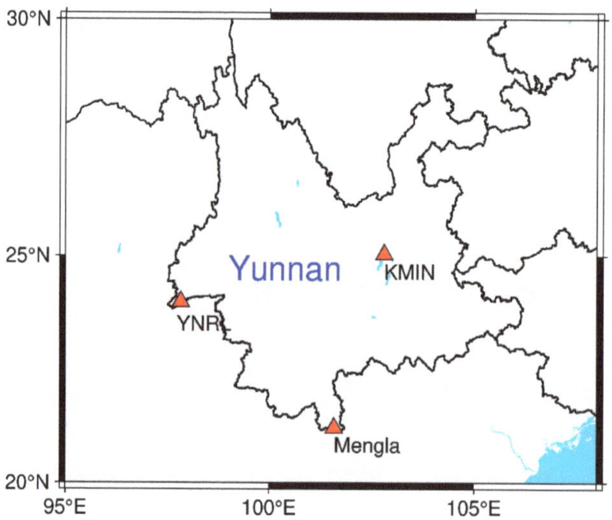

Figure A1. Locations of Station Mengla, Station YNRL, and Station KMIN. The red triangles mark the locations of the GNSS stations.

References

1. Anand, K.; Inamdar, V.; Ramanathan, N.; Loeb, G. Satellite observations of the water vapor greenhouse effect and column longwave cooling rates: Relative roles of the continuum and vibration-rotation to pure rotation bands. *J. Geophys. Res.-Atmos.* **2004**, *109*, 1–9.
2. Easterbrook, D. Greenhouse gases. In *Evidence-Based Climate Science: Data Opposing CO_2 Emissions as the Primary Source of Global Warming*, 2nd ed.; Easterbrook, D., Ed.; Elsevier: Oxford, UK, 2016; pp. 163–173.
3. Mills, E. *Weather Studies: Introduction to Atmospheric Science*, 6th ed.; American Meteorological Society: Boston, MA, USA, 2015.
4. Salby, M. *Fundamentals of Atmospheric Physics*; Academic Press: San Diego, CA, USA, 1996; pp. 25–29.
5. Wang, J.; Zhang, L.; Dai, A.; Immler, F.; Sommer, M.; Vomel, H. Radiation dry bias correction of Vaisala RS92 humidity data and its impacts on historical radiosonde data. *J. Atmos. Ocean. Technol.* **2013**, *30*, 197–214. [CrossRef]
6. Rocken, C.; Ware, R.; Van Hove, T.; Solheim, F.; Alber, C.; Johnson, J.; Bevis, M.; Businger, S. Sensing atmospheric water vapor with the Global Positioning System. *Geophy. Res. Lett.* **1993**, *20*, 2631–2634. [CrossRef]
7. Rocken, C.; Hove, T.; Johnson, J.; Solheim, F.; Ware, R.; Bevis, M.; Chiswell, S.; Businger, S. GPS/STORM—GPS sensing of atmospheric water vapor for meteorology. *J. Atmos. Ocean Tech.* **1995**, *12*, 468–478. [CrossRef]
8. Wang, M. The Assessment and Meteorological Applications of High Spatiotemporal Resolution GPS ZTD/PW Derived by Precise Point Positioning. Ph.D. Thesis, Tong University, Shanghai, China, 2019.
9. Bevis, M.; Businger, S.; Herring, T.; Rocken, C.; Anthes, R.; Ware, R. GPS meteorology: Remote sensing of atmospheric water vapor using the Global Positioning System. *J. Geophys. Res.* **1992**, *97*, 15787–15801. [CrossRef]
10. Askne, J.; Nordius, H. Estimation of tropospheric delay for microwaves from surface weather data. *Radio Sci.* **1987**, *22*, 379–386. [CrossRef]
11. Duan, J.; Bevis, M.; Fang, P.; Bock, Y.; Chiswell, S.; Businger, S.; Rochen, C.; Solheim, F.; Van Hove, T.; Ware, R.; et al. GPS meteorology: Direct estimation of the absolute value of precipitable water. *J. Appl. Meteorol.* **1996**, *35*, 830–838. [CrossRef]
12. Fang, P.; Bevis, M.; Bock, Y.; Gutman, S.; Wolfe, D. GPS meteorology: Reducing systematic errors in geodetic estimates for zenith delay. *Geophys. Res. Lett.* **1998**, *25*, 3583–3586. [CrossRef]
13. Cao, Y.; Fang, Z.; Xia, Q. Relationship between GPS precipitable water vapor and precipitation. *J. Appl. Meteorol. Sci.* **2005**, *16*, 54–59.
14. Van Baelen, J.; Reverdy, M.; Tridon, F.; Labbouz, L.; Dick, G.; Bender, M.; Hagen, M. On the relationship between water vapour field evolution and the life cycle of precipitation systems. *Q. J. R. Meteorol. Soc.* **2011**, *137*, 204–223. [CrossRef]
15. Huang, L.; Mo, Z.; Xie, S.; Liu, L.; Chen, J.; Kang, C.; Wang, S. Spatiotemporal characteristics of GNSS-derived precipitable water vapor during heavy rainfall events in Guilin, China. *Satell. Navig.* **2021**, *2*, 13. [CrossRef]
16. Brenot, H.; Neméghaire, J.; Delobbe, L.; Clerbaux, N.; De Meutter, P.; Deckmyn, A.; Deckmyn, A.; Deleloo, A.; Frappez, L.; Roozendael, M. Preliminary signs of the initiation of deep convection by GNSS. *Atmos. Chem. Phys.* **2013**, *13*, 5425–5449. [CrossRef]
17. Adams, D.; Barbosa, H.; Gaitán De Los Ríos, K. A spatiotemporal water vapor-deep convection correlation metric derived from the Amazon dense GNSS meteorological network. *Mon. Weather Rev.* **2017**, *145*, 279–288. [CrossRef]
18. Shi, C.; Zhou, L.; Fan, L.; Zhang, W.; Cao, Y.; Wang, C.; Xiao, F.; Lv, G.; Liang, H. Analysis of '21·7' extreme rainstorm process in Henan Province using BeiDou/GNSS observation. *Chin. J. Geophys.-CH* **2022**, *65*, 186–196.
19. Vedel, H.; Huang, X. Impact of ground based GPS data on numerical weather prediction. *J. Meteorol. Soc. JPN* **2004**, *82*, 459–472. [CrossRef]
20. Bennitt, G.; Jupp, A. Operational assimilation of GPS zenith total delay observations into the Met Office numerical weather prediction models. *Mon. Weather Rev.* **2012**, *140*, 2706–2719. [CrossRef]
21. Means, J. GPS precipitable water as a diagnostic of the north American monsoon in California and Nevada. *J. Clim.* **2013**, *26*, 1432–1444. [CrossRef]
22. Moore, A.; Small, I.; Gutman, S.; Bock, Y.; Dumas, J.; Fang, P.; Haase, J.; Jackson, M.; Laber, J. National weather service forecasters use GPS precipitable water vapor for enhanced situational awareness during the Southern California Summer Monsoon. *Bull. Am. Meteorol. Soc.* **2015**, *96*, 1867–1877. [CrossRef]
23. Wang, M.; Wang, J.; Bock, Y.; Liang, H.; Dong, D.; Fang, P. Dynamic mapping of the movement of landfalling atmospheric rivers over southern California with GPS data. *Geophys. Res. Lett.* **2019**, *46*, 3551–3559. [CrossRef]
24. Zhao, Q.; Ma, X.; Yao, W.; Liu, Y.; Yao, Y. A drought monitoring method based on precipitable water vapor and precipitation. *J. Clim.* **2020**, *33*, 10727–10741. [CrossRef]
25. Jin, S.; Li, Z.; Cho, J. Integrated water vapor field and multiscale variations over China from GPS measurements. *J. Appl. Meteorol. Clim.* **2008**, *47*, 3008–3015. [CrossRef]
26. Jin, S.; Luo, O. Variability and climatology of PWV from global 13-year GPS observations. *IEEE Trans. Geosci. Remote* **2009**, *47*, 1918–1924. [CrossRef]
27. Wang, J.; Zhang, L. Climate applications of a global 2-hourly atmospheric precipitable water dataset derived from IGS tropospheric products. *J. Geod.* **2009**, *83*, 209–217. [CrossRef]
28. Shi, C.; Zhang, W.; Cao, Y.; Lou, Y.; Liang, H.; Fan, L.; Satirapod, C.; Trakolkul, C. Atmospheric water vapor climatological characteristics over Indo-China region based on BeiDou/GNSS and relationships with precipitation. *Acta Geod. Cartogr. Sin.* **2020**, *49*, 1112–1119.

29. Wu, M.; Jin, S.; Li, Z.; Cao, Y.; Ping, F.; Tang, X. High-precision GNSS PWV and its variation characteristics in China based on individual station meteorological data. *Remote Sens.* **2021**, *13*, 1296. [CrossRef]
30. Fu, R.; Duan, X.; Liu, J.; Sun, J.; Wang, M.; Chen, X.; Liu, Y. Characteristics of ground-based GPS-retrieved PWV in Yunnan. *Meteorol. Sci. Technol.* **2010**, *38*, 456–462.
31. Hai, Y.; Sun, J.; Chen, X. The analysis of GPS-retrieved PWV characteristic in Yunnan from 2007–2010. *Yunnan Geogr. Environ. Res.* **2011**, *23*, 78–84.
32. Shen, Y.; Duan, W. Characteristics of temporal and spatial distribution of water vapor resource in Yunnan area. *Environ. Sci. Surv.* **2016**, *35*, 36–41.
33. Li, Y.; Xu, A.; Dong, B. Variation characteristics of precipitable water volume observed by GPS in Dali. *J. Meteorol. Res. Appl.* **2020**, *41*, 32–37.
34. Hu, H.; Cao, Y.; Shi, C.; Lei, Y.; Wen, H.; Liang, H.; Tu, M.; Wan, X.; Wang, H.; Liang, J.; et al. Analysis of the precipitable water vapor observation in Yunnan–Guizhou Plateau during the convective weather system in summer. *Atmosphere* **2021**, *12*, 1085. [CrossRef]
35. Wang, M.; Cao, Y.; Liang, H.; Tu, M.; Liu, Z. On the accuracy of regional weighted mean temperature linear models over China. *J. Nanjing Univ. Inf. Sci. Technol. (Nat. Sci. Ed.)* **2021**, *13*, 161–169.
36. Wang, M.; Chen, J.; Han, J.; Zhang, Y.; Fan, M.; Yu, M.; Sun, C.; Xie, T. Region-specific and weather-dependent characteristics of the relation between GNSS weighted mean temperature and surface temperature over China. *Remote Sens.* **2023**, *15*, 1538. [CrossRef]
37. Niell, A. Preliminary evaluation of atmospheric mapping functions based on numerical weather models. *Phys. Chem. Earth* **2001**, *26*, 475–480. [CrossRef]
38. Böhm, J.; Schuh, H. Vienna mapping functions in VLBI analyses. *Geophy. Res. Lett.* **2004**, *31*, L01603. [CrossRef]
39. Neill, A. Global mapping functions for the atmosphere delay at radio wavelengths. *J. Geophys. Res.* **1996**, *101*, 3227–3246. [CrossRef]
40. Böhm, J.; Niell, A.; Tregoning, P.; Schuh, H. Global mapping function (GMF): A new empirical mapping function based on data from numerical weather model data. *Geophys. Res. Lett.* **2006**, *33*, L07304. [CrossRef]
41. Davis, J.; Elgered, G.; Niell, A.; Kuehn, C. Ground based measurement of gradients in the "wet" radio refractivity of air. *Radio Sci.* **1993**, *28*, 1003–1018. [CrossRef]
42. Saastamoinen, J. Atmospheric correction for troposphere and stratosphere in radio ranging of satellites. In *The Use of Artificial Satellites for Geodesy*; Henriksen, S., Mancini, A., Chovitz, B., Eds.; William Byrd Press: Richmond, VA, USA, 1972; Volume 15, pp. 247–252.
43. Bevis, M.; Businger, S.; Chiswell, S.; Herring, T.; Anthes, R.; Rocken, C.; Ware, R. GPS meteorology: Mapping zenith wet delays onto precipitable water. *J. Appl. Meteorol.* **1994**, *33*, 379–386. [CrossRef]
44. Davis, J.; Herring, T.; Shapiro, I.; Rogers, A.; Elgered, G. Geodesy by radio interferometry: Effects of atmospheric modeling errors on estimates of baseline length. *Radio Sci.* **1985**, *20*, 1593–1607. [CrossRef]
45. Dach, R.; Lutz, S.; Walser, P.; Fridez, P. *Bernese GNSS Software Version 5.2*; User Manual; Astronomical Institute, University of Bern, Bern Open Publishing: Bern, Switzerland, 2015; ISBN 978-3-906813-05-9. [CrossRef]
46. Chen, G.; Herring, T. Effects of atmospheric azimuthal asymmetry on the analysis of space geodetic data. *J. Geophy. Res.* **1997**, *102*, 20489–20502. [CrossRef]

remote sensing MDPI

Article

Comparison and Synthesis of Precipitation Data from CloudSat CPR and GPM KaPR

Yanzhi Liang [1], Leilei Kou [1,2,*], Anfan Huang [1], Haiyang Gao [1,2], Zhengjian Lin [3], Yanqing Xie [4] and Liguo Zhang [4]

1 School of Atmospheric Physics, Nanjing University of Information Science and Technology, Nanjing 210044, China; 20211205006@nuist.edu.cn (Y.L.); gaohy@nuist.edu.cn (H.G.)
2 Collaborative Innovation Center on Forecast and Evaluation of Meteorological Disasters, Key Laboratory for Aerosol-Cloud-Precipitation of China Meteorological Administration, Nanjing University of Information Science and Technology, Nanjing 210044, China
3 Hainan Province Meteorological Detection Center, Haikou 570203, China
4 Shanghai Institute of Satellite Engineering, Shanghai 201109, China
* Correspondence: 002358@nuist.edu.cn

Abstract: Employing different bands of radar to detect precipitation information in identical regions enables the acquisition of a more comprehensive precipitation cloud structure, thereby refining the continuity and completeness of precipitation measurements. This study first compared the coincident data from CloudSat W-band cloud profiling radar (CPR) and Global Precipitation Measurement Mission (GPM) Ka-band precipitation radar (KaPR) from 2014 to 2017, and then a synthesis of the radar reflectivity from CPR and KaPR was attempted to obtain a complete cloud and precipitation structure. The findings of the reflectivity comparisons indicated that the echo-top height identified by CPR is on average 3.6 to 4.2 km higher than that from KaPR, due to the higher sensitivity. Because of strong attenuation of CPR by liquid-phase particles, the reflectivity below the height of the melting layer usually shows an opposite gradient to KaPR with decreasing altitude. The difference in the near-surface rain rates of CPR and KaPR was found to be related to reflectivity gradients in the vertical direction, and the larger the reflectivity gradients, the greater the differences in near-surface rain rates. For better representing the complete vertical structure of precipitation clouds and improving the consistency of the reflectivity and precipitation rate, the radar reflectivity was weighted, synthesized from CPR and KaPR based on the gradient difference of the reflectivity from the two radars. We presented the synthesis results for a stratiform cloud and a deep convective case, and Spearman's rank correlation coefficient (r_s) between the GPM combined radiometer precipitation rate and the radar reflectivity was utilized to analyze the performance of the synthesis. The consistency between synthesized reflectivity and precipitation rate in the non-liquid phase was improved compared with KaPR, and the r_s of the ice and mixed phases was increased by about 12% and 10%, respectively.

Keywords: CloudSat cloud profiling radar (CPR); Global Precipitation Measurement Mission (GPM) Ka-band precipitation radar (KaPR); precipitation data; comparison; synthesis

Citation: Liang, Y.; Kou, L.; Huang, A.; Gao, H.; Lin, Z.; Xie, Y.; Zhang, L. Comparison and Synthesis of Precipitation Data from CloudSat CPR and GPM KaPR. *Remote Sens.* **2024**, *16*, 745. https://doi.org/10.3390/rs16050745

Academic Editor: Kenji Nakamura

Received: 16 January 2024
Revised: 11 February 2024
Accepted: 18 February 2024
Published: 21 February 2024

1. Introduction

An accurate characterization of global precipitation is indispensable for improving global climate and weather forecasting models, as well as for evaluating the global energy budget and water resources [1]. Clouds play an important role in the hydrological cycle, redistributing water and generating precipitation on a global scale. To thoroughly comprehend the significance of clouds and precipitation, and to enhance their performance in numerical models, it is crucial to obtain global-scale measurements of these phenomena and conduct detailed investigations into their formation, development, and dissipation processes across all scales [2]. Microwave sensors outperform infrared and visible-light sensors in precipitation retrieval, since they exhibit higher sensitivity to rain and ice particles [3]. Moreover, active microwave cloud and precipitation radars have the advantages of strong penetration and high accuracy of precipitation estimation [4].

The CloudSat satellite equipped with a W-band cloud profiling radar (CPR) can quantitatively measure the vertical structure of clouds and precipitation [5]. The Global Precipitation Measurement Mission (GPM) core satellite carries a dual-frequency precipitation radar (DPR) operating in the Ku band and Ka band [6]. DPR can provide measurements of the 3D precipitation structure on a global scale. Owing to differences in orbital inclinations, the CloudSat satellite and the GPM satellite occasionally intersect, thereby affording an opportunity for multi-frequency precipitation measurement.

Conducting a comparative study on precipitation measurement using radar systems operating at different bands is of paramount importance in order to harness and integrate the unique strengths of each band. In conjunction with the CloudSat CPR, Berg et al. (2010) utilized the Tropical Rainfall Measuring Mission (TRMM) precipitation radar (PR) to assess rainfall distribution and intensity in tropical and subtropical oceans [7]. Their results indicated that CPR detected rainfall occurrences nearly 2.5 times more frequently than TRMM PR. Tang et al. (2017) conducted a global-scale evaluation by matching TRMM PR, GPM dual-frequency precipitation radar (DPR), and CloudSat CPR data [8]. Sindhu et al. (2017) compared the reflectivity derived from CloudSat CPR and TRMM PR by analyzing their coincident events [9]. The results demonstrated that the reflectivity from CloudSat CPR was approximately 10 dBZ lower than that of TRMM PR below a height of 4 km. Fall et al. (2013) analyzed the vertical structure of storms utilizing data from CloudSat CPR, TRMM PR, and ground-based radar [10]. They further conducted multi-frequency measurements of microphysical quantities within different regions of the melting layer and carried out comparisons among them. This study revealed the characteristics of the bright band in cold-season precipitation and its impact on radar-based quantitative precipitation estimation. CloudSat CPR exhibited excellent sensitivity in detecting snow, whereas GPM KaPR, as a supplemental payload designed to compensate for the limitations of the Ku band in detecting snow, ice, and light rain, also demonstrated robust snow detection capabilities. The ability of DPR to identify snowfall was evaluated by utilizing a dataset with observation intervals of less than 5 min between DPR and CPR [11]. Mroz et al. (2021) compared surface snowfall rates estimated by GPM DPR and CloudSat CPR with estimates derived from multi-radar and multi-sensor composite products covering the continental United States from November 2014 to September 2020 [12]. Skofronick-Jackson et al. (2019) compared three years of CPR and DPR snowfall data, taking into account differences in hardware, inversion algorithms, and radar bands. It was found that the global average near-surface snowfall accumulation in DPR was about 43% lower than that in CPR [13].

The Ka- and W-band spaceborne radars possess different advantages in the measurement of precipitation structures. The higher frequency of the W-band radar enables it to achieve superior spatial resolution and sensitivity, capturing detailed cloud structure and light precipitation. Conversely, the relatively lower frequency of the Ka-band radar allows for partial penetration of clouds and precipitation, facilitating the accurate measurement of light-to-heavy precipitation. The integration of these two radar bands yields complementary information, fostering a more comprehensive understanding of precipitation cloud characteristics. Notably, NASA's Aerosol/Cloud/Ecosystems (ACE) mission incorporates dual-band and dual-polarization modes in the W and Ka bands to enhance the functionality and performance of cloud radar [14,15]. Currently, China has successfully launched its first spaceborne precipitation radar, and the National Satellite Meteorological Center is planning the deployment of a cloud measurement satellite. The primary payload under consideration for this initiative is a W- and Ka-band measurement radar [16].

In this paper, we compare and synthesize precipitation data obtained by CPR-KaPR coincidence datasets. To begin with, we match the precipitation observations from both radars in terms of space and time. Next, we compare the differences in precipitation measurement between the two radar systems, including radar reflectivity, as well as the spatial distribution and intensity of the near-surface precipitation rate with different phases.

Finally, we synthesize the radar data from the two bands to obtain a more complete and comprehensive vertical structure of precipitation.

2. Data and Matching Methods

2.1. CloudSat CPR Data

The CloudSat satellite carries the most widely used spaceborne cloud radar, known as millimeter-wave cloud profiling radar (CPR). CPR is a W-band radar that is pointed toward the nadir and has a minimum detectable signal of approximately −29 dBZ. The radar footprint size is 1.7 km along-track and 1.4 km cross-track, while the vertical resolution is about 500 m [17,18].

In this study, we utilize radar bin data with a cloud mask value of 40, indicating a high confidence level in cloud identification [19,20]. The retrieval process for near-surface precipitation rates accounts for factors such as radar beam attenuation, multiple scattering effects, and precipitation identification in the ground clutter region (the five lowest radar bins above the surface, approximately 1.2 km) [21]. Retrieval of near-surface precipitation rates relies on radar reflectivity and the estimated value of path-integrated attenuation (PIA), which are determined based on surface reflectance characteristics. However, due to the complex nature of land surface reflectance compared to that of the ocean, retrieval of precipitation data over land surfaces is not feasible [22].

For snowfall products, the backscattering characteristics of snow particles are calculated using a combination of the original model for small particles and an aggregation model for large particles [23]. Assuming an exponential particle size distribution for snow, the slope and intercept parameters of the particle size distribution are obtained through the application of the optimal estimation method [24]. This optimization method minimizes a cost function that represents the weighted sum of differences between simulated and observed radar reflectance, as well as the disparities between prior and estimated microphysical properties of snow. Ultimately, the retrieved particle size distribution parameter curve allows for the derivation of snowfall estimates.

For this research, we used a variety of parameters, such as the radar reflectivity factor, cloud scene, near-surface rain rate, snowfall rate, and others. These parameters were derived from the 2B-GEOPROF, 2C-PRECIP-COLUMN, and 2C-SNOW-PROFILE products. These products were specifically extracted from the CloudSat Version 5 data release by the CloudSat data team in 2019.

2.2. GPM KaPR Data

The GPM is a new generation of precipitation observation program following the TRMM. Its core satellite carries two instruments for measuring precipitation: the GPM Microwave Imager (GMI), and the DPR. One goal of the DPR is to measure light-to-heavy precipitation using Ka-band (35.5 GHz) radar (KaPR) and Ku-band (13.6 GHz) radar (KuPR) [25]. This study uses KaPR HS-mode data because the Ka-band HS mode is more sensitive than that of the Ku-band. In addition to detecting more occurrences of light precipitation, it can also detect more clouds. The scanning angle range of the KaPR antenna is ±8.5°, the vertical resolution is 500 m for KaHS, the scanning width is 120 m, and the spatial resolution is 5 km. The minimum detectable reflectivity of KaPR is 10.40 dBZ, corresponding to the minimum detectable precipitation rate of 0.16 mm/h [26]. The technical indicators of KaPR and CPR are shown in Table 1.

The Ka-band data used in this paper were the standard 2A-DPR High-Sensitivity Mode (HS) product of GPM DPR L2 version 7, which were obtained by attenuation correction and non-uniform filling correction based on level L1 base data [26].

Table 1. Brief description of CPR and KaPR.

Instrument	CloudSat CPR	GPM KaPR
Launch time	18 April 2006	27 February 2014
Inclination angle (°)	98.23	65
Altitude (km)	705	407
Frequency (GHz)	94	35
Vertical resolution (m)	500	500 (KaHS)
Along-track resolution (km)	1.7	5
Cross-track resolution (km)	1.4	5
Minimum detectable Ze (dBZ)	−29	10.40 (KaHS)

2.3. Data Matching and Coincidence Data Processing

Before comparing the datasets generated by the two radar systems, a requirement lies in the alignment and time synchronization of their measurements. Within this context, the term "coincident event" assumes paramount importance, signifying a conjunctive occurrence of radar pixels. The critical criterion for a coincident event hinges upon time and location, stipulating that the coincidence between the footprints of the two radar systems must be within 15 min. To initiate this process, it is imperative to query the KaPR data associated with the footprint, within the specified 15-min window, predicated upon the transit time of the CPR. In instances where the trajectories of the footprint indeed coincide, recourse should be made to the nearest-neighbor methodology [27]. This approach facilitates the judicious amalgamation of the two radar datasets, thereby yielding the coveted CPR-KaPR coincidence data. The time difference threshold of 15 min has very little impact on the accuracy of the statistical results [28]. We chose 15 min by considering the trade-off between the number of samples and the time difference and statistical accuracy.

Following the spatiotemporal matching methodology delineated earlier, it can be discerned that, due to the superior footprint resolution exhibited by the CPR as compared to the KaPR, multiple observations of the footprint by the CPR are found to correspond to a single footprint observation by KaPR, as visually illustrated in Figure 1a. To rectify this disparity and establish a one-to-one correspondence between the CPR and KaPR pixel values, a resampling procedure was executed along the trajectory of the CPR orbit. Under the assumption of a uniform distribution of cloud features along the direction of the CPR orbit, the pixel values of the CPR data were spatially averaged within each corresponding KaPR pixel. In this manner, the mean values encompassing C2 to C4 correspond to C1′, while the mean values spanning C5 to C8 correspond to C2′. For the value of KaPR footprints, Kn = Kn′. KaPR footprints designated with primed symbols in Figure 1b serve the purpose of distinguishing data before and after the resampling process. The post-resampling dataset maintains parity with the KaPR dataset in terms of volume, and it is visually depicted in Figure 1b.

Following the rigorous data resampling process, a substantial dataset comprising a total of 310,095 sets of CloudSat CPR and GPM KaPR coincident events was successfully collated, spanning from March 2014 to December 2017. Within this illustrative representation, the dots serve to delineate the trajectory of the coincident orbit, with the depth of color encoding the corresponding coincident frequency, as shown in Figure 2. It is imperative to underscore that, owing to the disparate inclination angles governing the orbits of the two satellites, a notable variation in coincident frequency can be discerned across distinct latitudinal bands. Notably, within the geographical expanses spanning 55 to 67 degrees (north and south latitudes), the coincident frequency registers a noteworthy augmentation, ranging from 50% to an impressive 800% higher than observed at other latitudinal coordinates.

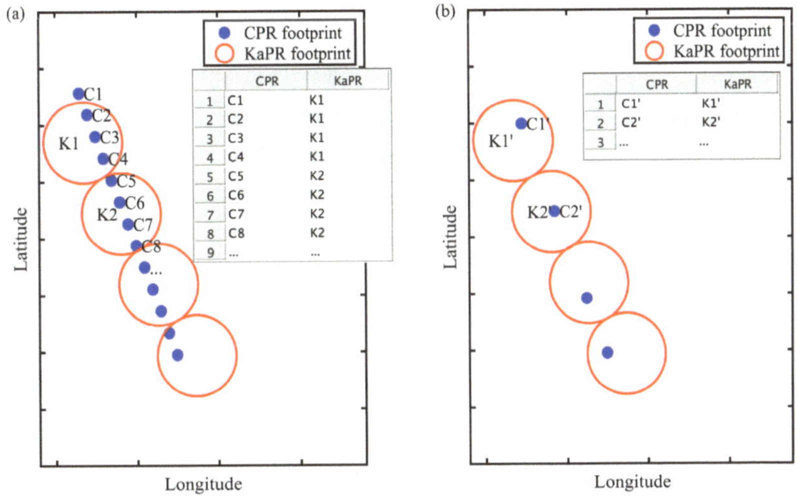

Figure 1. Resampling diagram ((**a**): data before resampling, (**b**): data after resampling).

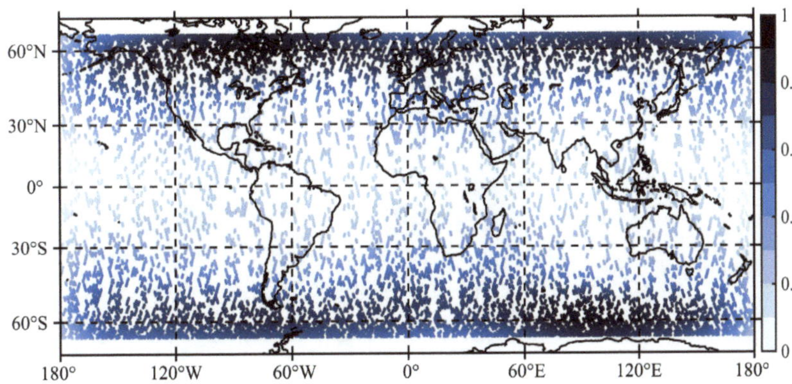

Figure 2. Footprints of CPR and KaPR coincident events between March 2014 and December 2017.

3. Results

3.1. Comparison of Reflectivity

Radar systems operating in distinct bands exhibit differences in their precipitation measurements. These disparities arise from the frequency-dependent nature of precipitation's microphysical characteristics, leading to differential scattering or absorption responses among radars operating in dissimilar frequency ranges. Through a comparative analysis of radar reflectivity across various frequency bands, it becomes possible to discern and investigate the sensitivity differentials exhibited by radars in response to different precipitation types.

In this study, the classification of precipitation phases within the study area relied upon the near-surface precipitation phase parameters from the KaPR. Specifically, footprints wherein both the CPR and the KaPR detected liquid-phase precipitation were designated as "rainfall footprints". Conversely, those footprints characterized by solid-phase precipitation were categorized as "snowfall footprints". When KaPR could not identify near-surface precipitation due to sensitivity reasons, we chose to trust CPR's precipitation phase identification. After identification, a total of 12,126 rainfall footprints and 14,737 snowfall footprints were selected. Given the greater precision of KaPR in discerning the bright band phenomenon, the classification of rain types within the rainfall footprints primarily hinges on the near-surface rain type parameters obtained from KaPR. Footprints identified by

KaPR as indicative of convective rainfall were designated as "convective rainfall footprints". Subsequently, those footprints identified by KaPR as representing stratiform rainfall were categorized as "stratiform rainfall footprints". This classification process yielded a final count of 4670 convective rainfall footprints and 7456 stratiform rainfall footprints.

Figure 3 presents a two-dimensional probability distribution diagram illustrating the radar reflectivity–height relationship for convection, stratiform rainfall, and snowfall during intersecting events observed by the CPR and the KaPR. Figure 3a–c depict the radar reflectivity–height profiles for CPR convection, stratiform rainfall, and snowfall, respectively. Subsequently, Figure 3d–f display the two-dimensional probability distribution images of radar reflectivity–height for KaPR convection, stratiform rainfall, and snowfall, respectively. In these representations, the abscissa denotes the radar reflectivity, the ordinate signifies the altitude, and the color mapping indicates the probability of observing a particular radar reflectivity at a given altitude in different precipitation types. The figure reveals distinctive patterns among the precipitation types. In convective rainfall, most CPR reflectivity between altitudes of 1.5 and 5 km exhibits a decreasing trend as altitude decreases, reflecting the presence of larger liquid particles with more pronounced attenuation of W-band radar waves. Conversely, most KaPR reflectivity gradually increases with decreasing altitudes during convective rainfall. For stratiform rainfall, both CPR and KaPR reflectivity between altitudes of 1 and 4 km remains relatively stable. CPR reflectivity primarily falls within the range of 7–12 dBZ, while KaPR reflectivity is predominantly distributed between 16–21 dBZ. This phenomenon arises from the stable intensity of stratiform rainfall and the larger size of liquid particles compared to convective rainfall.

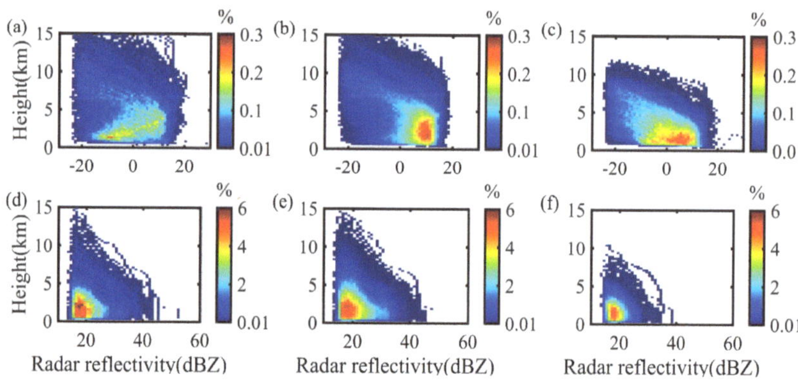

Figure 3. Reflectivity–height two-dimensional (2D) probability distribution of convective rain, stratiform rain, and snow in CPR-KaPR coincidence ((**a**): CPR convective rain; (**b**): CPR stratiform rain; (**c**): CPR snow; (**d**): KaPR convective rain; (**e**): KaPR stratiform rain; (**f**): KaPR snow).

Furthermore, the figure highlights variations in the two-dimensional probability distribution images between snowfall and rainfall, with CPR exhibiting more pronounced characteristics compared to KaPR. This divergence can be attributed to the higher frequency of the W band, which results in greater scattering efficiency in smaller water droplets and ice crystals. Such small particles are prevalent during snowfall, contributing to the broader radar reflectivity–height distribution exhibited by CPR during snowfall events. Consequently, W-band radar can offer more detailed structural information regarding snowfall events.

Regarding echo height, the average echo-top heights observed in convective rainfall, stratiform rainfall, and snowfall were as follows: for the CPR, they measured 7.56 km, 8.16 km, and 6.70 km, respectively, whereas for the KaPR, the average echo-top heights were 3.92 km, 4.11 km, and 3.08 km, respectively. Notably, the average echo-top heights detected by CPR exceeded those detected by KaPR, by approximately 3.6 to 4.2 km on average. This discrepancy arises due to CPR's superior resolution, which enables it to detect smaller cloud particles and provides a more comprehensive vertical profile of cloud particle information.

Figure 4 displays the average radar reflectivity profiles for convective rainfall, stratiform rainfall, and snowfall during the coincident event. The horizontal error bars in the figure are one standard deviation of the average radar reflectivity at each corresponding altitude. The observations reveal distinct performance in different regions:

1. In the lower region, below 1 km altitude, within convective rainfall, both CPR and KaPR exhibit a gradual increase in average reflectivity with decreasing altitude. This phenomenon arises due to the relatively lower attenuation of W-band radar waves in this region compared to the increase in the detection values.
2. In the altitude range of 1.5 to 5 km, the reflectivity for most CPR observations within convective rainfall exhibits variations with altitude, while KaPR observations show the opposite trend. This difference can be attributed to the substantial attenuation of W-band radar waves, which is more pronounced than the corresponding increase in detection values in this altitude range.
3. Within stratiform rainfall, the CPR average reflectivity initially increases and then decreases as the altitude rises between 1 and 3 km. In contrast, the KaPR reflectivity gradually decreases with increasing altitude in the same altitude range. This behavior can be attributed to the greater attenuation of W-band radar by liquid particles between 1 and 3 km, resulting in a reduction in the average reflectivity. While Ka-band electromagnetic waves also experience attenuation, their average attenuation is less pronounced than the corresponding increase in reflectivity. Hence, the average profile value for CPR gradually decreases with decreasing altitude, whereas the average profile value for KaPR increases with decreasing altitude.
4. In the case of snowfall, the average reflectivity profile of the CPR exhibits more pronounced variations with height compared to that of the KaPR, with a notable peak occurring near the altitude of 10 km. This divergence can be attributed to the heightened sensitivity of the W band, which enables the CPR to detect more detailed snow features that fall below the detection threshold of the KaPR. Consequently, the CPR reveals more gradients in reflectivity, offering a more comprehensive characterization of snowfall particle features.

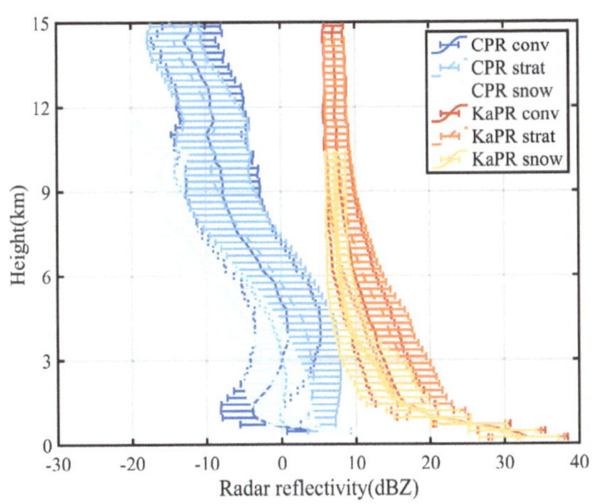

Figure 4. Average profiles of radar reflectivity for convective rain, stratiform rain, and snow in CPR and KaPR coincident events.

From Figures 3 and 4, we contend that the reflectivity measurements by cloud and precipitation radars operating in different bands can effectively complement one another. The W-band radar, with its higher spatial resolution and sensitivity, can provide light precipitation and cloud structure details, whereas the Ka-band radar excels in its ability to mitigate signal attenuation, allowing it to penetrate thick cloud cover and measure light-to-high precipitation. By comparing the radar reflectivity from the two bands, we found that the joint measurements are complementary and can expand the vertical structure range of precipitation clouds. Considering the continuity of precipitation, future advancements in precipitation algorithms could benefit from incorporating additional upper-level cloud information as constraints in the precipitation estimation process.

3.2. Comparison of Near-Surface Precipitation

In addition to assessing the vertical structure measurement of the two radars, it is essential to conduct a global-scale comparison of the near-surface precipitation data retrieved by them. This comprehensive evaluation will provide insights into their capabilities for assessing and monitoring precipitation across various regions and varying intensity levels.

3.2.1. Comparison of Rain Rates

In the comparison of near-surface rain rates, it is crucial to consider that the retrieval process for near-surface precipitation using the CPR takes into account factors such as attenuation and multiple scattering effects of radar beams, as well as the identification of precipitation in areas with ground clutter. This retrieval process involves utilizing data related to near-surface radar reflectivity and surface reflection characteristics. Path-integrated attenuation estimates are then employed to retrieve near-surface rain rates. Due to the inherent complexity of land surface albedo in contrast to the ocean, it is not feasible to retrieve precipitation data over land surfaces. Consequently, our comparison is limited to the assessment of ocean surface rainfall data retrieved by both radar systems.

For statistical purposes in the coincident event, all footprints reporting rain were selected, as it is important to note that the CPR can detect rainfall of lower intensity compared to the KaPR. Consequently, there may be instances where the CPR detects rainfall while the KaPR does not, resulting in cases where the CPR has a non-zero retrieval value while the KaPR records a value of 0 for near-surface rain rate.

To assess the performance of the two radars in rainfall estimation across diverse global regions, Figure 5 presents the average near-surface rain rate distribution based on coincident events observed by the two satellite-borne radars. The resolution for this analysis is 5 degrees * 5 degrees. Specifically, Figure 5a,b illustrate the average near-surface rain rate distributions for CPR and KaPR, respectively. Figure 5c depicts the distribution of the average near-surface rain rate differences between the two radars. The boxplot of the average near-surface rain rate between the two radars, with 40 degrees latitude as the boundary, is shown in Figure 5d.

The findings reveal that, within the latitudinal range of 40 degrees north and south, the average near-surface rainfall retrieved by KaPR is 0.4 mm/h greater than that retrieved by CPR. Conversely, above 40 degrees north and south, the average value retrieved by CPR surpasses KaPR by 0.1 mm/h. This variation suggests that, at low latitudes, the impact of heavy rainfall is more pronounced, leading to greater attenuation of CPR's electromagnetic waves. Consequently, CPR may struggle to accurately estimate rainfall exceeding its upper threshold, resulting in a lower average rainfall retrieval compared to KaPR. In addition to the fact that the probability of heavy precipitation events at high latitudes is smaller than that at low latitudes, as the latitude increases, the sampling frequency gradually increases, and the proportion of cases where there is a large difference in the retrieval values between the two also decreases. At high latitudes, there is a prevalence of medium- and small-intensity rainfall events compared to mid-latitudes. CPR's heightened sensitivity allows it to detect 15% more rainfall than KaPR in such conditions.

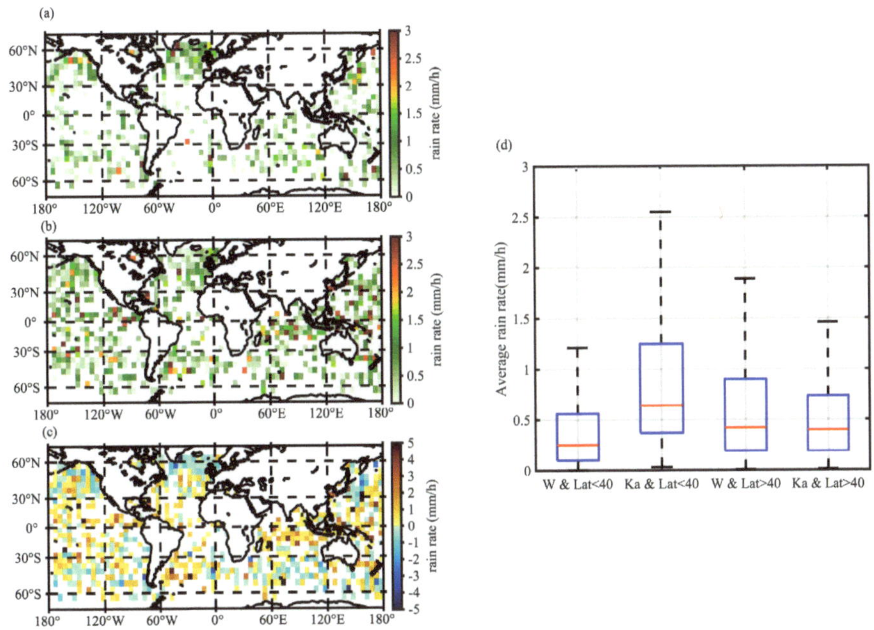

Figure 5. Average near-surface rain rate distribution based on coincident events ((**a**): CPR, (**b**): KaPR, (**c**): average near-surface rain rate difference between the two, (**d**): boxplot of the average near-surface rain rate between the two).

After comparing the rainfall detection capabilities of the two radars across different regions, we proceeded to evaluate their performance in near-surface rainfall of varying intensities. We selected footprints with rainfall retrieval values from both radar systems for this comparison. Figure 6a presents a scatter probability distribution of the near-surface rainfall rates observed by the CPR and the KaPR during coincident events. The majority of the rainfall events detected by both radar systems fall within the range of 0.2 to 1 mm/h.

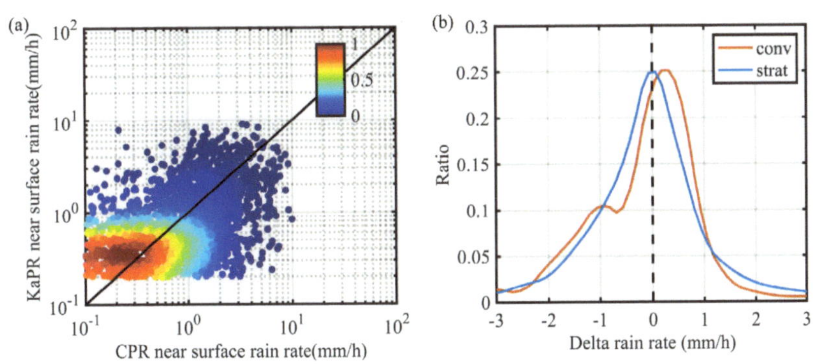

Figure 6. (**a**) CPR and KaPR near-surface rain rate scatter probability distribution and (**b**) convective and stratiform rain rate difference frequency distribution (Δrain rate = KaPR rain rate—CPR rain rate).

To quantify the agreement between the two radar systems in terms of near-surface rain rate, we calculated the correlation coefficient (CC) and the root-mean-square error (RMSE) for the inversion results. The overall correlation coefficient for the near-surface rain rate retrieved by both radars was 0.4, while the root-mean-square error was 2.18 mm/h.

These metrics suggest a moderate level of correlation and an average level of agreement between the near-surface rain rate estimates obtained from the two radar systems.

To analyze the differences in the near-surface rain rate values simultaneously retrieved by both radars, we categorized these differences based on whether the precipitation was convective or stratiform. Subsequently, we compiled and tabulated the frequency distribution of the differences between the near-surface rain rates derived from the CPR and the KaPR, as illustrated in Figure 6b.

Figure 6b reveals that the differences in near-surface rainfall between the CPR and KaPR follow a Gaussian distribution. Specifically, during convective rainfall events, the KaPR tends to produce higher values compared to the CPR. In contrast, during stratiform rainfall events, the differences between the two radar systems follow a standard normal distribution with a mean of zero.

We attribute these findings to the characteristics of convective and stratiform precipitation. In convective rainfall, precipitation cloud clusters are typically thicker, and near-surface precipitation particles tend to be larger. The CPR near-surface rain rate algorithm may underestimate the attenuation of W-band radar waves in such conditions, resulting in lower average near-surface rain rate estimates compared to KaPR.

On the other hand, stratiform rainfall tends to be more stable. In these conditions, the attenuation of W-band radar waves is less pronounced compared to convective rainfall. While W-band electromagnetic waves do experience substantial attenuation in cases of heavy stratiform rainfall, they may provide more accurate estimates of near-surface rain rates during periods of lower-intensity precipitation. Hence, for stratiform rainfall, differences between the two radars tend to conform to a standard normal distribution.

Numerous factors contribute to the disparities in near-surface rain rate retrieval between the CPR and the KaPR. These encompass distinct retrieval algorithms employed by the radars, as well as the influence of radar electromagnetic wave refraction and multiple scattering phenomena during the detection process. CPR and DPR employ different microphysical assumptions and retrieval algorithms, which is also an important factor influencing differences in rain rate. Based on the comparison of reflectivity and near-surface rain rate between the two radars, due to the attenuation of CPR, the difference in near-surface rain rate may also be related to the change in reflectivity in the vertical direction.

To illustrate the relationship between the change in reflectivity in the vertical direction and near-surface rainfall, we introduced a parameter linked to the gradient of the radar reflectivity. When the radar reflectivity within this frequency increases with decreasing altitude, a value of 1 is assigned to the bin, while a value of -1 is assigned when the reflectivity decreases with altitude. This procedure enables the derivation of the radar reflectivity gradient profile specific to the precipitation cloud within this frequency. The gradient profile difference across the entire precipitation cloud was computed by subtracting the radar reflectivity gradient profiles of KaPR and CPR. Figure 7a provides a schematic diagram depicting this process.

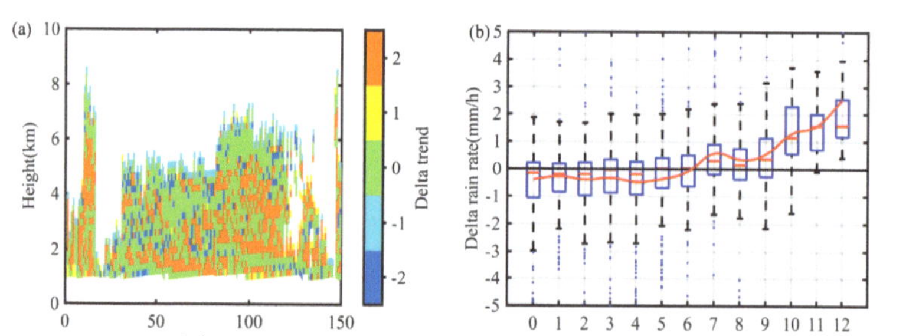

Figure 7. (**a**) Schematic diagram of the gradient profile difference of KaPR and CPR radar reflectivity of precipitation clouds, and (**b**) the boxplot of the delta near-surface rain rate and amount of bins whose gradient difference is 2 within the column.

When the radar reflectivity of both radars exhibits the same trend as altitude decreases, the resulting profile difference is 0. However, during heavy rainfall, the KaPR reflectivity gradually increases with decreasing altitude, while the CPR reflectivity undergoes attenuation due to raindrop particles, leading to a gradual decrease with decreasing altitude. Consequently, the profile difference reaches a value of 2. We proceeded to tally the number of data bins where the difference between the radar reflectivity gradient profiles of precipitation clouds derived from KaPR and CPR equaled "2". This count is presented in Figure 7b, in the form of a boxplot, alongside the near-surface rain rate difference.

The boxplot reveals that when the amount of bins exhibiting a difference of "2" surpasses or equals six within the column, the average near-surface rain rate retrieved by CPR is smaller than that obtained by KaPR. When the count of bins with a profile difference equal to "2" reaches or exceeds 10, CPR consistently yields smaller average near-surface rain rates compared to KaPR. Additionally, the average near-surface rain rate difference varies commensurately with the count of bins exhibiting a profile difference of "2". These findings underscore that, at a certain level of rainfall rate, the attenuation of W-band radar waves cannot be accurately estimated, resulting in the underestimation of the near-surface rain rate by CPR.

3.2.2. Comparison of Snowfall Rates

Snowfall is the meteorological phenomenon characterized by the descent of ice crystals from the atmosphere to the Earth's surface, forming snow. This process signifies the atmospheric transfer of water to the surface, constituting an integral component of the natural water cycle [29]. In the context of precipitation comparisons, it is important to conduct a global-scale evaluation of near-surface snow rates retrieved by the two radar systems.

The average distribution of near-surface snow rates retrieved by the CPR and the KaPR is illustrated in Figure 8a,b, respectively. Meanwhile, Figure 8c displays the distribution of the average near-surface snow rate difference between the two radar systems. The boxplot of the average near-surface snow rate between the two radars, with 50 degrees latitude as the boundary, is shown in Figure 8d. It is noteworthy that the CPR snow rate algorithm truncates the reflectivity profile at an altitude of approximately 1 km to mitigate the impact of ground clutter [30]. Consequently, in regions characterized by terrain-related factors between 30 and 50 degrees (north and south latitudes), CPR's inversion values tend to be lower than those of KaPR. Figure 8c illustrates this phenomenon, with specific regions such as the Tibetan Plateau in mainland China, the Rocky Mountains in the Western United States, and the Cordillera Mountains in South America exhibiting lower near-surface snow rate inversion values for CPR compared to KaPR. However, CPR's heightened sensitivity enables it to detect a greater number of snowfall events than KaPR in areas with latitudes exceeding 50 degrees, accounting for 86% of the total snowfall events.

The discrepancy in radar wave frequencies results in a deviation of several hundred meters in the selected sampling heights for the two near-surface snow rate algorithms near the ground. Therefore, KaPR tends to yield higher snow rates, attributed to terrain-related factors, compared to CPR. Nevertheless, CPR's advantage in terms of high sensitivity allows it to detect more snowfall events in high-latitude regions than KaPR.

Apart from terrain-related factors, as shown in Figure 8d, the average near-surface snow rate retrieved by KaPR in areas below 50 degrees (north and south latitudes) exceeds that of CPR by 0.2 mm/h. Conversely, in regions above 50 degrees (north and south latitudes), the average near-surface snow rate retrieved by CPR surpasses that of KaPR by 0.1 mm/h. This discrepancy arises because CPR detects 45% more light snow events (with intensities in the order of 0.01 mm/h) than KaPR in regions above 50 degrees (north and south latitudes).

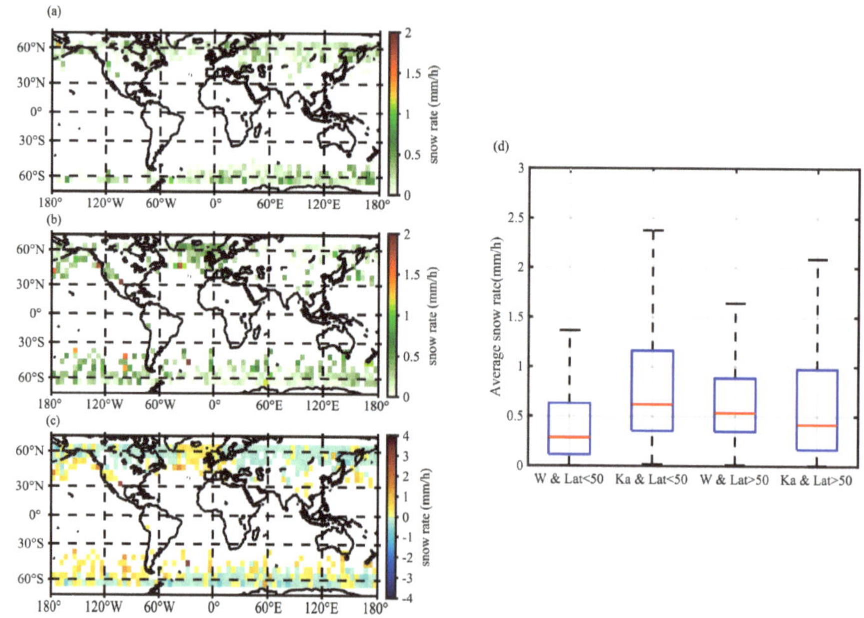

Figure 8. Average near-surface snow rate distribution based on coincident events of two spaceborne radars ((**a**): CPR, (**b**): KaPR, (**c**): difference in average near-surface snow rate, (**d**): boxplot of the average near-surface snow rate between the two).

4. Precipitation Synthesis from CPR and KaPR

The comparative analysis of near-surface precipitation between the two radar systems reveals that CPR excels in detecting weaker precipitation compared to KaPR. For heavier precipitation, the W-band radar waves experience attenuation, leading to an underestimation of precipitation by CPR. The KaPR has the advantage in detecting light-to-heavy precipitation.

Combining the near-surface precipitation data retrieved by both radar systems involves considering instances where one or both radars retrieve near-surface precipitation. In cases where both radars detect near-surface precipitation simultaneously, the maximum value is selected for inclusion in the statistics, resulting in a maximum near-surface rain rate dataset. The frequency distribution statistics of near-surface precipitation rates in coincident events for both radar systems are presented in Figure 9.

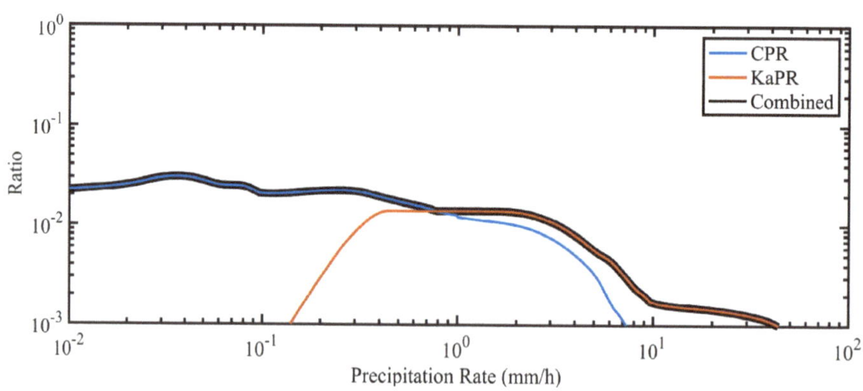

Figure 9. Frequency distribution of near-surface precipitation rates of CPR and KaPR coincident events.

In Figure 9, the x-axis indicates the precipitation rate, while the y-axis indicates the occurrence frequency. The blue and orange lines indicate the frequency of precipitation at different intensities detected by CPR and KaPR, respectively. The black lines indicate the near-surface precipitation rates obtained from the maximum combination of the two radar datasets. The observed trends in Figure 9 are as follows:

1. In the precipitation rate range of 0.01–0.2 mm/h, KaPR, due to its sensitivity limitations, is unable to detect precipitation within this range.
2. In the rate range of 0.2–8 mm/h, as precipitation rate increases, the near-surface rain rate retrieved by the CPR gradually underestimates, leading to a decrease in the proportion of CPR-detected precipitation and an increase in the proportion of KaPR-detected precipitation.
3. When the precipitation rate exceeds 7 mm/h, the values retrieved by CPR are smaller than those retrieved by KaPR.

This analysis demonstrates that the combined detection of near-surface precipitation by both radar systems allows for a more complete detection of precipitation, with rates ranging from 0.01 to 40 mm/h or higher levels.

In addition to the synthesis of near-surface precipitation, we also synthesized reflectivity cross-sections to better represent the vertical structures of precipitation clouds. Based on the previous analysis of the comparison of reflectivity, near-surface precipitation, and the reflectivity gradient, we adopted a weighted synthesis method associated with the reflectivity gradient difference to synthesize the reflectivity in the vertical cross-sections. Figure 10 illustrates the flowchart of synthesizing radar reflectivity from both radars. The following steps outline this synthesis procedure: Firstly, import both radar datasets and subject them to quality control measures. Secondly, calculate the vertical gradient of the radar reflectivity for each footprint. In this calculation, assign a value of 1 to the bin gradient where the radar reflectivity increases with decreasing altitude. Conversely, assign a value of −1 to the radar bin gradient where the radar reflectivity decreases with decreasing altitude. Thirdly, subtract the gradient value of the W band from the gradient value of the Ka band to derive the gradient difference for each bin. Since the reflectivity of the two radars changes differently as the altitude decreases, the value of the gradient difference of each bin may be any integer between the values of −2 and 2. When bins with a gradient difference equal to 2 appear continuously within a footprint, it means the occurrence of precipitation above the CPR measurement threshold. We assigned a synthetic weight factor "k" to each bin based on the total number of gradient differences equal to 2 in the vertical direction of the footprints. It can be seen from Figure 7b that when the total number of bins with a gradient difference equal to 2 in the footprints reaches 10, the precipitation rates estimated by CPR are smaller than those estimated by KaPR. When bins with a gradient difference equal to 2 appear continuously, the "k" of the top bin is assigned a value of 0.9. Then, "k" decreases in height by 0.1 with each successive bin featuring a gradient difference equal to 2, until reaching a value of 0. Finally, the composite reflectivity is obtained from both frequencies with a weighted average method incorporating the gradient information.

Due to scattering differences caused by different bands, the synthesis process involves simulating the scattering of electromagnetic waves in both the W and Ka bands when interacting with solid, mixed, and liquid particles. This step refers to the simulation method in Kou's [31] work, but the types of particles are not divided in detail in this synthesis. We simply classified particles into ice, mixed, and liquid. The "Ice" bins were referenced to cloud ice for scattering simulation, the "mix" bins were referenced to wet snow for scattering simulation, and the "liquid" bins were referenced to raindrop particles for scattering simulation. This simulation helped establish the conversion relationship of radar reflectivity between the W and Ka bands. Subsequently, based on the phase parameters provided by KaPR, the reflectivity factor Ze_W in the W band was converted to the Ka-band radar reflectivity factor $Ze_{W\text{-}Ka}$.

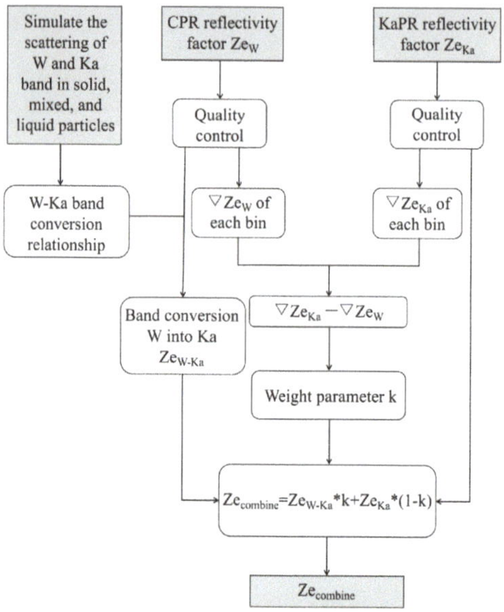

Figure 10. Flowchart of CPR and KaPR reflectivity synthesis.

The final step in this process is the calculation of the composite reflectivity $Ze_{combine}$ for each bin, incorporating the composite weight factor k. Specifically, when only a single radar detects clouds within a bin, the reflectivity factor value of that radar is used. Conversely, when both radars detect clouds simultaneously, the synthesized reflectivity $Ze_{combine}$ is employed.

Figure 11 depicts the radar reflectivity cross-sections from a CPR and KaPR coincident event that occurred on 28 July 2016, over the Baltic Sea. The weather background at that time was that the atmospheric stratification was stable, the humidity in the lower atmosphere was high, and the vertical movement of the atmosphere was weak. The figure includes the original and combined reflectivity cross-sections and their average profiles.

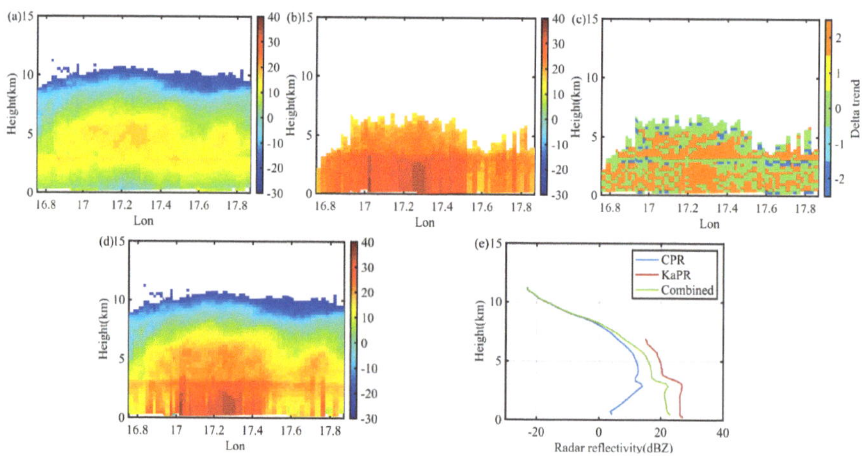

Figure 11. CPR and KaPR reflectivity and their synthesized cross-sections; average reflectivity of coincident case 1 ((**a**): CPR reflectivity cross-section, (**b**): KaPR reflectivity cross-section, (**c**): gradient difference in reflectivity of the two radars, (**d**): synthesis of two radar reflectivity cross-sections, (**e**): average profiles for CPR, KaPR, and synthesized reflectivity).

The observations from this precipitation case reveal a distinct bright band at an altitude of 3 km, indicating stratiform precipitation. The radar reflectivity values for CPR and KaPR are presented in Figure 11a,b, respectively. The precipitation cloud information detected by CPR is 48.86% more than that detected by KaPR. Both radars effectively measure precipitation in the 3–5 km altitude range. However, below 3 km, notable differences emerge in the reflectivity between the two radars. CPR experiences attenuation due to raindrop particles, hindering its ability to accurately indicate the location of strong echoes. Conversely, KaPR provides a clearer depiction of strong echoes below 3 km. The reflectivity gradient difference diagram in Figure 11c confirms that, below 3 km, the reflectivity of the two radars exhibits opposite trends as the altitude decreases. Figure 11d,e display the synthesized reflectivity and average profiles from both radars. Combining data from both radar frequencies allows for a comprehensive representation of the cloud/precipitation structure. The smaller ice particles at the top of the cloud system exhibit minimal attenuation of the W band, with the difference between the band conversion and the original reflectivity staying within 0.5 dB. As the moisture content increases below 5 km, the attenuation of the W band by precipitation particles gradually intensifies. Consequently, the weight of the KaPR reflectivity in the synthetic profile also increases gradually, bringing the synthesized reflectivity closer to the measured value in the Ka band.

Figure 12 illustrates the radar reflectivity cross-sections from a CPR and KaPR coincident event that took place on 13 June 2016, along the Gulf Coast of Louisiana, USA. The precipitation in this case extended up to 15 km in altitude, indicating a convective precipitation. The radar reflectivity values for CPR and KaPR are displayed in Figure 12a,b, respectively. CPR detected more precipitation cloud information than KaPR, accounting for 48.83%. Strong precipitation was observed in the area below 5 km, specifically between 93.5 and 93.6 degrees west. In this region, the CPR electromagnetic waves experience substantial interference and attenuation. Conversely, KaPR reflectivity provides a more accurate characterization in this region.

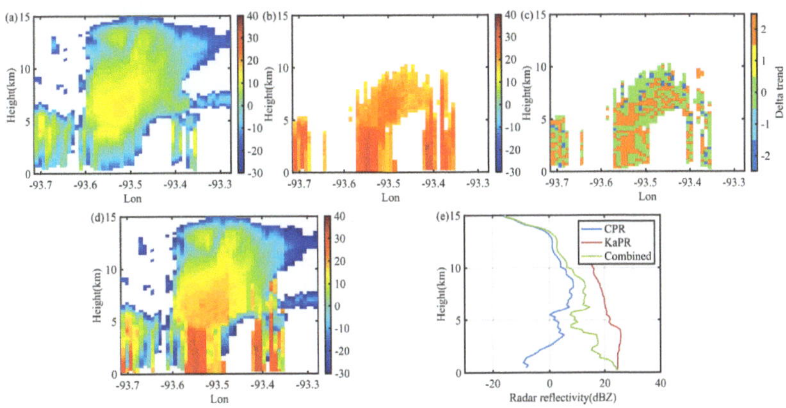

Figure 12. CPR and KaPR reflectivity cross-sections, and their synthesized cross-sections, for coincident case 2 ((**a**): CPR reflectivity, (**b**): KaPR reflectivity, (**c**): gradient difference in reflectivity of the two radars, (**d**): synthesized reflectivity cross-section, (**e**): average profiles for CPR, KaPR, and synthesized reflectivity).

A yellow area is evident in the W-band reflectivity near the 14 km altitude in Figure 12a. Above 10 km altitude, there may be a vigorous upward motion, causing the air within the precipitation system to rise rapidly to around 14 km and then cool rapidly. This process leads to the continuous condensation and growth of ice particles, resulting in larger ice crystals. These larger ice crystals generate strong radar wave reflections. A similar feature can also be observed near 14 km in the average radar reflectivity factor profile in Figure 12e, culminating in a peak.

Figure 12d shows the synthesized reflectivity from case 2, obtained by combining data from both radars. This synthesis yields a comprehensive representation of the cloud/precipitation structure, allowing for a more comprehensive assessment of the cloud characteristics associated with convective precipitation.

Data from the Level 2B GPM Combined Radar–Radiometer Precipitation Algorithm (2BCMB) were used to verify the synthesized results. The 2BCMB precipitation algorithm integrates Level 2 calibrated reflectivity profiles from the DPR and Level 1C GMI brightness temperatures. An a priori database of particle size distributions and corresponding environmental conditions was used to accurately retrieve precipitation data [32].

To demonstrate the performance of the synthesized reflectivity cross-section, we used the Spearman rank correlation coefficient between the 2BCMB precipitation rate and the radar reflectivity to analyze the performance of the synthesis. There is a relationship between the precipitation rate and radar reflectivity, because radar reflectivity is usually used to describe the intensity of scattering of radar waves by particles in the atmosphere (such as raindrops, snow, etc.). High radar reflectivity is generally associated with greater precipitation rates, and the Spearman rank correlation coefficient works for the monotonic relationship [33].

The Spearman rank correlation coefficient parameters (r_s) are defined as follows:

$$r_s = 1 - \frac{6\sum_{i=1}^{n} d_i^2}{n(n^2 - 1)}$$

where n is the number of bins with radar echo values, while d_i is the grade difference between the radar reflectivity and precipitation rate. The statistical quantitative comparison results are shown in Table 2.

Table 2. The statistical quantitative comparison results for case 1 and case 2.

		KaPR	Combined
Case 1	Ice phase	0.8035	0.8956
	Mixed phase	0.8153	0.8989
	Liquid phase	0.9231	0.9028
Case 2	Ice phase	0.6058	0.6750
	Mixed phase	0.5863	0.6453
	Liquid phase	0.4177	0.5110

From the comparison results of the stratiform and convective cases shown in Figure 13, we can see that the stratiform case has better consistency. The r_s for stratiform rain reached 0.8 for all three kinds of phase particles. Compared to stratiform precipitation, convective precipitation is usually generated under unstable atmospheric conditions. The r_s for convective rain reached only about 0.4~0.6 for the three kinds of phase particles. From the comparison of the r_s before and after synthesis, as shown in Table 2, whether in stratiform or convective precipitation, the results of the non-liquid-phase bin are better, and the r_s values of the ice phase and mixed phase are increased by 12% and 10%, respectively. The DPR data, combined with data from GMI, contribute to more accurate precipitation rate estimates. With the addition of the W band, the synthesized reflectivity in the non-liquid-phase bin has a better correlation with the more accurate precipitation rate retrieval in 2BCMB. The addition of W-band radar can play a positive role in the non-liquid-phase radar bin. The synthesis results are better in the ice-phase bin than in the mixed-phase bin. There are many types of precipitation particles in the mixed-phase bin, and the estimation of radar reflectivity is not accurate enough.

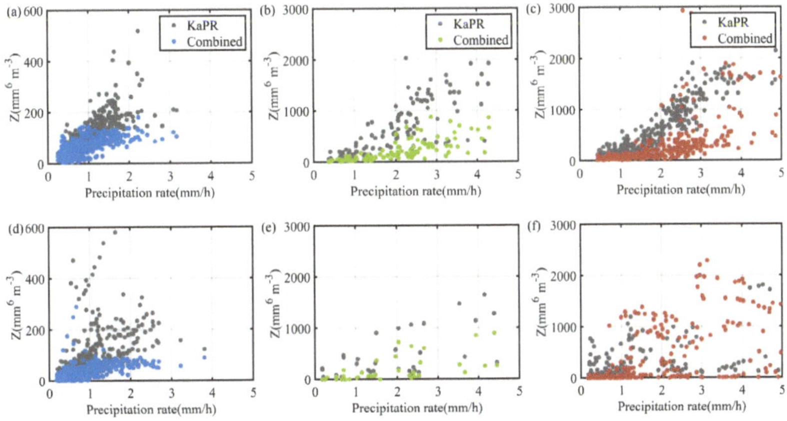

Figure 13. Scatterplots between the 2BCMB precipitation rate and KaPR reflectivity; synthesized reflectivity in different phases for cases 1 and 2 ((**a**): ice phase in case 1, (**b**): mixed phase in case 1, (**c**): liquid phase in case 1, (**d**): ice phase in case 2, (**e**): mixed phase in case 2, (**f**): liquid phase in case 2).

5. Discussion

This study compares and synthesizes the precipitation data from the coincidence of CPR and DPR observations, occurring within a ±15-min window, spanning from March 2014 to December 2017. The primary objective was to assess the disparities in radar reflectivity and precipitation rates between W- and Ka-band spaceborne radars, especially in diverse precipitation phases and varying intensities. Additionally, an attempt was made to synthesize the precipitation data from both radars to enhance the integrity and comprehensiveness of the precipitation information. The key findings of this investigation are as follows:

1. Radar reflectivity comparison: In convective rainfall, the presence of larger liquid particles leads to pronounced attenuation of W-band radar waves. Below the altitudes of the melting layer, CPR generally exhibits a decrease in reflectivity with decreasing altitude, while KaPR reflectivity tends to increase. In stratiform rainfall, where precipitation remains relatively stable, CPR reflectivity is predominantly found between 7 dBZ and 12 dBZ, whereas KaPR reflectivity is mainly distributed between 16 dBZ and 21 dBZ. Regarding snowfall, there is a peak around 10 km in the CPR snowfall average reflectivity profile, indicating that CPR can detect larger ice crystals formed during snowfall. Due to differences in radar sensitivity, the average echo-top height detected by CPR is approximately 3.6–4.2 km higher than that detected by KaPR, enabling CPR to capture more detailed vertical cloud structures.
2. Near-surface precipitation rate comparison: When comparing near-surface rain rates, the average near-surface rain rate of KaPR on the ocean surface is 0.4 mm/h higher than that of CPR within the range of 40 degrees north and south. This could be caused by the presence of large rainfall particles that lead to significant attenuation of CPR's electromagnetic waves, limiting its ability to accurately estimate rain rates beyond its upper threshold. However, at latitudes above 40 degrees, CPR's heightened sensitivity allows it to detect approximately 15% more near-surface rainfall events than KaPR, resulting in an average near-surface rain rate that is 0.1 mm/h greater than KaPR's. When comparing snowfall, it is important to note that for snowfall events influenced by terrain factors between 30 and 50 degrees north and south, CPR consistently retrieves lower values than KaPR. However, in areas with latitudes greater than 50 degrees, CPR detected more snowfall events than KaPR, accounting for 86% of the total snowfall events. In terms of near-surface snow rate comparisons, KaPR retrieves an average near-surface snow rate 0.2 mm/h higher than CPR's in regions below 50 degrees latitude. Conversely, in areas above 50 degrees latitude,

CPR's sensitivity leads to the detection of 45% more light snowfall events in the 0.01 mm/h class than KaPR, resulting in an average near-surface snowfall rate that is 0.1 mm/h higher than KaPR's.

3. Combining the near-surface precipitation rates of both radars allows for a more complete detection of precipitation, with rates ranging from 0.01 to 40 mm/h or higher levels. By synthesizing reflectivity from the CPR and KaPR, with consideration of their vertical gradient differences, a comprehensive cloud precipitation profile can be obtained. The precipitation cloud information detected by CPR but not KaPR accounted for 49% of the total cloud bodies on matching vertical sections. This reflectivity synthesis not only provides a complete view of the vertical structure of precipitation clouds, but also enables a more continuous reflectivity of precipitation cloud bodies. The Spearman rank correlation coefficient parameters for stratiform rain reached 0.8 for all three kinds of phase particles, while they only reached about 0.4~0.6 for the three kinds of phase particles for convective rain. This synthesis method works better in non-liquid-phase radar bins, whether in stratiform or convective precipitation, and the r_s of the ice phase and mixed phase are increased by about 12% and 10%, respectively, while the Spearman rank correlation coefficient of the precipitation rate and radar reflectivity is improved.

6. Conclusions

In this study, we compared the coincident data from CPR and KaPR from 2014 to 2017, and then a synthesis of the radar reflectivity from CPR and KaPR was attempted to obtain a complete cloud and precipitation structure. After comparing the near-surface precipitation rates, we found that the difference in near-surface precipitation rates between CPR and KaPR is related to the reflectivity gradient in the vertical direction. The greater the reflectivity gradient, the greater the difference in near-surface precipitation rates. Finally, the radar reflectivity was weighted and synthesized from CPR and KaPR based on the gradient difference of the reflectivity from the two radars. The results for a stratiform cloud and a deep convective case show that the synthesis can improve the consistency of reflectivity and precipitation rate measurements. In the future, the joint measurement of Ka and W bands holds the potential to establish a correlation between cloud properties and precipitation. This advancement is expected to deepen our comprehension of the cloud precipitation process, thereby contributing to an enhanced understanding of the intricacies involved in cloud-related precipitation phenomena. Additionally, the characteristics of the upper cloud system reflect the intensity and development trends of the lower-level rainfall system [34]. The combined use of the W and Ka bands offers a more objective and rational explanation of the connection between upper-level cloud structure and lower-level precipitation rates. Our next work will be to retrieve the precipitation rates of particles in different phases using combined W- and Ka-band radar data.

Author Contributions: Conceptualization, Y.L. and L.K.; methodology, Y.L, L.K. and Z.L.; software, H.G. and A.H.; validation, Y.L. and L.K.; formal analysis, L.K. and A.H.; investigation, Y.L.; resources, Y.X. and L.Z.; data curation, L.K. and Y.L.; writing—original draft preparation, Y.L.; writing—review and editing, L.K.; visualization, L.K.; supervision, L.K. and Y.X.; project administration, H.G.; funding acquisition, H.G. All authors have read and agreed to the published version of the manuscript.

Funding: This study was supported in part by the National Key Research and Development Program of China (2021YFC2802502), and in part by the Industry University Research Cooperation Fund of the Eighth Research Institute of China Aerospace Science and Technology Corporation and the National Natural Science Foundation of China (41975027).

Data Availability Statement: GPM data were downloaded from NASA PPS at https://storm.pps.eosdis.nasa.gov/storm/, accessed on 15 March 2022. CloudSat data were downloaded from the CloudSat data processing center at https://www.cloudsat.cira.colostate.edu/order/, accessed on 29 January 2022.

Acknowledgments: We acknowledge the data providers of the National Centers for the Environmental Prediction Final Operational Model Global Tropospheric Analyses. We would also like to thank the scientists and those who contributed to the CloudSat mission.

Conflicts of Interest: The authors declare no conflicts of interest.

References

1. Xie, P.; Arkin, P.A. Global Precipitation: A 17-Year Monthly Analysis Based on Gauge Observations, Satellite Estimates, and Numerical Model Outputs. *Bull. Am. Meteorol. Soc.* **1997**, *78*, 2539–2558. [CrossRef]
2. Bony, S.; Colman, R.; Kattsov, V.M.; Allan, R.P.; Bretherton, C.S.; Dufresne, J.-L.; Hall, A.; Hallegatte, S.; Holland, M.M.; Ingram, W.; et al. How Well Do We Understand and Evaluate Climate Change Feedback Processes? *J. Clim.* **2006**, *19*, 3445–3482. [CrossRef]
3. Kidd, C.; Kniveton, D.; Todd, M.C.; Bellerby, T. Satellite Rainfall Estimation Using Combined Passive Microwave and Infrared Algorithms. *J. Hydrometeorol.* **2003**, *4*, 1088–1104. [CrossRef]
4. Kidd, C.; Levizzani, V. Status of Satellite Precipitation Retrievals. *Hydrol. Earth Syst. Sci.* **2011**, *15*, 1109–1116. [CrossRef]
5. Kulie, M.S.; Milani, L.; Wood, N.B.; Tushaus, S.A.; Bennartz, R.; L'Ecuyer, T.S. A Shallow Cumuliform Snowfall Census Using Spaceborne Radar. *J. Hydrometeorol.* **2016**, *17*, 1261–1279. [CrossRef]
6. Hou, A.Y.; Kakar, R.K.; Neeck, S.; Azarbarzin, A.A.; Kummerow, C.D.; Kojima, M.; Oki, R.; Nakamura, K.; Iguchi, T. The Global Precipitation Measurement Mission. *Bull. Am. Meteorol. Soc.* **2014**, *95*, 701–722. [CrossRef]
7. Berg, W.; L'Ecuyer, T.; Haynes, J.M. The Distribution of Rainfall over Oceans from Spaceborne Radars. *J. Appl. Meteorol. Climatol.* **2010**, *49*, 535–543. [CrossRef]
8. Tang, G.; Wen, Y.; Gao, J.; Long, D.; Ma, Y.; Wan, W.; Hong, Y. Similarities and Differences between Three Coexisting Spaceborne Radars in Global Rainfall and Snowfall Estimation. *Water Resour. Res.* **2017**, *53*, 3835–3853. [CrossRef]
9. Sindhu, K.D.; Bhat, G.S. Comparison of CloudSat and TRMM Radar Reflectivities. *J. Earth Syst. Sci.* **2013**, *122*, 947–956. [CrossRef]
10. Fall, V.M.; Cao, Q.; Hong, Y. Intercomparison of Vertical Structure of Storms Revealed by Ground-Based (NMQ) and Spaceborne Radars (CloudSat-CPR and TRMM-PR). *Sci. World J.* **2013**, *2013*, 270726. [CrossRef]
11. Casella, D.; Panegrossi, G.; Sanò, P.; Marra, A.C.; Dietrich, S.; Johnson, B.E.; Kulie, M.S. Evaluation of the GPM-DPR Snowfall Detection Capability: Comparison with CloudSat-CPR. *Atmos. Res.* **2017**, *197*, 64–75. [CrossRef]
12. Mroz, K.; Montopoli, M.; Battaglia, A.; Panegrossi, G.; Kirstetter, P.-E.; Baldini, L. Cross-Validation of Active and Passive Microwave Snowfall Products over the Continental United States. *J. Hydrometeorol.* **2021**, *22*, 1297–1315. [CrossRef]
13. Skofronick-Jackson, G.; Kulie, M.; Milani, L.; Munchak, S.J.; Wood, N.B.; Levizzani, V. Satellite Estimation of Falling Snow: A Global Precipitation Measurement (GPM) Core Observatory Perspective. *J. Appl. Meteorol. Climatol.* **2019**, *58*, 1429–1448. [CrossRef]
14. Durden, S.L.; Tanelli, S.; Epp, L.W.; Jamnejad, V.; Long, E.M.; Perez, R.M.; Prata, A. System Design and Subsystem Technology for a Future Spaceborne Cloud Radar. *IEEE Geosci. Remote Sens. Lett.* **2016**, *13*, 560–564. [CrossRef]
15. Tanelli, S.; Haddad, Z.S.; Im, E.; Durden, S.L.; Sy, O.O.; Peral, E.; Sadowy, G.; Sanchez-Barbetty, M. Radar Concepts for the next Generation of Spacebome Observations of Cloud and Precipitation Processes. In Proceedings of the 2018 IEEE Radar Conference (RadarConf18), Oklahoma City, OK, USA, 23–27 April 2018; IEEE: Piscataway, NJ, USA, 2018. [CrossRef]
16. Wu, Q.; Yang, M.; Dou, F.; Guo, Y.; An, D. A Study of Cloud Parameters Retrieval Algorithm for Spaceborne Millimeter Wavelength Cloud Radar. *Acta Meteorol. Sin.* **2018**, *76*, 160–168. (In Chinese) [CrossRef]
17. Stephens, G.L.; Vane, D.G.; Tanelli, S.; Im, E.; Durden, S.; Rokey, M.; Reinke, D.; Partain, P.; Mace, G.G.; Austin, R.; et al. CloudSat Mission: Performance and Early Science after the First Year of Operation. *J. Geophys. Res.* **2008**, *113*, D00A18. [CrossRef]
18. Tanelli, S.; Durden, S.L.; Im, E.; Pak, K.S.; Reinke, D.G.; Partain, P.; Haynes, J.M.; Marchand, R.T. CloudSat's Cloud Profiling Radar After Two Years in Orbit: Performance, Calibration, and Processing. *IEEE Trans. Geosci. Remote Sens.* **2008**, *46*, 3560–3573. [CrossRef]
19. Bacmeister, J.T.; Stephens, G.L. Spatial Statistics of Likely Convective Clouds in CloudSat Data. *J. Geophys. Res.* **2011**, *116*, D04104. [CrossRef]
20. Riley, E.M.; Mapes, B.E. Unexpected Peak near −15 °C in CloudSat Echo Top Climatology. *Geophys. Res. Lett.* **2009**, *36*, L09819. [CrossRef]
21. Marchand, R.; Mace, G.G.; Ackerman, T.; Stephens, G. Hydrometeor Detection Using Cloudsat—An Earth-Orbiting 94-GHz Cloud Radar. *J. Atmos. Ocean. Technol.* **2008**, *25*, 519–533. [CrossRef]
22. Behrangi, A.; Christensen, M.; Richardson, M.; Lebsock, M.; Stephens, G.; Huffman, G.J.; Bolvin, D.; Adler, R.F.; Gardner, A.; Lambrigtsen, B.; et al. Status of High-Latitude Precipitation Estimates from Observations and Reanalyses. *J. Geophys. Res. Atmos.* **2016**, *121*, 4468–4486. [CrossRef]
23. Wood, N.B.; L'Ecuyer, T.S.; Heymsfield, A.J.; Stephens, G.L.; Hudak, D.R.; Rodriguez, P. Estimating Snow Microphysical Properties Using Collocated Multisensor Observations. *J. Geophys. Res. Atmos.* **2014**, *119*, 8941–8961. [CrossRef]
24. Rodgers, C.D. Retrieval of Atmospheric Temperature and Composition from Remote Measurements of Thermal Radiation. *Rev. Geophys.* **1976**, *14*, 609. [CrossRef]

25. Hamada, A.; Takayabu, Y.N. Improvements in Detection of Light Precipitation with the Global Precipitation Measurement Dual-Frequency Precipitation Radar (GPM DPR). *J. Atmos. Ocean. Technol.* **2016**, *33*, 653–667. [CrossRef]
26. Iguchi, T.; Seto, S.; Meneghini, R.; Yoshida, N.; Awaka, J.; Kubota, T. *GPM/DPR Level-2 Algorithm Theoretical Basis Document*; NASA Goddard Space Flight Center: Washington, DC, USA, 2023.
27. Joseph Turk, F.; Ringerud, S.; Camplani, A.; Casella, D.; Chase, R.J.; Ebtehaj, A.M.; Gong, J.; Kulie, M.S.; Liu, G.; Milani, L.; et al. Applications of a CloudSat-TRMM and CloudSat-GPM Satellite Coincidence Dataset. *Remote Sens.* **2021**, *13*, 2264. [CrossRef]
28. Yin, M.; Liu, G.; Honeyager, R.; Joseph Turk, F. Observed Differences of Triple-Frequency Radar Signatures between Snowflakes in Stratiform and Convective Clouds. *J. Quant. Spectrosc. Radiat. Transf.* **2017**, *193*, 13–20. [CrossRef]
29. Dong, C. Remote Sensing, Hydrological Modeling and in Situ Observations in Snow Cover Research: A Review. *J. Hydrol.* **2018**, *561*, 573–583. [CrossRef]
30. Chen, S.; Hong, Y.; Kulie, M.S.; Behrangi, A.; Stepanian, P.M.; Cao, Q.; You, Y.; Zhang, J.; Hu, J.; Zhang, X. Comparison of Snowfall Estimates from the NASA CloudSat Cloud Profiling Radar and NOAA/NSSL Multi-Radar Multi-Sensor System. *J. Hydrol.* **2016**, *541*, 862–872. [CrossRef]
31. Kou, L.; Lin, Z.; Gao, H.; Liao, S.; Ding, P. Simulation and Sensitivity Analysis for Cloud and Precipitation Measurements via Spaceborne Millimeter-Wave Radar. *Atmos. Meas. Tech.* **2023**, *16*, 1723–1744. [CrossRef]
32. Olson, W. *GPM Combined Radar-Radiometer Precipitation Algorithm Theoretical Basis Document (Version 5)*; NASA: Washington, DC, USA, 2018.
33. Rosenblad, A. The Concise Encyclopedia of Statistics. *J. Appl. Stat.* **2011**, *38*, 502. [CrossRef]
34. Tang, L.; Gao, W.; Xue, L.; Zhang, G.; Guo, J. Climatological Characteristics of Hydrometeors in Precipitating Clouds over Eastern China and Their Relationship with Precipitation Based on ERA5 Reanalysis. *J. Appl. Meteorol. Climatol.* **2023**, *62*, 625–641. [CrossRef]

remote sensing

MDPI

Article

The Sensitivity of Polar Mesospheric Clouds to Mesospheric Temperature and Water Vapor

Jae N. Lee [1,2,*], Dong L. Wu [2], Brentha Thurairajah [3], Yuta Hozumi [2,4] and Takuo Tsuda [5]

1 Goddard Earth Sciences Technology and Research II, University of Maryland, Baltimore County, Baltimore, MD 20742, USA
2 NASA Goddard Space Flight Center, Greenbelt, MD 20771, USA; yuta.hozumi@nasa.gov (Y.H.)
3 Center for Space Science and Engineering Research, Bradley Department of Electrical and Computer Engineering, Virginia Tech, Blacksburg, VA 24061, USA
4 Department of Physics, Catholic University of America, Washington, DC 20064, USA
5 Department of Computer and Network Engineering, University of Electro-Communications, Chofu 182-8585, Japan; takuo.tsuda@uec.ac.jp
* Correspondence: jae.n.lee@nasa.gov

Abstract: Polar mesospheric cloud (PMC) data obtained from the Aeronomy of Ice in the Mesosphere (AIM)/Cloud Imaging and Particle Size (CIPS) experiment and Himawari-8/Advanced Himawari Imager (AHI) observations are analyzed for multi-year climatology and interannual variations. Linkages between PMCs, mesospheric temperature, and water vapor (H_2O) are further investigated with data from the Microwave Limb Sounder (MLS). Our analysis shows that PMC onset date and occurrence rate are strongly dependent on the atmospheric environment, i.e., the underlying seasonal behavior of temperature and water vapor. Upper-mesospheric dehydration by PMCs is evident in the MLS water vapor observations. The spatial patterns of the depleted water vapor correspond to the PMC occurrence region over the Arctic and Antarctic during the days after the summer solstice. The year-to-year variabilities in PMC occurrence rates and onset dates are highly correlated with mesospheric temperature and H_2O. They show quasi-quadrennial oscillation (QQO) with 4–5-year periods, particularly in the southern hemisphere (SH). The combined influence of mesospheric cooling and the mesospheric H_2O increase provides favorable conditions for PMC formation. The global increase in mesospheric H_2O during the last decade may explain the increased PMC occurrence in the northern hemisphere (NH). Although mesospheric temperature and H_2O exhibit a strong 11-year variation, little solar cycle signatures are found in the PMC occurrence during 2007–2021.

Keywords: polar mesospheric clouds; water vapor; temperature; mesosphere

Citation: Lee, J.N.; Wu, D.L.; Thurairajah, B.; Hozumi, Y.; Tsuda, T. The Sensitivity of Polar Mesospheric Clouds to Mesospheric Temperature and Water Vapor. *Remote Sens.* **2024**, *16*, 1563. https://doi.org/10.3390/rs16091563

Academic Editors: Haobo Li, Suelynn Choy, Yuriy Kuleshov, Mayra I. Oyola-Merced and Xiaoming Wang

Received: 19 March 2024
Revised: 22 April 2024
Accepted: 23 April 2024
Published: 28 April 2024

1. Introduction

Polar mesospheric clouds (PMCs), or noctilucent clouds (NLCs), are Earth's highest clouds that occur in the summertime, at high latitudes in a narrow altitude region near ~82 km. These clouds consist of ice crystals in nanometers that are nucleated onto meteoric particles with an ice water content (IWC) of less than 100 g/km^2 [1]. The occurrence of PMCs in the summer polar mesosphere indicates that PMC formation requires favorable atmospheric conditions, i.e., low temperature (T) and sufficient water vapor (H_2O) abundance. While the mesospheric H_2O supply is largely from the slow ascent of polar air in the mean upper-atmospheric circulation, the regional temperature can be modulated by waves, i.e., planetary waves, gravity waves (GWs), and tides originated from the lower atmosphere. Different from the lower atmosphere, the wave-driven dynamic polar mesosphere is coldest during the summer solstice. The cold mesosphere in the summer hemisphere is a manifestation of strong dynamic forcing [2–6]. Gravity wave momentum deposition induces a mean meridional flow from the summer to winter hemisphere, resulting in an upward flow in the summer mesosphere and downward flow in the winter mesosphere.

Adiabatic cooling and warming, associated with this seasonal vertical mean flow, drive strong seasonal variations in mesospheric temperature and atmospheric tracers. Vertical thermal structures and their relations to PMCs have been extensively studied in numerous publications [7–10].

Upper atmosphere cooling is likely to increase from doubling CO_2 radiative forcing [11]. Observational and modeling studies have linked PMC changes to enhanced ice particle production by increased CO_2 cooling [10,12,13]. Linkages between increases in PMC occurrence and brightness to climate change on different time scales have been discussed by a number of authors [14–18]. On a centennial timescale, it has been suggested that a H_2O increase of about ~1 ppmv due to methane oxidation in the mesosphere during the industrialization era could be one of the major contributing factors to the more frequent occurrence of PMCs observed in recent years [19]. Since 11-year solar cycle variations can also induce changes in PMC occurrence by modulating mesospheric T and H_2O, the sensitivity of PMC occurrence to the solar cycle has also been extensively investigated [20–25]. The micro-physical properties of PMCs, i.e., vapor pressure over ice and the nucleation of mesospheric ice particles in the relevant temperature range, have also been explored with laboratory observations in combination with detailed microphysical modeling of cloud processes under the conditions of the polar summer mesopause [26–33].

Although PMCs have been observed for more than 120 years [34–36], observation techniques have significantly improved in recent decades due to advanced new instruments from the ground and rockets [37–46]. However, ground-based and rocket observations are limited in terms of spatiotemporal coverage and unable to make consistent daily measurements because they are often obscured by low clouds and local weather conditions. Thus, sampling bias is a fundamental limitation in ground-based PMC observations.

Recent satellite observations of PMCs from low-Earth orbit (LEO) [47–56] and geostationary-Earth-orbit (GEO) [57–60] have provided the needed global coverage of PMCs with consistent temporal and spatial samplings. Sun synchronous satellite instruments observe PMCs at two fixed local times from the ascending and descending orbits. As the orbit can shift over the lifetime of the satellite, a local time shift in the observation can affect the multi-year time series of the PMCs due to the interference of the tidal amplitudes and phases with the observation times. Stable and continuous observations of PMCs and the middle atmosphere environment are critical for understanding the variations in PMCs, especially with their implication for long-term and/or solar cycle influences on the mesosphere.

Beyond all the difficulties in diagnosing changes in the mesosphere, observational evidence of the cooling trend in the mesospheric temperature can be found from Rayleigh lidars [61] and satellite records [62–64]. The estimated decadal variability in the mesosphere is often hampered by other large interannual fluctuations such as stratospheric sudden warming and warming due to volcanic eruptions. Nevertheless, the observed temperature in the middle atmosphere appears to exhibit a general cooling trend of 1–2 K/decade in response to CO_2 increase [65].

Water vapor (H_2O), a highly variable greenhouse gas in the lower atmosphere, is an important tracer of the middle atmosphere dynamics [66] and an indicator of climate change. In contrast to its increasing trend in the past, a ~10% decrease in stratospheric water vapor since 2000 has been reported as a potential contributing factor to the global warming hiatus, which may have caused a slow increase in the global surface temperature [67]. A recent analysis from Microwave Limb Sounder (MLS) and Sounding of the Atmosphere using Broadband Emission Radiometry (SABER) observations suggests an increasing H_2O trend of 0.2–0.3 ppmv/0.1–0.2 ppmv during the last decade in the stratosphere/lower mesosphere, respectively [13]. An anthropogenic increase in methane (CH_4) [68] could possibly drive a steady increase in H_2O, since methane oxidation is a principal source of stratospheric water vapor.

In this paper, we conduct a comprehensive study of PMC occurrence and the mesospheric environment for cloud formation to quantify the sensitivity of PMCs to environmental conditions by analyzing two independent satellite PMC measurements. We characterize the monthly and seasonal climatologies as well as year-to-year variations in PMC occurrence rates, as observed by the Aeronomy of Ice in the Mesosphere (AIM)/Cloud Imaging and Particle Size (CIPS) experiment and Himawari-8/Advanced Himawari Imager (AHI).

To evaluate the PMC occurrence changes with respect to the warming climate and altered mesospheric environment, we correlate the MLS T and H_2O variations with the PMC occurrence rates and PMC onset dates for the periods when AIM/CIPS and Himawari-8/AHI observations are available. We also examine the interannual variations in PMCs across 4–5-year periods (quasi-quadrennial oscillation) from 2007 to the 2021/2022 PMC season, in conjunction with mesospheric T and H_2O.

2. Data

2.1. MLS T and H_2O

We use V5.0 MLS T and H_2O in the present study. MLS is an instrument onboard the NASA/Aura satellite, launched in July 2004 into a Sun-synchronous polar orbit. MLS measures thermal microwave emissions from the Earth's limb from 82°S to 82°N. The daily temperature and H_2O fields are mapped onto a 4° (latitude) × 8° (longitude) grid for daytime (ascending) and nighttime (descending) orbits. The daily mean is estimated as an average of the data from ascending and descending orbits. MLS observations provide nearly global coverage (82°S–82°N) by passing the same location two times every day: ascending orbit at 1:45 ± 15 PM local time and descending orbit at 1:45 ± 15 AM local time (https://aura.gsfc.nasa.gov/scinst.html, accessed on 31 January 2023). There are differences in the MLS T and H_2O data from ascending and descending orbits [69,70], however, averaging the data from the ascending and descending orbits does not affect the results of the paper.

Inferred from the emissions of molecular oxygen [O_2] at 118 GHz and its isotope at 239 GHz, MLS temperature measurements have a vertical resolution of ~6 km and precision of ~2.3 K in the mesosphere [71,72]. The uncertainty due to noise and a priori information is ~2.5 K above the mesosphere. MLS temperature measurements are recommended for scientific use up to 0.0005 hPa (~104 km), as updated in the V5.0 data quality and description document (https://mls.jpl.nasa.gov/data/v5-0_data_quality_document.pdf, accessed on 31 January 2023).

The H_2O volume mixing ratio (VMR) is derived from the 183.3 GHz line using a 190 GHz radiometer. The typical precision values of the MLS V5.0 water vapor vary between 0.3 and 2 ppmv over the range from 68 to 0.001 hPa with vertical resolution of 12–16 km in the mesosphere. The highest vertical level recommended for scientific use is 316–0.001 hPa. In the comparison with the version 3.0 data from the Sub-Millimetre Radiometer (SMR) aboard the Odin satellite, the profiles of the MLS T and H_2O mixing ratio are consistent with those of the SMR with relative differences within 5% in the lower mesosphere [73].

2.2. MLS Relative Humidity with Respect to Ice (RHI)

The MLS relative humidity with respect to ice (RHI) product is only validated and used for tropospheric studies [74]. The analysis of mesospheric RHI in this study is preliminary, in which we focus mainly on its relative variations with time and altitude. As in an earlier MLS algorithm [75], the RHI algorithm is based on the Goff–Gratch function [76] and is recommended for use by the World Meteorological Organization (WMO). In essence, this formula is fundamentally based on the Clausius–Clapeyron equation,

$$\frac{\mathrm{d}(\ln e_i)}{\mathrm{d}T} = \frac{L_{ice}(T)}{RT^2} \quad (1)$$

where e_i is the ice saturation pressure, $L_{ice}(T)$ is the latent heat of the sublimation of ice as a function of temperature, and R is the molar gas constant. If the H_2O VMR is measured RHI (in %) can be expressed as

$$\text{RHI} = \frac{p \cdot \text{VMR}_{H2O}}{e_i} \times 100 \quad (2)$$

where p is the atmospheric pressure. Therefore, extension to the mesosphere would depend primarily on the knowledge of $L_{ice}(T)$ in a cold environment (130 K–180 K), which is derived from the molar heat capacities of ice and water vapor. The water vapor capacity is approximately constant for these temperatures, whereas the ice capacity as a function of temperature is known quite well down to 15 K. A more comprehensive review on ice and water vapor properties can be found in [77].

In summary, the MLS RHI should contain scientifically useful measurements of H_2O saturation in the mesosphere and warrant an in-depth analysis in PMC studies. Figure 1 shows two examples of the averaged RHI profiles in the NH and SH from a 30-day period after the 2011 solstice. Despite the coarse vertical resolution in the mesosphere, the MLS RHI appears to reveal a high saturation level in the upper mesosphere (0.01–0.002 hPa) with a peak near 0.005 hPa. As discussed in later sections, these periods represent the enhanced occurrence of PMC formation in the summer polar region. This high saturation with an RHI > 100% is consistent with the PMC-forming potential at this time of year.

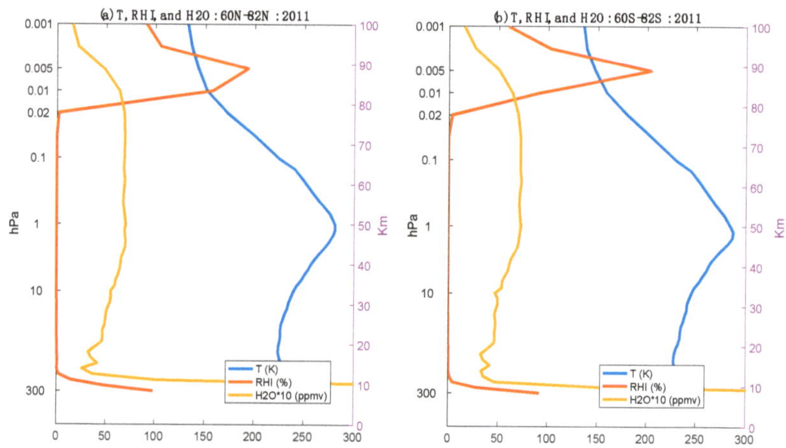

Figure 1. Profiles of MLS T (K), RHI (%), and H_2O (ppmv) averaged over 30 days after the 2011 summer solstice for the (**a**) NH and (**b**) SH.

2.3. AIMS/CIPS

The CIPS instrument [78] onboard the AIM satellite has been operational since May 2007. CIPS uses four UV imagers to measure the 265 nm radiation scattered by the atmosphere from a ~30–85° latitude and completes about 15 orbits per day, crossing the equator at 12 AM/PM local times. CIPS is a nadir-imaging instrument that utilizes a unique four-camera design to discriminate the scattering of ultraviolet (UV) solar photons from PMC ice particles against the background sunlit atmosphere. From 2007 through to February of 2016, images were acquired in the summer hemisphere between the terminator and a dayside latitude of about 40 degrees. Since 12 February 2016, images have been acquired at all sunlit latitudes. The fundamental measurement is albedo, defined as the ratio of the scattered radiance to the incoming solar irradiance.

In this study, we use the version 5.20, level 3c data set, which consists of orbit-by-orbit albedos binned in 1° latitude bins for the entire season. We use an albedo threshold of 5×10^{-6} sr^{-1} to reduce false detections and determine the PMC occurrence rate. The occurrence rate (OR) is calculated as the ratio of the number of clouds (with albedo values $\geq 5 \times 10^{-6}$ sr^{-1}) to the number of observations. Note that PMC retrievals are not available for the NH 2017 and SH 2017–2018 seasons due to the difficulties in accounting for the changing orbital parameters. Detailed descriptions of the observations, algorithms, data products, and uncertainties can be found in [79,80].

2.4. *Himawari-8/AHI*

Himawari-8 is a Japanese GEO meteorological satellite, operated since July 2015. Himawari-8/AHI has 16 observation bands, including 3 visible bands: blue (0.47 μm), green (0.51 μm), and red (0.64 μm). These visible bands can be used for PMC observations [59]. In this work, we use Himawari-8/AHI PMC occurrence rate data from 2016 to 2021 for the NH and from 2016/2017 to 2021/2022 for the SH. The PMC data sets are produced by the PMC detection method, which was developed in [60]. To perform the PMC detection, the resampled optical intensity data are calculated from the full-disk HSD band-1 (blue) images, which have 11,000 × 11,000 pixels with a time resolution of 10 min. The spatial resolutions (or the resampled grid) of the resampled data are 1 degree in latitude and 1 km in height. Then, the PMC detections are performed for the height profiles of the resampled intensity data at each latitude with a 10-min interval. We calculate the daily PMC occurrence rates from the detected PMC data. The Himawari-8/AHI PMC OR data used in this work are available with resolutions of 1 day in time and 1 degree in latitude. More detailed information for the AHI PMC measurement can be found in [60].

3. Results and Discussion

3.1. *Climatology of Mesospheric T and H_2O*

In this section, we investigate the climatology of the MLS T and H_2O mixing ratio. PMCs are expected to alter and to be altered by T and H_2O in the summer mesopause region. Figure 2 shows the zonal mean monthly climatology calculated from the MLS T (left panel) and H_2O (right panel) during 2005–2021 for each month of the year. The climatology is shown for 82°S–82°N over the altitude range from 5 to 0.001 hPa (~37 to ~100 km). The climatology of the middle atmosphere T from MLS clearly shows the hemispheric asymmetry of the cold temperature minimum over high latitudes in both summer hemispheres in the upper mesosphere (between 0.01 hPa and 0.001 hPa). Because the UV absorption by ozone in the Hartley band near 240 nm is a major source of solar heating in the mesosphere, the mean temperature monotonically decreases with altitude as the ozone VMR decreases. However, additional radiative, chemical, and dynamical processes are involved in the determination of the temperature structure near the mesopause [81]. A large amplitude of the annual oscillation in the high latitude implies that dynamic forcing is stronger there, compared to low-latitude zones. The summertime cold region is obvious during summer months, extending from the upper mesosphere to below 0.01 hPa. The temperature range for the summer mesosphere extends from 140 to 180 K in both hemispheres.

The climatology of H_2O (right panel) shows a general decrease with altitude to near the mesopause. A meridional tilt is obvious during the solstice season and the H_2O mixing ratio decreases from the summer to winter hemisphere. For a given altitude, H_2O increases with an increasing latitude, reflecting the summertime upwelling of water vapor following the mean meridional circulation, which maximizes at high latitudes. The H_2O amount decreases with latitudes, so a H_2O maximum appears at the equator during equinox. A noticeable perturbation in the H_2O field in summer high latitudes can be found in the upper mesosphere. At a high latitude above 0.01 hPa, the H_2O meridional increase slows down and does not increase with latitude beyond 50°N/S during the summer months. In contrast, the summertime H_2O reaches a maximum below the 0.01 hPa level. This perturbation indicates that the summertime high-latitude H_2O depletion above the 0.01 hPa level may be linked with the formation of PMCs. The decrease in H_2O above this level in the upper polar mesosphere could be due to the photodissiation of the H_2O where it is brought up from a lower altitude by the overturning circulation as it reaches the mesopause [82]. The

H_2O below PMC level can be enhanced by the sublimation of PMCs when they release H_2O back into the surrounding environment.

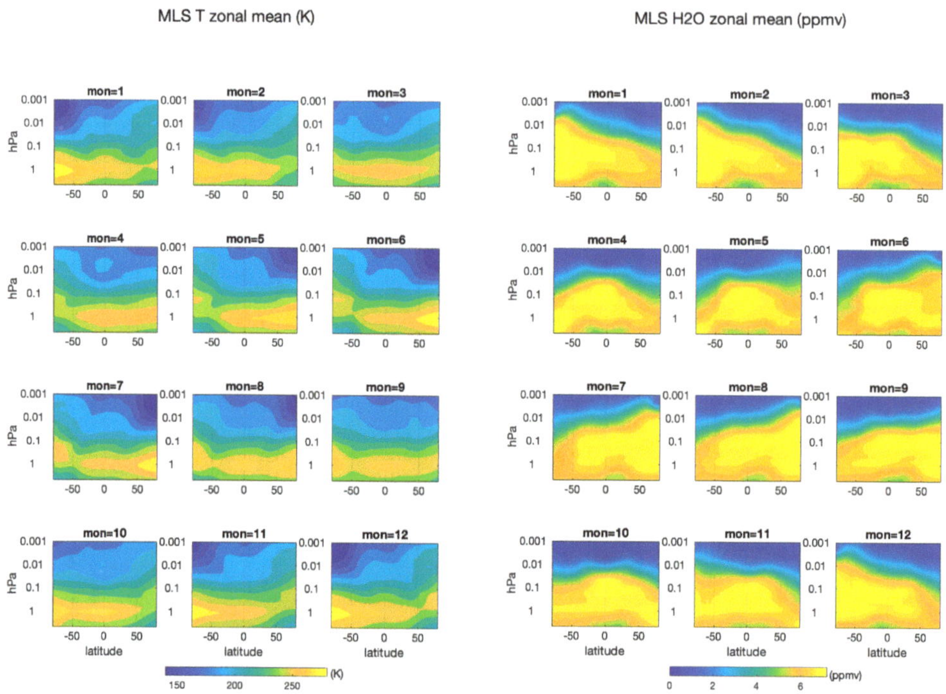

Figure 2. Latitude height distribution of MLS T (K) (**left** panel) and H_2O (ppmv) (**right** panel) monthly zonal mean climatology calculated during 2005–2021.

3.2. *Upper-Mesospheric Dehydration*

We investigate this dehydration process in further detail with the MLS T and H_2O measurements to examine its temporal evolution and spatial distribution in connection to PMC formation. In Figure 3, the zonal mean MLS H_2O climatology (2005–2021) near the mesopause region above 0.02 hPa is shown at a given time and latitude for the NH (left panel) and SH (right panel). At the 0.02 hPa level, a high amount of H_2O VMR i.e., a H_2O VMR greater than 6 ppmv, appeared approximately 10 to 15 days before the solstice in both hemispheres. This high H_2O concentration continues to grow and peaks at greater than 7 ppmv at 80°N/S during the 'core' of PMC seasons around the summer solstice. This high H_2O amount at 0.02 hPa in high latitudes persists over the next 90–100 days and shows equatorward propagating signals. The extent of the equatorward propagation signal is slightly more prolonged in the SH relative to the NH, but the structure of the hemispheric distribution of H_2O VMR is quite similar.

Significant H_2O depletion in the upper mesosphere (above 0.005 hPa) is evident in the MLS climatology. This perturbation suggests that the summertime high-latitude H_2O above the PMCs is depleted by the formation of PMCs, while the H_2O below the PMC altitudes is enhanced by the sublimation of PMCs when they release H_2O back into the surrounding environment. This is consistent with the WACCM model simulations, which indicates that PMC formation is expected to decrease the distribution of atomic hydrogen and H_2O around PMCs [83].

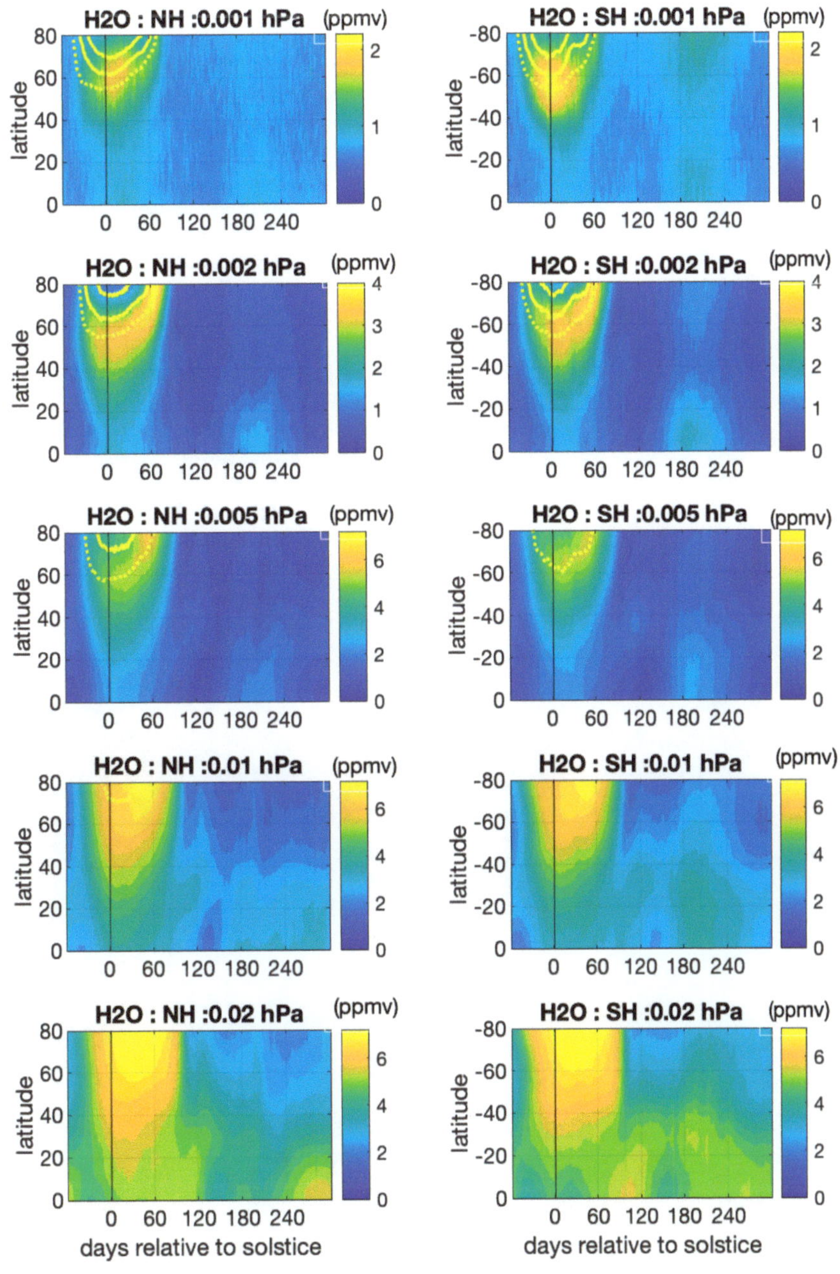

Figure 3. MLS zonal daily mean H_2O VMR climatology (ppmv) in the mesosphere at 0.02, 0.01, 0.005, 0.002, and 0.001 hPa for the NH (**left** column) and SH (**right** column) (2005–2021). Black lines indicate local summer solstices. Above 0.01 hPa, white dotted contour lines represent T at 150 K, and solid lines represent T at 140 K (**lower** latitudes) and 130 K (**higher** latitudes), respectively.

In the lower mesosphere at 0.01 hPa and 0.02 hPa, Figure 3 (lower panels) shows the composites of the H_2O maximum at 30–40 days after solstice, not near the solstice. The wave-driven ascent is known to be responsible for the cold summer polar temperature and high H_2O in the mesosphere. However, delayed H_2O maxima (i.e., delayed 30–40 days from solstice) at these altitudes indicate that the wave-driven ascent in the summer pole may not occur simultaneously with the temperature minimum at the summer solstice, but is a delayed response due to additional subsequent dynamic processes.

The MLS H_2O measurement is based on its 183 GHz emission feature from the limb atmosphere. There is a two-order magnitude of sensitivity difference between emissions from gas and the condensed phase of water. MLS is very sensitive to water vapor depletion as it freezes into a condensed phase because it is insensitive to cloud ice. Similar studies with these MLS data were carried out for the dehydration process near the tropical tropopause layer (TTL) where water vapor is frozen to cirrus [84], and for the denitrification in the polar stratosphere where nitric acid gas (HNO_3) is taken up onto ice surfaces [85].

The spatial patterns of the H_2O hole, a region of exceptionally depleted H_2O in the upper mesosphere at 0.002 hPa over the Arctic and Antarctic regions, are shown in Figure 4 for NH (Figure 4a,c) and SH (Figure 4b,d) for the 2007 (NH) and 2007/2008 (SH) summers The H_2O VMRs are averaged for 0 to 30 days from the summer solstice. The depleted H_2O region appears at the upper mesosphere above 0.002 hPa from the beginning of the summer solstice and lasts more than 30 days. We infer that the H_2O holes in these regions are probably caused by the formation of PMCs, not by atmospheric dynamics, since this kind of perturbation is not found in the MLS CO distribution. The H_2O bound in ice particles is coupled with the background environment by freeze drying and sublimation The formation of clouds can dehydrate the surrounding atmosphere, and they release the H_2O back to the surrounding atmosphere within ~60 days from the solstice. The presence of PMCs presumably causes an obstruction to the upwelling of H_2O and may have caused the H_2O holes above the PMCs.

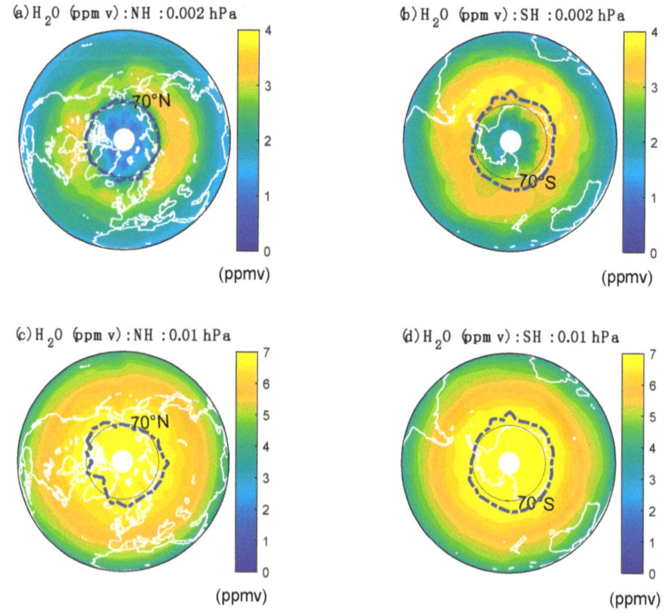

Figure 4. The H_2O dehydration (in ppmv) at 0.002 hPa due to PMC formation in 2007 NH and 2007/2008 SH summer seasons. H_2O VMRs are averaged for 0 to 30 days from summer solstice (**a**) for the NH and (**b**) for the SH, respectively. Similarly, the H_2O (in ppmv) at 0.01 hPa is shown (**c**) for the NH and (**d**) for the SH. Overlaid blue lines represent a temperature contour of 140 K for 0.002 hPa and 150 K for 0.01 hPa.

In both hemispheres, these depleted H_2O regions are nearly concurrent with low-temperature regions, as shown by the 140 K temperature blue contour lines in Figure 4. At

0.002 hPa, regions of H_2O holes are evident where the temperature is lower than 140 K. Note that the 140 K temperature contour lines are almost coincident with the 70°N/S latitude line. The ice particles in the upper mesosphere are transported according to background winds, eddy diffusion, and sedimentation. Ice particle nucleation and growth are well-described with microphysical processes and saturation [86,87].

3.3. *Climatology of PMCs*

To show the daily variability in the frequency of PMC occurrence at different latitudes, the average PMC occurrence rate (OR) from the two instruments, AIM/CIPS and Himawari-8/AHI, are compared in Figure 5. The upper panel represents AIM/CIPS-measured PMC ORs for the NH (Figure 5a) and SH (Figure 5b) and the lower panel represents those from Himawari-8/AHI for the NH (Figure 5c) and SH (Figure 5d). The PMC ORs are calculated from the daily occurrence rates during 2007–2021 in the NH and 2007/2008–2021/2022 in the SH from AIM/CIPS and 2016–2021 in the NH and 2016/2017–2021/2022 from the Himawari-8/AHI measurements. The AIM/CIPS PMC measurements cover 15+ years from 2007, while Himawari-8/AHI only covers 7 years since 2016. Therefore, the climatologies from the two instruments represent the temporal and spatial distribution of an average PMC OR from two different periods. A comparison of the two data sets also shows different latitudinal coverages, with AHI observations limited to below 81°N/S.

Despite differences in the sampling and algorithms of the measurements, the daily variations in the PMC ORs from two independent measurements show remarkable similarities in their overall spatial extent, timing, and the duration of the cloud occurrence. The latitudinal extent of the PMC occurrence starts increasing from ~30 days before the summer solstice to ~30 days after the solstice, until they reach the peak of occurrence. In both hemispheres, PMCs occur most frequently above 70°N/S in latitude and for a ~60 to 90 day duration centered around the peak, which occurs about 30 days after the summer solstice. The OR from AHI reaches its maximum within 80°N/S latitudes about 20 days after the solstice, which is ~10 days earlier than that from CIPS.

To examine the impact of atmospheric conditions on PMC formation, mesospheric T and H_2O climatology at 0.02 hPa during the same periods of each PMC observation are also overlaid with PMC climatology. Solid and dotted contour lines in magenta represent high-latitude (60°N/S–82°N/S) H_2O VMRs. Similarly, the white contour line represents the high-latitude mean T at the same level. During the summer solstice, the average temperature is ~5 K higher in the SH compared to the NH within the high-latitude region. Hence, the different threshold values of T are shown in Figure 5 for the two hemispheres. The T threshold values for the SH and NH are 178 K and 173 K, respectively.

In both hemispheres, the maximum PMC OR is not coincident with the date of the minimum temperature, but lags up to 30 days after the summer solstice. The mesospheric temperature at 0.02 hPa and the near-mesopause temperature reaches its local minimum in the PMC region around the summer solstice, but the PMC occurrence peak from CIPS lags the temperature minimum, reaching its maximum ~30 days after the solstice in both hemispheres. The date of the seasonal high-latitude mesospheric H_2O maxima seen from MLS is up to 30 days later than the date of the corresponding local T minima. The lagged days between the PMC OR and T are reduced in AHI-measured PMC ORs. In both cases, however, the PMC OR peak occurs between the T minima and H_2O maxima.

The two magenta contour lines in Figure 5 represent the daily zonal mean H_2O VMR over 60°N/S–82°N/S. On average, the CIPS NH PMC fields greater than 20% of the OR are mostly confined to humid regions where the H_2O VMR is higher than 7.1 ppmv (solid magenta lines). The PMC ORs greater than 80% are mostly coincident with the region of a high H_2O VMR greater than 7.5 ppmv (dotted magenta lines). On average, high H_2O VMRs of >7.1 ppmv and H_2O VMRs of >7.5 ppmv last ~60 days and ~30 days, respectively. In the case of AHI, regions of the same high H_2O VMR are broader and days of the same high H_2O VMR extend as long as ~70 days. The extended days of the high H_2O VMR in Figure 5c,d reflect the increase in mesospheric H_2O during recent years.

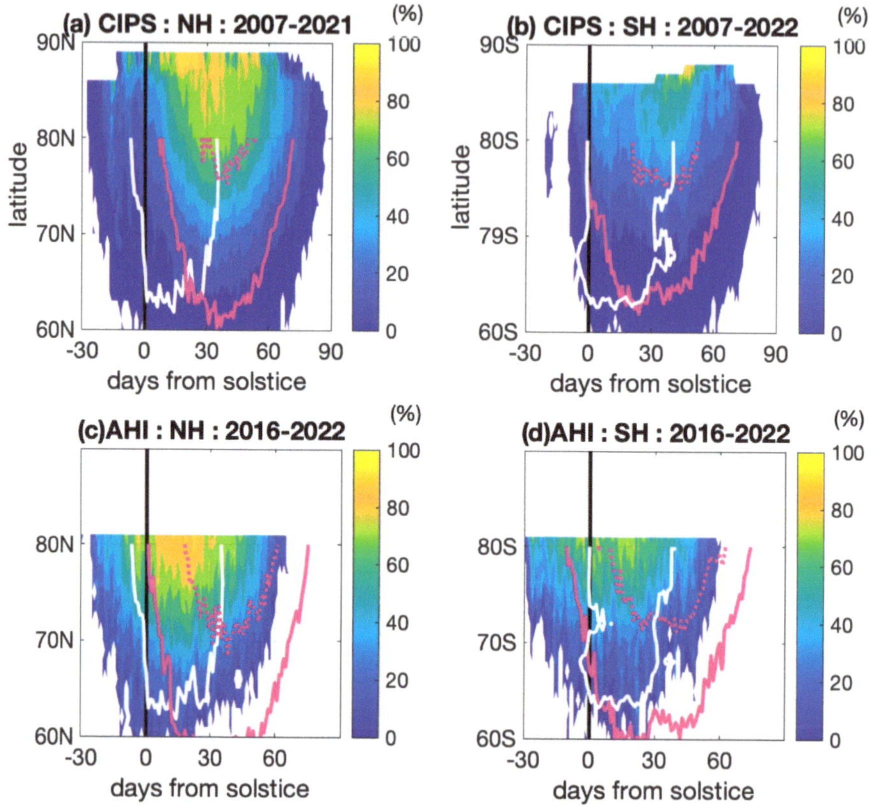

Figure 5. Climatology of the daily PMC occurrence rate (OR) in % as observed from AIM/CIPS (upper panel: 2007–2021) and Himawari-8/AHI (lower panel: 2016–2021) in the NH and SH at each latitude bin. Black lines indicate the day of local summer solstice. Solid and dotted contour lines in magenta represent high latitude (60°N/S–82°N/S) H_2O VMR of 7.1 ppmv and 7.5 ppmv, respectively, at 0.02 hPa (near PMC altitude). White lines represent high latitude mean T = 173 K (NH) and T = 178 K (SH) at the same level.

In general, during the MLS observations (2005–2021), the NH high latitude (70°N–82°N) H_2O VMR at 0.02 hPa increased by ~0.4 ppmv (~7%). These results are consistent with the current understanding of the conditions of T and H_2O needed for forming PMCs. A combination of the recent mesospheric cooling and increments in the mesospheric H_2O concentration would provide favorable conditions for the formation of PMCs in the coming years. Detailed analyses of the year-to-year variability in PMCs, T, and H_2O are discussed in Section 3.4.

The climatology of the PMC OR exhibits a hemispheric asymmetry, as the multiyear averaged PMC occurrence is more frequent in the NH than in the SH in both the AHI and CIPS observations. The multiyear average of the daily PMC OR can reach 100% in NH high latitudes, but it is rarely higher than 80% in the SH. Several studies have noted the hemispheric asymmetry in PMC occurrence, with PMCs more frequently observed with a greater extent in the NH relative to the SH [27,39,54,87]. The smaller extent of PMCs in the SH compared to the NH is also obvious in recent years (2015–2021), as clearly seen from the Himawari-8/AHI observations [60].

Besides the frequency of their occurrence, distinct interhemispheric differences in PMC altitudes and micro-physical properties have also been reported. For example, it was shown that the mean cloud altitude is higher and they are brighter with a greater mean particle size in the SH compared to the NH [49]. Interhemispheric teleconnections between the two hemispheres via stratospheric wind reversal have been proposed as a dynamic control mechanism [12,21,88–90]. In addition to the cold temperature, the upward transport of H_2O to the summer polar mesopause region is another important factor in PMC formation. The higher-altitude PMCs in the SH are indicators of the stronger mesospheric upwelling compared to that in the NH. Different solar forcing in the two hemispheres is suggested as a main cause of the interhemispheric differences in the altitudes of the supersaturation region and the upwelling vertical wind, as the solar flux in January is 6% greater than the solar flux in July due to the Earth's orbital eccentricity [39]. More precise comparison of the daily PMC ORs for latitudes between 38° and 81°, obtained from simultaneous Himawari-8/AHI and AIM/CIPS data, have already been presented in [60].

3.4. Dependence of Year-to-Year Variations of PMC on T and H_2O

3.4.1. Onset Time of PMC

To examine the relationship between PMC formation and atmospheric conditions, the PMC onset dates from CIPs are compared to the timing of the high H_2O and cold temperature at 0.01 hPa (Figure 6). The high H_2O onset date is estimated as the first day of the PMC season, when the MLS high-latitude (60–82°N/S) zonal mean H_2O VMR at 0.01 hPa is greater than 4.2 ppmv and 6.5 ppmv for the NH and SH, respectively. Similarly, the cold temperature onset date is estimated as the first day of the PMC season when the MLS high-latitude (60–82°N/S) zonal mean mesospheric temperature at 0.01 hPa is lower than 165 K and 185 K for the NH and SH, respectively. Besides the frequency of their occurrence, distinct interhemispheric differences in the onset time of PMCs are a good indicator of the wave activity involved in each hemisphere's PMC season [90].

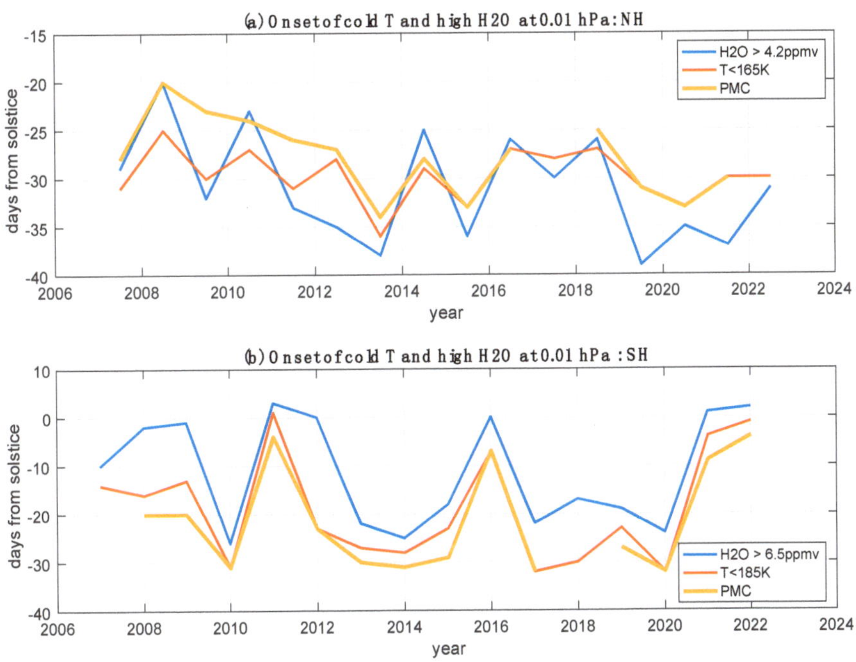

Figure 6. Interannual variations in the CIPS PMC onset date (in yellow) and onset date of the high H_2O (in blue) VMR and low temperature (in red) with respect to the summer solstice in the (**a**) NH and (**b**) SH. Note the different y axes for each hemisphere's PMC season.

The PMC onset date, as seen by CIPS, is estimated from the version 5.20 level 3c data. The onset date is the first day of the PMC season, when CIPS observed an orbit-average cloud albedo of at least 5×10^{-6} sr^{-1} in the latitude range of 70°–80°N [91]. A good consistency in the onset dates between AHI and CIPS are already shown [60], where the detection threshold for the CIPS albedo is 3×10^{-6} sr^{-1}. Because the estimate of the onset date is sensitive to the threshold values, the results from AHI may be a bit later if the threshold is increased to 5×10^{-6} sr^{-1}.

In general, the PMC onset dates occur 20–30 days prior to the summer solstice. During the 15 years of the CIPS measurements, the earliest onset in the NH was 34 days before the solstice in 2013 and the latest onset was 20 days before the solstice in 2008. The NH PMC onset date was from 10 to 20 days earlier than that of the SH. In the NH, the PMC onset date occurs between 20 and 34 days prior to the summer solstice, while it occurs even after the summer solstice in the SH. Nevertheless, this shows that the variability in the SH season onset during CIPS periods is 25 days, significantly greater than that in the NH. The earliest onset in the SH was 30 days before solstice in 2011 and the latest onset was 5 days before the solstice in 2022. The PMC onset dates in the NH tend to show an increasing trend towards an early onset of the season. Interestingly, the onset dates in the SH show dramatic interannual variability with 4- to 5-year oscillation. Furthermore, they show the latest onset during the last year of the observation in 2021/2022.

This year-to-year variability in PMC onset time has been reported in earlier observations. In the SH, solar cycle variation and stratospheric control (inter-hemispheric coupling) have been suggested as the mechanisms which drive the interannual variation in PMC onset [21,90,92]. It was proposed that the SH PMC onset is mainly controlled by the timing of the SH stratospheric polar vortex breaking [93]. An early reversal of winter (eastward) to summer (westward) stratospheric wind causes the early onset of a net eastward GW drag, which can lead to an early deceleration of the mesospheric zonal mean flow. Accordingly, the early onset of an eastward GW drag would accelerate equatorward meridional drift and mesospheric upwelling. In the NH, inter-hemispheric coupling has been suggested as one of the drivers of the variation in PMC season onset [22,92]. In this mechanism, planetary-wave-induced changes in the winter stratosphere modulate the zonal wind, leading to changes in the GW filtering, the mean meridional circulation, and the summer mesospheric temperature. The 11-year solar cycle [21], early springtime PW breaking in the summer stratosphere [43,90], and enhanced quasi 5-day PW activity [91] have also been suggested as drivers of the interannual variation in PMC onset time in the NH.

Our analysis clearly shows that the summer polar mesospheric cold temperature and high H_2O VMR onsets are correlated with the interannual variability in PMC onset. In the NH, the mesospheric T and H_2O VMR onset time show 2–3-year oscillations and the PMC onset time tracks these variations well. In the SH, the mesospheric T and H_2O VMR show unique 4–5-year oscillations and the PMC onset time is highly correlated with these atmospheric oscillations. There are four cases in the SH (2010/2011, 2015/2016, 2020/2021, and 2021/2022) where the PMC onset times are late. Negative ENSO events associated with the westerly Quasi Biennial Oscillation (QBO) phase have been suggested as the potential cause of early SH stratospheric polar vortex breaking around the solstice [92]. Although early PMC onset has occurred during a strong positive ENSO year, anomalous zonal wind filtering may induce anomalous eastward GW forcing in the upper mesosphere. We can infer that the early onset of high mesospheric H_2O and cold T at the PMC level, achieved by the enhanced mesospheric upwelling due to the early onset of a GW drag, would lead to an early onset time of SH PMC. More on this oscillation is discussed in Section 3.4.3.

3.4.2. Occurrence of PMC

Many studies have discussed the linkages between the increasing trends in the occurrence and brightness of PMCs and climate change on different time scales [14,16,17]. Substantial trends in PMC occurrence have been consistently reported since a review paper by [50]. A more frequent, brighter, and broader extent of PMCs in the NH compared to those in the SH has been emphasized [39,54,60,88]. A centennial-scale H_2O increase of about ~1 ppmv due to CH_4 oxidation in the mesosphere during the industrialization era has been proposed as one of the major contributing factors to the increased occurrence of PMCs [19].

To examine the recent trends and hemispheric asymmetry in the PMC occurrence rate (OR), we compare the interannual variations in PMC ORs from CIPS and AHI in Figure 7. To address the atmospheric environment related to these PMC ORs, a comparison of MLS H_2O, RHI, and temperature is also shown for both the NH and SH. For each year, the daily occurrences of PMCs in each latitude bin from 60°–85°N/S are averaged. Similarly, MLS H_2O (ppmv) within the H_2O hole beyond 72°N/S at 0.01 hPa, MLS RHI, and T at near mesopause (0.001 hPa) beyond 62 N/S are averaged for 0 to 30 days from the summer solstice for each hemisphere. The error bars indicate the standard deviation of the daily time series of each variable during 30 days from the summer solstice.

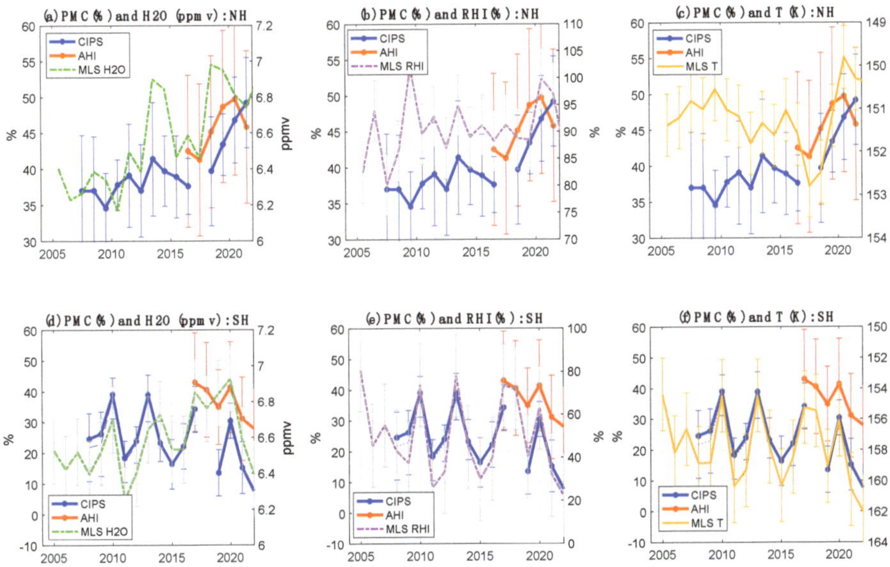

Figure 7. Year-to-year variations in seasonal mean PMC occurrence Rate (%) for NH (**a**–**c**) and SH (**d**–**f**) from CIPS (in blue) and AHI (in red). Similarly, year-to-year variations in MLS H_2O (ppmv), RHI (%), and T beyond 62°N/S at 0.01 hPa are given on the right axis. H_2O VMR, RHI, and T are averaged for 0 to 30 days from summer solstice for each hemisphere. The *y*-axis for T is reversed for a better presentation. The vertical bars indicate standard deviations for the MLS variables.

In the NH (Figure 7, upper panel), the PMC ORs as observed by CIPS increased by 0.75%/year during the observation periods (2007–2021). Even though large discrepancies between CIPS and AHI are shown at the end of the time series in 2021, the ORs, as observed by AHI, also increased from 2017. The MLS H_2O VMR within the H_2O hole also generally increased by ~0.4 ppmv (~7%) during the analysis periods within 2005–2021. An increase in mesospheric H_2O, which is greater in the NH than in the SH, can be the most plausible candidate for the PMC formation process which leads to a consistent positive PMC trend in the NH.

Different from the NH, the PMC occurrence rates in the SH are evident during 2016 and 2022 from both observations. While PMC occurrences consistently increase in the NH (0.75%/year), the SH PMC ORs show a decrease (−0.96%/year). The ORs are relatively low during recent years, i.e., 2020–2022. In the SH, the MLS H_2O VMR within the H_2O hole only increased by ~0.25 ppmv (~4%) and showed similar low local minima in 2016/2017 and

2021/2022, consistent with the PMC occurrences. PMCs can be indicators of anthropogenic H_2O increases, because increasing H_2O significantly enhances the visibility of PMCs.

More interestingly, the PMC occurrence rate variations in the SH are highly correlated with the mesospheric (0.01 hPa) RHI (γ = 0.93) and T (γ = 0.97) variations. The PMC, RHI, and T show 4–5 years of mesospheric oscillations. The correlation in the year-to-year variability between the PMC occurrence and H_2O, RHI, and temperature in the PMC region can be investigated further, as the PMC and middle atmosphere observations continue. The PMC ORs in the SH show unique features, decoupled from the NH variations. The perturbations in the NH stratosphere and mesosphere, i.e., stratospheric sudden warming, do not impact the SH to change the gravity wave forcing, since the midsummer jet in the SH is strong enough not to be coupled with anomalies in the NH [94].

3.4.3. Quasi Quadrennial Oscillation

In the SH, the PMC and summer mesospheric high-latitude T, H_2O, and RHI time series contains an oscillation over an approximately 4–5-year periods (Figure 7). A peak-to-peak amplitude of the QQO feature in PMC is 20–25%. The amplitude of mesospheric T, H_2O, and RH oscillation at 0.01 hPa is 4–5 K, 0.3–0.5 ppmv, and 40–50%, respectively. The distinct features of Quasi Quadrennial Oscillation (QQO) and the unusual periodic 4–5-year oscillation in the high-latitude near-mesopause region T and meridional wind are reported [95,96] with multiple satellite and ground observations.

Considering the significant negative correlation between the high-latitude T and H_2O at 0.01 hPa, we can infer that the QQO variation in the T and H_2O is an adiabatic response. In the SH summer, the equatorward circulation leads to upwelling and adiabatic cooling, while poleward circulation leads to downwelling and warming. Similarly, a lower H_2O mixing ratio is expected with the downwelling, and a higher H_2O mixing ratio is expected with the upwelling. In the same way, a higher T and lower H_2O mixing ratio are still expected with descent in the SH winter. A significant anticorrelation between meridional wind and temperature has been found [96] with wintertime radar measurements at Davis station, Antarctica. They suggest that the wintertime QQO is linked to adiabatic heating and cooling driven by the meridional flow. As a mechanism behind the QQO, their analysis supports tidal and planetary waves, which may filter GWs to drive an adiabatic response in the mesosphere. However, their explanation is a more plausible explanation for the winter hemisphere, since GW is relatively weak in the summer hemisphere.

4. Discussion

Quantifying the relationship between PMC, T, and water vapor (H_2O) is complicated due to the interaction of PMC with the surrounding H_2O. When the surrounding atmosphere is cold enough to form PMCs, PMCs can dehydrate the surrounding environment. When the surrounding environment is warmed, the clouds sublimate and release the water vapor back into the atmosphere. This process is similar in many aspects to that near the tropical tropopause, where the in situ formed cirrus interacts with local water vapor [84]. In other words, the total water (vapor + ice) is approximately conserved over a short (months) period of time. Although the degree of saturation (S) is commonly used by the PMC community [30], RHI (standard product from the MLS retrieval) is an equivalent variable to S and widely used in Earth and planetary sciences. RHI is simply the ratio of water vapor pressure over the local atmospheric saturation pressure.

4.1. Relationships between T, H_2O, and RHI

The MLS mesospheric temperature (T) data have been validated extensively against other measurements [71,72,97]. In the polar regions, they generally agree within ~5 K, with SABER temperature data and improvements have been further made with the version 5 data. As shown in [45], the MLS T is close (within 5 K) to the frost point at the altitude where PMCs form. As shown in Figure 6, at 0.01 hPa, the MLS T is ~150 K, in line with the conditions for PMC formation. The 170 K contour used at 0.02 hPa is used to illustrate the sharp vertical gradient in the atmosphere where PMCs are likely to form at a higher altitude. In addition, it is imperative to recognize that PMC formation can be intermittent, wave-like, or spotty in space and time. The MLS T results shown in this study reflect a daily or monthly mean condition. Transient wave perturbations can occur and induce a colder temperature condition for PMCs to form intermittently.

Atmospheric T and H_2O are intimately related through the Clausius Clapeyron (CC) equation. However, the following conditions would be required to determine/validate one from the other: (1) simultaneous PMC detection at the same location as the T and H_2O measurements; (2) either T or H_2O is precisely measured; and (3) the supersaturation ratio is well known for PMC formation. The MLS RHI uncertainty is typically ~50% [75], which is largely determined by its T error (1 K ~ 16% in the CC equation). The larger MLS T error in the mesosphere is likely to increase the RHI uncertainty to 200%, which may help to constrain PMC supersaturation.

Given the large uncertainties about the range of PMC supersaturation, which is associated with their formation process and ice and nuclei microphysics, the MLS RHI data serve as a surrogate for evaluating the validity of the intimate relationship between T and H_2O. Like the situation near the tropopause, the RHI measurement is derived fundamentally from the CC equation. In this study, we presented a basis of how this principle is applicable to a wider temperature and pressure range, including very cold conditions near the mesopause.

4.2. *Influence of PMC on H_2O and RHI*

The impact of the PMC on the atmospheric environment, via the processes of dehydration resulting from ice growth and H_2O enhancement resulting from ice sublimation, is examined with MLS RHI. In Figure 8, a comparison of MLS RHI (%) profiles is shown for different time periods of the summer solstice. MLS RHIs are averaged for 0 to 30 days before the solstice, from 0 to 30 days after the solstice, and from 30 to 60 days after the solstice, respectively, for 60°N/S–82°N/S during 2005–2022. The 30 days after the solstice (in red curves) can be considered as the PMC formation phase and the 30 to 60 days after the solstice (in yellow curves) can be considered as the sublimation period. The 30 days before the solstice (in blue curves) can be considered as a background state profile which is not influenced by the PMC formation.

As seen in Figure 8a,b, the RHI profiles show a rapid increase in RHI with an increasing altitude from 0.02 hPa. In both hemispheres, enhanced RHI profiles above 0.02 hPa indicate the increase in RHI from altitudes near the bottom of the PMC layers from 0 to 30 days after the solstice. After 30 days from the solstice, the RHI profiles tend to go back to the background states in both hemispheres. In the SH, the RHI from 30 to 60 days after the solstice is slightly lower than the RHI before the solstice. It is noticeable that the hydration layers in the SH are elevated to above 0.01 hPa in comparison to those in the NH, and that this follows from SH PMC being located ~1 km higher than that in the NH [62].

To show the differences in the RHI distribution among different periods around the summer solstice, the probability density functions (PDFs) from the same temporal and spatial conditions of RHI at 0.01 hPa are calculated (Figure 8c,d). The bin size is 5%, and the total integration of the area under each PDF is normalized to be one. The PDFs of RHI (%) during different periods indicate that supersaturation (above 100% of RHI) happens more frequently at 30 days after solstice, when PMCs are forming.

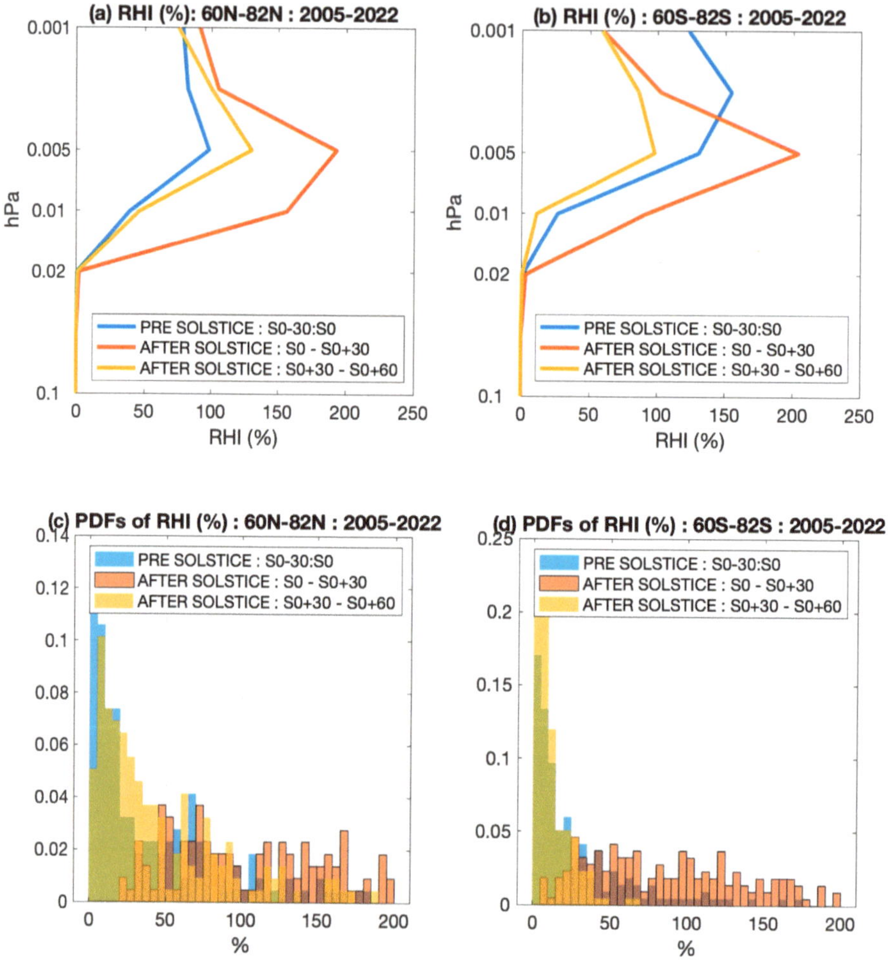

Figure 8. Comparison of profiles of MLS RHI (%) among different periods around summer solstice in (**a**) NH and (**b**) SH. Comparison of probability density function (PDF) of RHI in (**c**) NH and (**d**) SH. RHI data at 0.01 hPa are averaged for 0 to 30 days before solstice (in blue), 0 to 30 days after solstice (in red), and 30 to 60 days after solstice (in yellow), respectively, for 60°–82°N/S latitude band during 2005–2021.

4.3. Diminishing Solar Cycle Variations in PMC

Discerning the near-term trends in the atmospheric tracers from the influence of the solar cycle variation is a challenging task. When the variation associated with the solar cycle is not thoroughly accounted for, there can be biases for any remaining trends. The variations in the annual mean near-global (82°S–82°N) area weighted T and H_2O near the mesopause to the 11-year solar cycle are investigated by analyzing MLS observations from 2005 to 2021. In this study, the V19 Total Solar Irradiance (TSI) data from the SOlar Radiation and Climate Experiment (SORCE) Total Irradiance Monitor (TIM) [98] are used as a solar irradiance index, which can indicate the solar activity level over the solar cycle. With an absolute accuracy, the TSI can be the preferred reference data, compared to F10.7 flux and Ly-α flux, to determine the amplitude of solar cycle and difference between two minima. The TIM TSI data have a three times higher accuracy (0.035%) than previous TSI measurements with a long-term stability of 0.001% per year [99].

In Figure 9, the annual mean and standard deviations of T and H_2O near the mesopause (0.001 hPa) are shown for the interannual variations related to the solar cycle. At a given level at 0.001 hPa, the annual mean temperature (Figure 9, left panel) is maximum at mid-latitudes (30°N/S–40°N/S) and minimum at high latitudes above 60°N/S. The annual mean temperature clearly shows interannual variations associated with the 11-year solar cycle (SC). The near-global mean temperature shows an in-phase correlation with the SORCE-measured TSI without lag (γ = 0.81) with a 95% confidence level [100], as the T is higher during the solar maximum (2012–2014) and lower during the solar minima (SC 24; 2007–2009 and SC25; 2018–2020). The amplitude of temperature variations associated with the solar cycle is ~5 K during SC 24–25.

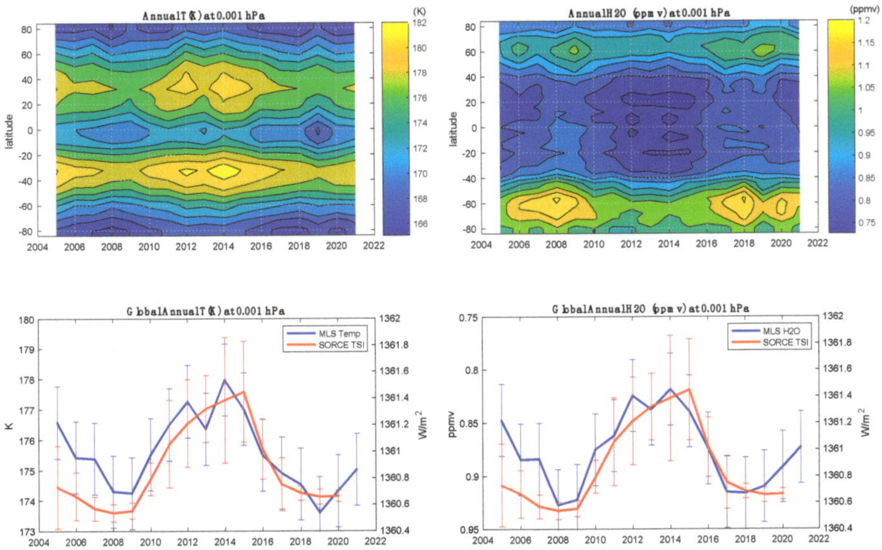

Figure 9. Latitudinal distribution of annual mean temperature (K) and H_2O (ppmv) at 0.001 hPa is shown in the upper panel. The near-global averaged (82°S–82°N) annual mean time series and their standard deviations are shown in the lower panel with SORCE TSI (W/m^2) variation. The *y*-axis for H_2O plot (**lower right**) is reversed for a better presentation.

The inferred annual mean temperature response to solar forcing from the Hamburg Model of Neutral and Ionized Components (HAMMONIA) is 0.5–1 K/100 sfu [101]. The solar cycle variation amplitude of the global mesopause temperature estimated from the Sounding of the Atmosphere using Broadband Emission Radiometry (SABER) is ~5 K [102] and 3–5 K [103,104] per 100 solar flux unit (sfu) depending on latitude. Since the F10.7 flux changes by roughly ~100 sfu over the 11-year solar cycle, our estimate corroborates with the previous results from the SABER measurements.

The analysis from satellite observations and model simulations suggests a decrease in mesospheric H_2O with solar cycle [14,105] which may be less favorable for PMC formation. The Ly-α flux plays an important role in the chemical composition of H_2O and the mesosphere [106], by the two major photolysis of H_2O at the Ly-α line, i.e.,

$$H_2O + h\nu \rightarrow H + OH \tag{3}$$

$$H_2O + h\nu \rightarrow H_2 + O(1) \tag{4}$$

For a given altitude at 0.001 hPa, the annual mean H_2O VMR (Figure 9, right panel) generally increases with an increasing latitude in both hemispheres, reflecting the summertime upwelling of H_2O following the mean meridional circulation, which maximizes at high latitudes. It shows a maximum amount over the SH high latitudes, 60 S–80 S. Similar to temperature, the near-global mean H_2O VMR time series also clearly shows an 11-year solar cycle signal without lag ($\gamma = -0.85$ with 95% confidence level) with TSI. The response process of water vapor to solar forcing and the pathways of solar signals are complex, because the H_2O VMR in the mesosphere is driven by two processes: transport and chemistry. The "top–down" pathway considers solar variations in the Lyman-α (121.6 nm) decreasing the mesospheric H_2O, which then couples down to the stratosphere via a wintertime polar vortex [69]. In "bottom-up" pathway, the tropospheric H_2O response may play a role because both H_2O and CH_4 in the stratosphere originate from here. With an increased solar flux during the solar maximum condition, a lower H_2O in the upper mesosphere is expected due to the enhanced H_2O photolysis process.

The amplitude of the solar-cycle-driven near-global annual mean H_2O VMR is about 0.1 ppmv at near mesopause (0.001 hPa). This 0.1 ppmv solar variation in H_2O is consistent with [106], shown with the composite data at 68°S and 68°N. The solar cycle signal is significant above 0.005 hPa. But below that level, H_2O is insensitive to solar irradiance variation. While the solar maximum is generally expected to induce lower water vapor in the upper mesosphere, estimates of the solar cycle variation from H_2O measurements often depend on the analysis period. For example, the relatively large amplitude of the H_2O solar cycle variations over 1 ppmv obtained from the Halogen Occultation Experiment (HALOE) (1993–2005) is diminished after 2004. It was suggested that HALOE H_2O measurements near the summer mesopause can be contaminated by PMCs [106]. Enhanced PMC extinction during the solar minimum can falsely increase H_2O, and a higher H_2O concentration during the solar minimum could have included this artifact.

Solar cycle variations are expected to disturb PMC formation by modulating the temperature and humidity in the middle atmosphere. A higher temperature [107] and H_2O reduction by Ly-α flux-driven photolysis during the solar maximum period should provide less favorable conditions for PMC formation. However, quantifying solar cycle signals from PMCs from the data record covering less than a few solar cycles is a complicated exercise. Despite an expectation of the high sensitivity of PMCs to solar irradiance, little solar cycle signatures are found in the PMC occurrence during 2007–2021. The linear correlation between yearly PMC ORs and TSI from 2007 to 2021 is not significant ($\gamma < 0.1$) for both hemispheres, as can be seen from Figures 6 and 7 in Section 3.4. The steady global increase in the mesospheric H_2O during the last decade may have overwhelmed the H_2O decrease driven by solar activity. Mesospheric H_2O increases due to anthropogenic methane increases probably play a dominant role in weakening the solar cycle response in PMCs.

5. Summary

The sensitivities of PMCs to the MLS-measured T and H_2O were analyzed to investigate the characteristics of PMC occurrence in relation to the middle atmosphere conditions. The AIM/CIPS and Himawari-8/AHI measurements of the PMC cover 16 years from 2007 and 6 years from 2016, respectively. The measurements of the middle atmosphere from MLS fully cover more than 18 years from August 2004.

1. We compared the climatology and year-to-year variability in the daily PMC ORs from AIM/CIPS and Himawari-8/AHI. Despite differences in the sampling and algorithms of the measurements, daily variations in the PMC occurrence rate in two independent measurements showed a remarkable similarity in their overall spatial extent, timing, and the duration of the cloud occurrence. The OR from AHI reached its maximum, within 80 N/S latitudes, about 20 days after the solstice, which was ~10 days earlier than that from CIPS. The climatologies of the two PMC ORs exhibited a hemispheric asymmetry between the two hemispheres, as the multiyear averaged PMC occurrence was more frequent in the NH than the SH in both observations.

2. The climatologies of the two PMC ORs were compared with the climatology of the summer solstice H_2O VMR and T at the near-PMC level (0.02 hPa) at a high latitude (60°N/S–82°N/S). The PMC occurrence above 60°N/S was directly related to T and H_2O variations, and the combination of these two determined the PMC's seasonal development. In the CIPS case, the high frequency of the occurrence nearly co-occurred with a high H_2O, but it lagged up to 30 days from the temperature drop. The date of the seasonal high latitude mesospheric H_2O maxima seen from the MLS was up to 30 days later than the date of the corresponding local T minima. The lagged days between the PMC and T were reduced in the AHI-measured PMC ORs. In both cases, the PMC OR peak occurred between the T minima and H_2O maxima.
3. We showed the spatial patterns of the H_2O hole, a region of exceptionally depleted H_2O in the upper mesosphere at 0.002 hPa during the PMC season over the Arctic and Antarctic beyond 70°N/S centered at the poles. We inferred that the H_2O depletion in this region was probably caused by the formation of PMCs. These clouds dehydrate the surrounding atmosphere when they are formed. The H_2O amount integrated below these PMC regions generally increased during the last 17 years in the NH. It showed consistent interannual variations similar to the CIPS- and AHI-measured PMC occurrence variations in the NH.
4. Our analysis estimated that the 11-year solar cycle signals in the near-global annual mean T and H_2O were ~1 K and ~0.1 ppmv at mthe esopause level (0.001 hPa) with TSI variations without a lag ($\gamma = 0.81$ for T and $\gamma = 0.85$ for H_2O with 95% confidence level), respectively. However, there was no significant anti-correlation between the PMC occurrence and solar cycle.
5. In the NH, the increases in PMC during recent years were correlated with the positive trend of the mesospheric H_2O, as observed from MLS. Abundant H_2O can significantly enhance the PMC formation. In the NH, the PMC onset dates also became 5–10 days earlier during the last decade. The NH PMC OR increased ~0.75%/year during 2007–2021. The significant increase in mesospheric H_2O in the NH due to anthropogenic forcing during the last decade may explain the diminished solar cycle signals in PMC occurrences during recent years.
6. While PMC occurrences consistently increased in the NH, the SH PMC ORs showed a decrease (0.96%/year) during 2008–2022 with relatively low rates during recent years i.e., 2020–2022. In the SH, the summer mesospheric high-latitude T and H_2O VMR time series showed a unique 4–5 years of quasi-quadrennial oscillations (QQOs). Similarly, the PMC OR and onset time also showed these distinct oscillations. The peak-to-peak amplitude of the QQO feature in PMC was 20–25%. The amplitude of mesospheric T and H_2O at 0.01 hPa were 4–5 K and 0.3–0.5 ppmv, respectively. The cold years in the mesosphere coincided with the humid years with abundant H_2O.
7. Solar cycle variations are expected to disturb PMC formation by modulating the temperature and humidity in the middle atmosphere. Higher temperature and H_2O reduction by Ly-α flux-driven photolysis during solar maximum period should provide less favorable conditions for PMC formation. Despite an expectation of the high sensitivity of the PMC to the solar irradiance, little solar cycle signatures were found in the PMC occurrence during the analysis period of 2007–2021. The steady global increase in the mesospheric H_2O due to anthropogenic methane increase during the last decade may have overwhelmed the H_2O decrease driven by solar activity.

Author Contributions: Conceptualization, J.N.L. and D.L.W.; methodology, J.N.L. and D.L.W.; software, J.N.L., D.L.W., B.T., Y.H. and T.T.; validation, J.N.L., D.L.W., B.T., Y.H. and T.T.; formal analysis, J.N.L. and D.L.W.; investigation, J.N.L., D.L.W., B.T., Y.H. and T.T.; resources, B.T., Y.H. and T.T.; data curation, J.N.L., D.L.W., B.T., Y.H. and T.T.; writing—original draft preparation, J.N.L., D.L.W. and B.T.; writing—review and editing, B.T., Y.H. and T.T.; visualization, J.N.L.; supervision, D.L.W.; project administration, D.L.W.; funding acquisition, D.L.W. All authors have read and agreed to the published version of the manuscript.

Funding: This work is supported by NASA's Sun-Climate research at Goddard Space Flight Center. BT was supported by NSF AGS award 2149483.

Data Availability Statement: MLS Level 2, version 5 temperature data can be downloaded from https://disc.gsfc.nasa.gov/datasets/ML2T_005/summary (accessed on 31 January 2023). MLS Level 2, version 5 water vapor data can be downloaded from https://disc.gsfc.nasa.gov/datasets/ML2H2O_005/summary (accessed on 31 January 2023). CIPS Level 3c, version 5.20, revision 05 data can be downloaded from https://lasp.colorado.edu/aim/download/pmc/l3c (accessed on 31 January 2023). The PMC data derived from the Himawari-8/AHI data are available from http://ttt01.cei.uec.ac.jp/himawari/ (accessed on 31 January 2023). The SORCE V19 TSI data can be downloaded from https://disc.gsfc.nasa.gov/datasets/SOR3TSID_019/summary (accessed on 31 January 2023).

Acknowledgments: We thank the MLS, CIPS, Himawari, and SORCE Science Team for making the data available on their websites. We also thank editors and anonymous reviewers for their suggestions that helped to greatly improve this manuscript.

Conflicts of Interest: The authors declare no conflict of interest.

References

1. Siskind, D.E.; Merkel, A.W.; Marsh, D.R.; Randall, C.E.; Hervig, M.E.; Mlynczak, M.G.; Russell, J.M., III. Understanding the effects of polar mesospheric clouds on the environment of the upper mesosphere and lower thermosphere. *J. Geophys. Res. Atmos.* **2018**, *123*, 11705–11719. [CrossRef]
2. Dunkerton, T. On the Mean Meridional Mass Motions of the Stratosphere and Mesosphere. *J. Atmos. Sci.* **1978**, *35*, 2325–2333. [CrossRef]
3. Holton, J.R. The dynamics of large scale atmospheric motions. *Rev. Geophys.* **1983**, *21*, 1021–1027. [CrossRef]
4. Garcia, R.R.; Solomon, S. The effect of breaking gravity waves on the dynamics and chemical composition of the mesosphere and lower thermosphere. *J. Geophys. Res.* **1985**, *90*, 3850–3868. [CrossRef]
5. McLandress, C. On the importance of gravity waves in the middle atmosphere and their parameterization in general circulation models. *J. Atmos. Sol.-Terr. Phys.* **1998**, *60*, 1357–1383. [CrossRef]
6. Shepherd, T.G. Transport in the middle atmosphere. *J. Meteorol. Soc. Jpn.* **2007**, *85B*, 165–191. [CrossRef]
7. Witt, G. The nature of noctilucent clouds. *Space Res. IX* **1969**, 157–169. Available online: https://ui.adsabs.harvard.edu/abs/1969spre.conf..157W/abstract (accessed on 21 April 2024).
8. Lübken, F.; Fricke, K.; Langer, M. Noctilucent clouds and the thermal structure near the Arctic mesopause in summer. *J. Geophys. Res.* **1996**, *101*, 9489–9508. [CrossRef]
9. Lübken, F.-J. Thermal structure of the Arctic summer mesosphere. *J. Geophys. Res.* **1999**, *10*, 9135–9149. [CrossRef]
10. Lübken, F.-J.; Latteck, R.; Becker, E.; Höffner, J.; Murphy, D. Using polar mesosphere summer echoes and stratospheric/mesospheric winds to explain summer mesopause jumps in Antarctica. *J. Atmos. Sol.-Terr. Phys.* **2017**, *162*, 106–115. [CrossRef]
11. Roble, R.G.; Dickinson, R.E. How will changes in carbon dioxide and methane modify the mean structure of the mesosphere and thermosphere? *Geophys. Res. Lett.* **1989**, *16*, 1441–1444. [CrossRef]
12. Lübken, F.-J.; Höffner, J.; Viehl, T.P.; Becker, E.; Latteck, R.; Kaifler, B.; Murphy, D.J.; Morris, R.J. Winter/summer transition in the Antarctic mesopause region. *J. Geophys. Res. Atmos.* **2015**, *120*, 12394–12409. [CrossRef]
13. Yue, J.; Russell, J.; Gan, Q.; Wang, T.; Rong, P.; Garcia, R.; Mlynczak, M. Increasing Water Vapor in the Stratosphere and Mesosphere After 2002. *Geophys. Res. Lett.* **2019**, *46*, 13452–13460. [CrossRef]
14. Dalin, P.; Perminov, V.; Pertsev, N.; Romejko, V. Updated long-term trends in mesopause temperature, airglow emissions, and noctilucent clouds. *J. Geophys. Res. Atmos.* **2020**, *125*, e2019JD030814. [CrossRef]
15. Dalin, P.; Suzuki, H.; Pertsev, N.; Perminov, V.; Shevchuk, N.; Tsimerinov, E.; Zalcik, M.; Brausch, J.; McEwan, T.; McEachran, I.; et al. The strong activity of noctilucent clouds at middle latitudes in 2020. *Polar Sci.* **2022**, *35*, 100920. [CrossRef]
16. DeLand, M.T.; Thomas, G.E. Updated PMC trends derived from SBUV randeldata. *J. Geophys. Res. Atmos.* **2015**, *120*, 2140–2166. [CrossRef]
17. Shettle, E.P.; DeLand, M.T.; Thomas, G.E.; Olivero, J.J. Long term variations in the frequency of polar mesospheric clouds in the Northern Hemisphere from SBUV. *Geophys. Res. Lett.* **2009**, *36*, L02803. [CrossRef]
18. Thomas, G.E.; Olivero, J.J.; Jensen, E.J.; Schröder, W.; Toon, O.B. Relation between increasing methane and the presence of ice clouds at the mesopause. *Nature* **1989**, *338*, 490–492. [CrossRef]
19. Lübken, F.; Berger, U.; Baumgarten, G. On the Anthropogenic Impact on Long-Term Evolution of Noctilucent Clouds. *Geophys. Res. Lett.* **2018**, *45*, 6681–6689. [CrossRef]
20. Thomas, G.E.; McPeters, R.D.; Jensen, E.J. Satellite observations of polar mesospheric clouds by the solar backscattered ultraviolet spectral radiometer: Evidence of a solar cycle dependence. *J. Geophys. Res.* **1991**, *96*, 927–939. [CrossRef]

21. Benze, S.; Randall, C.E.; Karlsson, B.; Harvey, V.L.; DeLand, M.T.; Thomas, G.E.; Shettle, E.P. On the onset of polar mesospheric cloud seasons as observed by SBUV. *J. Geophys. Res. Atmos.* **2012**, *117*, D07104. [CrossRef]
22. Fiedler, J.; Baumgarten, G.; Lübken, F.-J. NLC observations during one solar cycle above ALOMAR. *J. Atmos. Sol.-Terres. Phys.* **2008**, *71*, 424–433. [CrossRef]
23. Hervig, M.; Siskind, D. Decadal and inter-hemispheric variability in polar mesospheric clouds, water vapor, and temperature. *J. Atmos. Sol.-Terr. Phys.* **2006**, *68*, 30–41. [CrossRef]
24. Hervig, M.E.; Siskind, D.E.; Bailey, S.M.; Russell, J.M. The influence of PMCs on water vapor and drivers behind PMC variability from SOFIE observations. *J. Atmos. Sol.-Terr. Phys.* **2015**, *132*, 124–134. [CrossRef]
25. Hervig, M.E.; Berger, U.; Siskind, D.E. Decadal variability in PMCs and implications for changing temperature and water vapor in the upper mesosphere. *J. Geophys. Res. Atmos.* **2016**, *121*, 2383–2392. [CrossRef]
26. Turco, R.P.; Toon, O.B.; Whitten, R.C.; Keesee, R.G.; Hollenback, D. Noctilucent clouds: Simulation studies of their genesis, properties, and global influences. *Planet. Space Sci.* **1982**, *30*, 1147–1181. [CrossRef]
27. Olivero, J.J.; Thomas, G.E. Climatology of polar mesospheric clouds. *J. Atmos. Sci.* **1986**, *43*, 1263–1274. [CrossRef]
28. Jensen, E.; Thomas, G.E. A growth-sedimentation model of polar mesospheric clouds: Comparison with SME measurements. *J. Geophys. Res.* **1988**, *93*, 2461–2473. [CrossRef]
29. Aschbrenner, R.; Griffin, J.; Snell, H.E. Polar Mesospheric Cloud Modeling and Seasonal Forecasts with Satellite Data. In Proceedings of the IGARSS 2008—2008 IEEE International Geoscience and Remote Sensing Symposium, Boston, MA, USA, 6–11 July 2008; pp. IV-1006–IV-1009. [CrossRef]
30. Rapp, M.; Thomas, G.E. Modeling the microphysics of mesospheric ice particles: Assessment of current capabilities and basic sensitivities. *J. Atmos. Sol.-Terr. Phys.* **2006**, *68*, 715–744. [CrossRef]
31. Nachbar, M.; Wilms, H.; Duft, D.; Aylett, T.; Kitajima, K.; Majima, T.; Plane, J.M.C.; Rapp, M.; Leisner, T. The impact of solar radiation on polar mesospheric ice particle formation. *Atmos. Meas. Technol.* **2019**, *19*, 4311–4322. [CrossRef]
32. Nachbar, M.; Duft, D.; Leisner, T. The vapor pressure of liquid and solid water phases at conditions relevant to the atmosphere. *J. Chem. Phys.* **2019**, *151*, 064504. [CrossRef]
33. Duft, D.; Nachbar, M.; Leisner, T. Unravelling the microphysics of polar mesospheric cloud formation. *Atmos. Chem. Phys.* **2019**, *19*, 2871–2879. [CrossRef]
34. Leslie, R.C. Sky glows. *Nature* **1885**, *32*, 245.
35. Gadsden, M.; Schröder, W. *Noctilucent Clouds*; Springer: New York, NY, USA, 1989.
36. Gordley, L.L.; Hervig, M.E.; Fish, C.; Russell, J.M., III; Bailey, S.; Cook, J.; Hansen, S.; Shumway, A.; Paxton, G.; Deaver, L.; et al. The solar occultation for ice experiment. *J. Atmos. Sol.-Terres. Phys.* **2009**, *71*, 300–315. [CrossRef]
37. Baumgarten, G.; Fiedler, J.; Lübken, F.; von Cossart, G. Particle properties and water content of noctilucent clouds and their interannual variation. *J. Geophys. Res.* **2008**, *113*, D06203. [CrossRef]
38. Baumgarten, G.; Chandran, A.; Fiedler, J.; Hoffmann, P.; Kaifler, N.; Lumpe, J.; Merkel, A.; Randall, C.E.; Rusch, D.; Thomas, G. On the horizontal and temporal structure of noctilucent clouds as observed by satellite and lidar at ALOMAR (69N). *Geophys. Res. Lett.* **2012**, *39*, L01803. [CrossRef]
39. Chu, X.; Gardner, C.S.; Roble, R.G. Lidar studies of interannual, seasonal, and diurnal variations of Polar Mesospheric Clouds at the south pole. *J. Geophys. Res. Atmos.* **2003**, *108*. [CrossRef]
40. Chu, X.; Huang, W.; Fong, W.; Yu, Z.; Wang, Z.; Smith, J.A.; Gardner, C.S. First lidar observations of polar mesospheric clouds and Fe temperatures at McMurdo (77.8°S, 166.7°E), Antarctica. *Geophys. Res. Lett.* **2011**, *38*, L16810. [CrossRef]
41. Fiedler, J.; Baumgarten, G.; von Cossart, G. Mean diurnal variations of noctilucent clouds during 7 years of lidar observations at ALOMAR. *Ann. Geophys.* **2005**, *23*, 1175–1181. [CrossRef]
42. Fiedler, J.; Baumgarten, G.; Berger, U.; Hoffmann, P.; Kaifler, N.; Lübken, F.-J. NLC and the background atmosphere above ALOMAR. *Atmos. Meas. Tech.* **2011**, *11*, 5701–5717. [CrossRef]
43. Fiedler, J.; Baumgarten, G.; Berger, U.; Gabriel, A.; Latteck, R.; Lübken, F.-J. On the early onset of the NLC season 2013 as observed at ALOMAR. *J. Atmos. Sol.-Terr. Phys.* **2015**, *127*, 73–77. [CrossRef]
44. Gumbel, J.; Stegman, J.; Murtagh, D.P.; Witt, G. Scattering phase functions and particle sizes in noctilucent clouds. *Geophys. Res. Lett.* **2001**, *28*, 1415–1418. [CrossRef]
45. Klekociuk, A.R.; Morris, R.J.; Innis, J.L. First Southern Hemisphere common-volume measurements of PMC and PMSE. *Geophys. Res. Lett.* **2008**, *35*, L24804. [CrossRef]
46. Seele, C.; Hartogh, P. Water vapor of the polar middle atmosphere: Annual variation and summer mesosphere Conditions as observed by ground-based microwave spectroscopy. *Geophys. Res. Lett.* **1999**, *26*, 1517–1520. [CrossRef]
47. Donahue, T.M.; Guenther, B.; Blamont, J.E. Noctilucent clouds in daytime: Circumpolar particulate layers near the summer mesopause. *J. Atmos. Sci.* **1972**, *30*, 515–517. [CrossRef]
48. Bailey, S.M.; Merkel, A.W.; Thomas, G.E.; Carstens, J.N. Observations of polar mesospheric clouds from the SNOE satellite. *J. Geophys. Res.* **2005**, *110*, D13203. [CrossRef]
49. von Savigny, C.; Petelina, S.V.; Karlsson, B.; Llewellyn, E.J.; Degenstein, D.A.; Lloyd, N.D.; Burrows, J.P. Vertical variation of NLC particle sizes retrieved from Odin/OSIRIS limb scattering observations. *Geophys. Res. Lett.* **2005**, *32*. [CrossRef]
50. DeLand, M.T.; Shettle, E.P.; Thomas, G.E.; Olivero, J.J. A quarter-century of satellite PMC observations. *J. Atmos. Sol.-Terres. Phys.* **2006**, *68*, 9–29. [CrossRef]

51. DeLand, M.T.; Thomas, G.E. Extending the SBUV polar mesospheric cloud data record with the OMPS NP. *Atmos. Meas. Tech.* **2019**, *19*, 7913–7925. [CrossRef]
52. DeLand, M.T.; Gorkavyi, N. PMC observations from the OMPS Limb Profiler. *J. Atmos. Sol.-Terr. Phys.* **2021**, *213*, 105505. [CrossRef]
53. Hervig, M.E.; Pagan, K.L.; Foschi, P.G. Analysis of polar stratospheric cloud measurements from AVHRR. *J. Geophys. Res.* **2001**, *106*, 10363–10374. [CrossRef]
54. Hervig, M.E.; Gordley, L.L.; Stevens, M.; Russell, J.M.; Bailey, S.; Baumgarten, G. Interpretation of SOFIE PMC measurements: Cloud identification and derivation of mass density, particle shape, and particle size. *J. Atmos. Sol.-Terr. Phys.* **2009**, *71*, 316–330. [CrossRef]
55. Hervig, M.E.; Stevens, M.H. Interpreting the 35 year SBUV PMC record with SOFIE observations. *J. Geophys. Res. Atmos.* **2014**, *119*, 12689–12705. [CrossRef]
56. Russell, J.M., III; Bailey, S.M.; Gordley, L.L.; Rusch, D.W.; Horányi, M.; Hervig, M.E.; Merkel, A.W. The Aeronomy of Ice in the Mesosphere (AIM) mission: Overview and early science results. *J. Atmos. Sol.-Terr. Phys.* **2009**, *71*, 289–299. [CrossRef]
57. Gadsden, M. Polar mesospheric clouds seen from geostationary orbit. *J. Atmos. Sol.-Terr. Phys.* **2000**, *62*, 31–36. [CrossRef]
58. Hozumi, Y.; Tsuda, T.T.; Hosokawa, K.; Ando, Y.; Suzuki, H.; Murata, T.; Nakamura, T. Horizontal Movement of Polar Mesospheric Clouds Observed from the Himawari-8 Geostationary Meteorological Satellite. *J. Geophys. Res. Atmos.* **2021**, *126*, e2021JD035081. [CrossRef]
59. Tsuda, T.T.; Hozumi, Y.; Kawaura, K.; Hosokawa, K.; Suzuki, H.; Nakamura, T. Initial report on polar mesospheric cloud observations by Himawari-8. *Atmos. Meas. Tech.* **2018**, *11*, 6163–6168. [CrossRef]
60. Tsuda, T.T.; Hozumi, Y.; Kawaura, K.; Tatsuzawa, K.; Ando, Y.; Hosokawa, K.; Suzuki, H.; Murata, K.T.; Nakamura, T.; Yue, J.; et al. Detection of polar mesospheric clouds utilizing Himawari-8/AHI full-disk images. *Earth Space Sci.* **2022**, *9*, e2021EA002076. [CrossRef]
61. Kishore, P.; Ratnam, M.V.; Velicogna, I.; Sivakumar, V.; Bencherif, H.; Clemesha, B.R.; Simonich, D.M.; Batista, P.P.; Beig, G. Long-term trends observed in the middle atmosphere temperatures using ground based LIDARs and satellite borne measurements. *Ann. Geophys.* **2014**, *32*, 301–317. [CrossRef]
62. Bailey, S.M.; Merkel, A.W.; Thomas, G.E.; Rusch, D.W. Hemispheric differences in Polar Mesospheric Cloud morphology observed by the Student Nitric Oxide Explorer. *J. Atmos. Sol.-Terr. Phys.* **2007**, *69*, 1407–1418. [CrossRef]
63. Beig, G.; Keckhut, P.; Lowe, R.P.; Roble, R.G.; Mlynczak, M.G.; Scheer, J.; Fomichev, V.I.; Offermann, D.; French, W.J.R.; Shepherd, M.G.; et al. Review of mesospheric temperature trends. *Rev. Geophys.* **2003**, *41*. [CrossRef]
64. Ratnam, M.V.; Raj, S.A.; Qian, L. Long-term trends in the low-latitude middle atmosphere temperature and winds: Observations and WACCM-X model simulations. *J. Geophys. Res. Space Phys.* **2019**, *124*, 7320–7331. [CrossRef]
65. Bailey, S.M.; Thurairajah, B.; Hervig, M.E.; Siskind, D.E.; Russell, J.M.; Gordley, L.L. Trends in the polar summer mesosphere temperature and pressure altitude from satellite observations. *J. Atmos. Sol.-Terr. Phys.* **2021**, *220*, 105650. [CrossRef]
66. Lee, J.N.; Wu, D.L.; Manney, G.L.; Schwartz, M.J.; Lambert, A.; Livesey, N.J.; Minschwaner, K.R.; Pumphrey, H.C.; Read, W.G. Aura Microwave Limb Sounder observations of the polar middle atmosphere: Dynamics and transport of CO and H_2O. *J. Geophys. Res. Atmos.* **2011**, *116*, D05110. [CrossRef]
67. Solomon, S.; Rosenlof, K.H.; Portmann, R.W.; Daniel, J.S.; Davis, S.M.; Sanford, T.J.; Plattner, G.-K. Contributions of Stratospheric Water Vapor to Decadal Changes in the Rate of Global Warming. *Science* **2010**, *327*, 1219–1223. [CrossRef]
68. Grygalashvyly, M.; Sonnemann, G. Trends of mesospheric water vapor due to the increase of methane—A model study particularly considering high latitudes. *Adv. Space Res.* **2006**, *38*, 2394–2401. [CrossRef]
69. Salinas, C.C.J.H.; Chang, L.C.; Yue, J.; Qian, L.; Gan, Q.; Russell, J.; Mlynczak, M. Estimating the migrating diurnal tide component of mesospheric water vapor. *J. Geophys. Res. Space Phys.* **2022**, *127*, e2021JA030187. [CrossRef]
70. Salinas, C.C.J.H.; Wu, D.L.; Lee, J.N.; Chang, L.C.; Qian, L.; Liu, H. Aura/MLS observes and SD-WACCM-X simulates the seasonality, quasi-biennial oscillation and El Niño–Southern Oscillation of the migrating diurnal tide driving upper mesospheric CO primarily through vertical advection. *Atmos. Chem. Phys.* **2023**, *23*, 1705–1730. [CrossRef]
71. Schwartz, M.J.; Lambert, A.; Manney, G.L.; Read, W.G.; Livesey, N.J.; Froidevaux, L.; Ao, C.O.; Bernath, P.F.; Boone, C.D.; Cofield, R.E.; et al. Validation of the Aura Microwave Limb Sounder temperature and geopotential height measurements. *J. Geophys. Res.* **2008**, *113*, D15S11. [CrossRef]
72. Schwartz, M.; Livesey, N.; Read, W. *MLS/Aura Level 2 Temperature V005*; Goddard Earth Sciences Data and Information Services Center (GES DISC): Pasadena, CA, USA, 2020. [CrossRef]
73. Grieco, F.; Pérot, K.; Murtagh, D.; Eriksson, P.; Rydberg, B.; Kiefer, M.; Garcia-Comas, M.; Lambert, A.; Walker, K.A. Improvement of Odin/SMR water vapour and temperature measurements and validation of the obtained data sets. *Atmos. Meas. Tech.* **2001**, *14*, 5823–5857. [CrossRef]
74. Read, W.G.; Lambert, A.; Bacmeister, J.; Cofield, R.E.; Christensen, L.E.; Cuddy, D.T.; Daffer, W.H.; Drouin, B.J.; Fetzer, E.; Froidevaux, L.; et al. Aura Microwave Limb Sounder upper tropospheric and lower stratospheric H_2O and relative humidity with respect to ice validation. *J. Geophys. Res. Atmos.* **2007**, *112*, D24S35. [CrossRef]
75. Read, W.G.; Shippony, Z.; Snyder, W.V. *Microwave Limb Sounder Forward Model Algorithm Theoretical Basis Document*; Technical Report; Jet Propulsion Laboratory: Pasadena, CA, USA, 2004; JPL D-18130.

76. Goff, J.A.; Gratch, S. Low-pressure properties of water from −160 to 212 F, in Transactions of the American society of heating and ventilating engineers. In Proceedings of the 52nd Annual Meeting of the American Society of Heating and Ventilating Engineers, New York, NY, USA, 10–13 June 1946; pp. 95–122.
77. Murphy, D.M.; Koop, T. Review of the vapour pressures of ice and supercooled water for atmospheric applications. *Q. J. R. Meteorol. Soc.* **2005**, *131*, 1539–1565. [CrossRef]
78. McClintock, W.; Rusch, D.W.; Thomas, G.E.; Markel, A.W.; Lankton, M.R.; Drake, V.A.; Bailey, S.M.; Russel, J.M., III. The cloud imaging and particle size experiment on the aeronomy of ice in the mesosphere mission: Instrument concept, design, calibration, and on-orbit performance. *J. Atmos. Sol.-Terres. Phys.* **2009**, *71*, 340–355. [CrossRef]
79. Carstens, J.N.; Bailey, S.M.; Lumpe, J.D.; Randall, C.E. Understanding uncertainties in the retrieval of polar mesospheric clouds from the cloud imaging and particle size experiment in the presence of a bright Rayleigh background. *J. Atmos. Sol. -Terr. Phys.* **2013**, *104*, 197–212. [CrossRef]
80. Lumpe, J.D.; Bailey, S.M.; Carstens, J.N.; Randall, C.E.; Rusch, D.W.; Thomas, G.E.; Nielsen, K.; Jeppesen, C.; McClintock, W.E.; Merkel, A.W.; et al. Retrieval of polar mesospheric cloud properties from CIPS: Algorithm description, error analysis and cloud detection sensitivity. *J. Atmos. Sol.-Terres. Phys.* **2013**, *104*, 167–196. [CrossRef]
81. States, R.J.; Gardner, C.S. Thermal Structure of the Mesopause Region (80–105 km) at 40°N Latitude. Part I: Seasonal Variations. *J. Atmos. Sci.* **2000**, *57*, 66–77. [CrossRef]
82. Christensen, O.M.; Eriksson, P.; Urban, J.; Murtagh, D.; Hultgren, K.; Gumbel, J. Tomographic retrieval of water vapour and temperature around polar mesospheric clouds using Odin-SMR. *Atmos. Meas. Technol.* **2015**, *8*, 1981–1999. [CrossRef]
83. Merkel, A.W.; Marsh, D.R.; Gettelman, A.; Jensen, E.J. On the relationship of polar mesospheric cloud ice water content, particle radius and mesospheric temperature and its use in multi-dimensional models. *Atmos. Meas. Tech.* **2009**, *9*, 8889–8901. [CrossRef]
84. Flury, T.; Wu, D.L.; Read, W.G. Correlation among cirrus ice content, water vapor and temperature in the TTL as observed by CALIPSO and Aura/MLS. *Atmos. Meas. Tech.* **2012**, *12*, 683–691. [CrossRef]
85. James, A.D.; Brooke, J.S.A.; Mangan, T.P.; Whale, T.F.; Plane, J.M.C.; Murray, B.J. Nucleation of nitric acid hydrates in polar stratospheric clouds by meteoric material. *Atmos. Meas. Tech.* **2018**, *18*, 4519–4531. [CrossRef]
86. Berger, U.; von Zahn, U. Icy particles in the summer mesopause region: Three-dimensional modeling of their environment and two-dimensional modeling of their transport. *J. Geophys. Res.* **2002**, *107*, 1366. [CrossRef]
87. Berger, U.; Lübken, F. Trends in mesospheric ice layers in the Northern Hemisphere during 1961–2013. *J. Geophys. Res. Atmos.* **2015**, *120*, 11277–11298. [CrossRef]
88. Russell, J.M., III; Rong, P.; Bailey, S.M.; Hervig, M.E.; Petelina, S.V. Relationship between the summer mesopause and polar mesospheric cloud heights. *J. Geophys. Res. Atmos.* **2010**, *115*. [CrossRef]
89. Shettle, E.P.; Nedoluha, G.E.; DeLand, M.T.; Thomas, G.E.; Olivero, J.J. SBUV observations of polar mesospheric clouds compared with MLS temperature and water vapor measurements. *Geophys. Res. Lett.* **2010**, *37*, L18810. [CrossRef]
90. Karlsson, B.; Randall, C.E.; Shepherd, T.G.; Harvey, V.L.; Lumpe, J.; Nielsen, K.; Bailey, S.M.; Hervig, M.; Russell, J.M. On the seasonal onset of polar mesospheric clouds and the breakdown of the stratospheric polar vortex in the Southern Hemisphere. *J. Geophys. Res.* **2011**, *116*, D18107. [CrossRef]
91. Thurairajah, B.; Bailey, S.M.; Harvey, V.L.; Randall, C.E.; France, J.A. The Role of the Quasi 5-Day Wave on the Onset of Polar Mesospheric Cloud Seasons in the Northern Hemisphere. *J. Geophys. Res. Atmos.* **2023**, *128*, e2022JD037982. [CrossRef]
92. Li, H.; Wu, J.; Zhou, Z. The formation of multiple layers of ice particles in the polar summer mesopause region. *Ann. Geophys.* **2016**, *34*, 117–122. [CrossRef]
93. Yang, C.; Li, T.; Lai, D.; Wang, X.; Xue, X.; Dou, X. The delayed response of the troposphere-stratosphere-mesosphere coupling to the 2019 southern SSW. *Geophys. Res. Lett.* **2022**, *49*, e2022GL101759. [CrossRef]
94. Siskind, D.E.; Allen, D.R.; Randall, C.E.; Harvey, V.L.; Hervig, M.E.; Lumpe, J.; Thurairajah, B.; Bailey, S.M.; Russell, J.M. Extreme stratospheric springs and their consequences for the onset of polar mesospheric clouds. *J. Atmos. Sol.-Terr. Phys.* **2015**, *132*, 74–81. [CrossRef]
95. French, W.; Mulligan, F.; Klekociuk, A. Analysis of 24 years of mesopause region OH rotational temperature observations at Davis, Antarctica—Part 1: Long-term trends. *Atmos. Chem. Phys.* **2020**, *20*, 6379–6394. [CrossRef]
96. French, W.J.R.; Klekociuk, A.R.; Mulligan, F.J. Analysis of 24 years of mesopause region OH rotational temperature observations at Davis, Antarctica—Part 2: Evidence of a quasi-quadrennial oscillation (QQO) in the polar mesosphere. *Atmos. Chem. Phys.* **2020**, *20*, 8691–8708. [CrossRef]
97. Chen, Z.; Schwartz, M.J.; Bhartia, P.K.; Schoeberl, M.; Kramarova, N.; Jaross, G.; DeLand, M. Mesospheric and upper stratospheric temperatures from OMPS-LP. *Earth Space Sci.* **2023**, *10*, e2022EA002763. [CrossRef]
98. Kopp, G. *SORCE Level 3 Total Solar Irradiance Daily Means V019*; Goddard Earth Sciences Data and Information Services Center (GES DISC): Greenbelt, MD, USA, 2020. [CrossRef]
99. Kopp, G. Science Highlights and Final Updates from 17 Years of Total Solar Irradiance Measurements from the SOlar Radiation and Climate Experiment/Total Irradiance Monitor (SORCE/TIM). *Sol. Phys.* **2021**, *296*, 133. [CrossRef] [PubMed]
100. Angell, J.K.; Korshover, J. Comparison between sea surface temperature in the equatorial eastern Pacific and United States surface temperature. *J. Appl. Meteorol.* **1981**, *20*, 1105–1110. [CrossRef]
101. Beig, G.; Fadnavis, S.; Schmidt, H.; Brasseur, G.P. Inter-comparison of 11-year solar cycle response in mesospheric ozone and temperature obtained by HALOE satellite data and HAMMONIA model. *J. Geophys. Res. Atmos.* **2012**, *117*. [CrossRef]

102. Tang, C.; Liu, D.; Wei, H.; Wang, Y.; Dai, C.; Wu, P.; Zhu, W.; Rao, R. The response of the temperature of cold-point mesopause to solar activity based on SABER data set. *J. Geophys. Res. Space Phys.* **2016**, *121*, 7245–7255. [CrossRef]
103. Forbes, J.M.; Zhang, X.; Marsh, D.R.; Al, F.E.T. Solar cycle dependence of middle atmosphere temperatures. *J. Geophys. Res. Atmos* **2014**, *119*, 9615–9625. [CrossRef]
104. Remsberg, E.; Damadeo, R.; Natarajan, M.; Bhatt, P. Observed responses of mesospheric water vapor to solar cycle and dynamical forcings. *J. Geophys. Res. Atmos.* **2018**, *123*, 3830–3843. [CrossRef] [PubMed]
105. Vardavas, I.M.; Carver, J.H.; Taylor, F.W. The role of water-vapour photodissociation on the formation of a deep minimum in mesopause ozone. *Ann. Geophys.* **1998**, *16*, 189–196. [CrossRef]
106. Hervig, M.E.; Siskind, D.E.; Bailey, S.M.; Merkel, A.W.; DeLand, M.T.; Russell, J.M. The Missing Solar Cycle Response of the Polar Summer Mesosphere. *Geophys. Res. Lett.* **2019**, *46*, 10132–10139. [CrossRef]
107. Lee, J.N.; Wu, D.L.; Ruzmaikin, A.; Fontenla, J. Solar cycle variations in mesospheric carbon monoxide. *J. Atmos. Sol.-Terr. Phys* **2018**, *170*, 21–34. [CrossRef]

remote sensing

MDPI

Article

Evaluatingthe Polarimetric Radio Occultation Technique Using NEXRAD Weather Radars

Antía Paz [1,2,†,*], Ramon Padullés [1,2,†] and Estel Cardellach [1,2,†]

1 Institute of Space Sciences, 08193 Barcelona, Spain; padulles@ice.csic.es (R.P.); estel@ice.csic.es (E.C.)
2 Institute for Space Studies of Catalonia, 08034 Barcelona, Spain
* Correspondence: a.paz@csic.es
† Current address: ICE-CSIC/IEEC, Campus UAB, C/Can Magrans S/N, Cerdanyola del Vallès, 08193 Barcelona, Spain

Abstract: Currently, it remains a challenge to effectively monitor areas experiencing intense precipitation and the associated atmospheric conditions on a global scale. This challenge arises due to the limitations on both active and passive remote sensing methods. Apart from the lack of observations in remote areas, the quality of some observations deteriorates when heavy precipitation is present, making it difficult to obtain highly accurate measurements of the thermodynamic parameters driving these weather events. However, there is a promising solution in the form of the Global Navigation Satellite System (GNSS) Polarimetric Radio Occultation (PRO) technique. This approach provides a way to assess the large-scale bulk-hydrometeor characteristics of regions with heavy precipitation and the meteorological conditions associated with them. PRO offers vertical profiles of atmospheric variables, including temperature, pressure, water vapor pressure, and information about hydrometeors, all in a single fine-vertical resolution observation. To continue validating the PRO technique, we make use of polarimetric weather data from Next Generation Weather Radars (NEXRAD), focusing on comparing specific differential phase shift (K_{dp}) structures to PRO observable differential phase shift ($\Delta\Phi$). We have seen that PAZ and NEXRAD exhibit a good agreement on the vertical structure of the observable $\Delta\Phi$ and that their combination could be useful for enhancing our understanding of the microphysics underlying heavy precipitation events.

Keywords: Global Navigation Satellite Systems (GNSS); Polarimetric Radio Occultations (PRO); Next Generation Weather Radars (NEXRAD)

Citation: Paz, A.; Padullés, R.; Cardellach, E. Evaluating the Polarimetric Radio Occultation Technique Using NEXRAD Weather Radars. *Remote Sens.* **2024**, *16*, 1118. https://doi.org/10.3390/rs16071118

Academic Editors: Yuriy Kuleshov, Xiaoming Wang, Suelynn Choy, Haobo Li and Mayra I. Oyola-Merced

Received: 30 January 2024
Revised: 19 March 2024
Accepted: 19 March 2024
Published: 22 March 2024

1. Introduction

The Radio Occultacion (RO) technique, originally developed for planetary sciences to study other planets' atmospheres (e.g., [1]), consists on tracking the signals emitted by a Global Positioning System (GPS) satellite from a Low Earth Orbit (LEO) satellite as it rises or sets behind the Earth's limb. The technique measures the delay and bending caused by the refractivity of radio signals during propagation through the atmosphere. This delay can be utilized to derive radio refractivity profiles and ionospheric total electron content. From these radio refractivities, valuable vertical profiles of thermodynamic variables, including atmospheric pressure, temperature, and water vapor pressure, can be extracted from the stratosphere down to the surface with a vertical resolution ranging from 100 to 300 m (e.g., [2]). These products, derived from the standard RO technique, are currently assimilated operationally into various global Numerical Weather Prediction (NWP) models (e.g., [3]).

On 22 February 2018, the Spanish Earth Observation satellite PAZ was successfully launched, carrying a Global Navigation Satellite System (GNSS) Polarimetric Radio Occultation (PRO) payload. This mission, known as the Radio Occultation and Heavy Precipitation (ROHP) experiment, is led by the Institut de Ciéncies de l'Espai-Consejo Superior de Investigaciones Científicas/Institut de Estudis Espacials de Catalunya (ICE-CSIC/IEEC) in collaboration with NOAA, UCAR, and the NASA/Jet Propulsion Laboratory. The novelty of the PAZ mission lies in the acquisition of RO measurements at two linear polarizations for the first time [4].

The primary objective of the PRO technique is to detect heavy precipitation by measuring the difference in the phase delay between the two polarizations, horizontal (H) and vertical (V), of the GPS signals. Operating at L-band frequencies (L1 at 1.57542 GHz; L2 at 1.22760 GHz), these signals propagate through the atmosphere and reach the LEO satellite, which in the case of PAZ is equipped with a modified Integrated GPS Occultation Receiver (IGOR+) advanced GPS receiver [4]. The measurements of H and V polarizations are conducted independently, yet synchronously, employing a dual linearly polarized antenna directed towards the Earth's limb in the anti-velocity direction of the LEO satellite. As the signals traverse deeper into denser atmospheric layers, they experience bending curvature induced by the refractive index vertical gradients. By acquiring the incoming electromagnetic field at the two linear and orthogonal polarizations, valuable information can be extracted concerning targets that introduce a differential phase shift ($\Delta\Phi$) between the H and V components of the propagating signals. These targets are primarily hydrometeors that undergo flattening due to air drag during their descent or that are naturally asymmetric (e.g., snowflakes, graupel, etc.). In the presence of heavy precipitation events, the large droplets stand out for being oblate–spheroid-like. Therefore, the PRO technique offers the additional benefit of inferring vertical information about precipitation, enabling the retrieval of both the standard thermodynamic state of the surrounding area and vertical precipitation information within the same measurement.

The validation of the $\Delta\Phi$ observable with two-dimensional data has been assessed in a statistical way using merged precipitation products like the Integrated Multi-satellitE Retrievals for GPM (IMERG) (e.g., [5,6]). Additionally, passive microwave radiometers have been used to help with the interpretation of the vertical structure [7], but these also provide limited vertical resolution retrievals. Furthermore, coincidences with the GPM Dual frequency Precipitation Radar (DPR) are sparse due to the limited swath of the space-based radar.

The initial hypothesis of the ROHP experiment, namely, that PRO observations are sensitive to heavy precipitation, was already demonstrated [5]. Furthermore, it has also been shown that the PRO observable $\Delta\Phi$ increases with higher precipitation rain rates, indicating also sensitivity to precipitation intensity. Moreover, subsequent studies have shown that PRO is not only sensitive to precipitation, but also to horizontally oriented frozen hydrometeors found in various vertical layers of convective clouds [8,9]. This sensitivity is particularly pronounced for snow, where aggregated ice crystals produce relatively large hydrometeor particles. The combined sensitivity of PRO to both heavy precipitation and the associated cloud structures, along with its inherent capacity to provide colocated thermodynamic profiles, makes the PRO technique highly advantageous for studying heavy precipitation events. Other space-based observing systems focused on obtaining vertical thermodynamic profiles often have lower vertical resolution and encounter challenges when attempting measurements within deep clouds (e.g., [10]). On the other hand, space-based precipitation radars lack the capability to provide information about the thermodynamics. Overall, the demonstrated capabilities of PRO render it an interesting choice for investigating heavy precipitation phenomena, and the microphysics underlying these events.

Given that prior validations relied on two-dimensional data, this study seeks to enhance validation by employing three-dimensional data. The specific focus is on validating the vertical structure, and this can be achieved with the Next Generation Weather Radars (NEXRAD) dataset. This entails comparing the observable $\Delta\Phi$ retrieved from PAZ with the vertical structures of K_{dp} derived from NEXRAD, which will permit us to calculate the equivalent differential phase shift observable from these radars.

This article is distributed as follows: Section 2 describes the data used in the analysis in order to make the comparison between NEXRAD and PAZ, and it also explains the methodology that has been followed; Section 3 shows the results of such comparison and the corresponding statistical analysis; finally, Section 4 accounts for the conclusions.

2. Materials and Methods

2.1. Polarimetric Radio Occultation Data

As it was already mentioned, the PRO technique allows us to compare the phase delay (Φ) associated with the two measured polarizations. Since situations of heavy precipitation are characterized by large horizontally oriented raindrops, the accumulated differential phase shift in these regions, $\Delta\Phi = \Phi_H - \Phi_V$, will be positive due to the depolarization effect [8]. $\Delta\Phi$ takes a value of $\pm\pi/2$ when the polarization is purely circular, and therefore, we will analyze $\Delta\Phi$ with respect to the value it would have if the received field was purely circular: $\Delta\Phi = \Delta\Phi - \pi/2$. More specifically, the Φ registered in both ports of PAZ's antenna contains the following terms:

$$\Phi_p = \omega t + \phi^{range} + \phi_p^{atm} + \phi_p^{ins} \tag{1}$$

where ω represents the carrier frequency, ϕ^{range} is the signature of the phase related to changes in the range between the transmitter and receiver, ϕ_p^{atm} denotes the signatures of the phase due to atmospheric effects in the p-polarization (where p can be H or V), and ϕ_p^{ins} indicates the signatures in the p-polarization phase induced by instrumental and platform environment effects [4].

As the first two terms in Equation (1) are independent of polarization, they cancel out when obtaining $\Delta\Phi(t)$:

$$\Delta\Phi(t) = \Phi_H(t) - \Phi_V(t) = \Delta\Phi^{atm}(t) + \Delta\Phi^{ins}(t) \tag{2}$$

where $\Delta\Phi(t)$ should remain constant with time if no differential shift is introduced across the ray path. The instrumental term, $\Delta\Phi^{ins}$, can be corrected with calibration [6]. Detailed theoretical analysis of other systematic effects can be found in [11].

The specific contribution to $\Delta\Phi$ at each point of the propagation path is defined as the specific differential phase, K_{dp}. The values of K_{dp} are expressed in units of length (mm-shift/km-rain) instead of radians because these are the general units used in the GNSS community [4], and that is why it is multiplied by $\lambda/2\pi$. The expression is the following:

$$K_{dp} = \frac{\lambda^2}{2\pi} \int \Re\{f_H(D) - f_V(D)\} N(D) dD \tag{3}$$

where the wavelength, λ, corresponds to the GNSS; $\Re$ represents the real part; $f_H(D)$ and $f_V(D)$ are the forward scattering amplitudes describing the effect of scattering of the GNSS propagating waves by hydrometeors for the horizontal and vertical components, respectively; the variable D refers to the equivalent diameter of the drops; and $N(D)$ is the particle size distribution (PSD). The terms accounting for the type of particle (liquid or solid) and its shape are the scattering amplitudes.

The total hydrometeors' contribution along the ray path is therefore described by the following expression:

$$\Delta\Phi = \int_L K_{dp}(l) dl \tag{4}$$

where the units of $\Delta\Phi$ are in *mm*, K_{dp} is formulated in Equation (3), and L is the ray path length. As we deduce from Equation (4), there is an intrinsic ambiguity between the extension, L, and the intensity, K_{dp} in the final $\Delta\Phi$ measurement.

The Calibrated PRO profiles from PAZ are available from May 2018 to the present [12]. Each file represents a PAZ observation and contains the vertical profile of the observable differential phase shift, $\Delta\Phi$, expressed in units of length (mm) and in terms of the tangential height of each PRO ray. As the PRO rays traverse the atmosphere from GPS to LEO, they bend due to refractivity gradients, eventually becoming tangential to the surface at their lowest height point, defined as the tangent point of the ray, h_t. The locations (i.e., latitude and longitude) representative of each PRO observation are defined at the tangent point of the ray with h_t = 4 km. Even though each ray is linked to its tangential height, hydrometeors that potentially present along the points in that ray are contributing to the value of $\Delta\Phi$ regardless of its height.

The PRO files also contain the locations (latitude, longitude and height) of each ray trajectory between the GPS and PAZ, obtained through ray-tracing techniques. These ray-path locations are calculated and re-gridded, so that only rays whose tangent height coincides with a regular grid between 0 and 20 km with a vertical resolution of 0.1 km are provided. This ensures a collection of ray trajectories with the same vertical resolution as $\Delta\Phi$, and that represent the trajectories that contribute to each $\Delta\Phi$ measurement. Only the portion of these trajectories traveling below 20 km from the surface are considered, since it is assumed that no clouds nor hydrometeors are present above those heights.

2.2. NEXRAD Data

NEXRAD is a 160 network of dual-polarized weather radars distributed across the United States (US) and its territories, developed and deployed by the National Weather Service (NWS) of the US. NEXRAD radars operate at S-band (2–4 GHz) and their main advantage is that they are equipped with polarimetric capabilities, enabling them to provide valuable data on precipitation characteristics, such as size, shape and type of hydrometeor present in the atmosphere. These data are crucial for understanding severe weather phenomena like thunderstorms, tornadoes and heavy rainfall events. Additionally, dual-polarization allows for better discrimination between different types of precipitation, aiding in the identification of potential hazards. With a wide coverage across the US territory, NEXRAD radars offer real-time, high-resolution imagery of atmospheric conditions The radars scan the sky in a 360-degree rotation, providing continuous updates on weather patterns, storm movements, and the evolution of weather systems. For more information about NEXRAD, see [13,14].

The NEXRAD data used here are obtained from the NEXRAD Level II dataset [15]. Each NEXRAD file contains fields for various variables, such as reflectivity (Z), differential reflectivity (Z_{dr}), total differential phase (Ψ_{dp}), cross-correlation ratio (ρ_{dp}), among others. These fields are provided as a function of azimuth, range, and elevation angle. For each radar, a 3D file is generated approximately every 8 minutes.

Typically, the radar scans have a range of elevation angles between 0.5° and 19.5°. Different variables have a different spatial resolution. For example, Z is provided at 1.0° azimuthal resolution and 1 km in range gate resolution, to a range of 460 km. Doppler velocity and spectrum width are provided at 1.0° azimuthal and 0.25 km in range gate resolution, to a range of 300 km.

2.3. Coincident Observations between PAZ and NEXRAD

The PAZ satellite is providing around 150/200 occultations per day, globally distributed, that pass quality control (QC). In the process of selecting the coincident observations, we have implemented a filtering criterion that includes all observations within a range of 250 km to a NEXRAD radar. The total number of observations between May 2018 and December 2022 that meet the colocation criteria is 3208.

To achieve the objectives of this study, we have carefully selected coincident observations from PAZ satellite and NEXRAD weather radars, ensuring they are colocated in both space and time. The time difference between the occultation and the NEXRAD data does not exceed 8 min, as we select the closest radar file to each observation. Typically, an occultation lasts for about 2 min. In terms of spatial alignment, we have chosen multiple radars for each PAZ observation to ensure a good coverage of the area sensed by the PRO rays below 20 km. To select the appropriate radars for each observation, we calculate the distances between all points along the PRO ray trajectories and the radar locations. Radars that do not reach any ray point within 250 km are discarded. Furthermore, the percentage of ray points that have at least one radar within a distance of 250 km is computed and stored as the *covered area* of each PRO.

Figure 1 shows the coverage of the NEXRAD radars over the continental U.S., panel (a), and an example of one coincident PRO observation, panels (b) and (c). Specifically, Figure 1b,c represents the observation in 3D and 2D, respectively. In Figure 1c we see, in grey, the projection on the surface of the portion of PRO rays below 20 km. This area is not the same for all observations since the geometry of the rays when they propagate through the atmosphere from the GPS to the LEO satellite will depend on the relative movement of both satellites. Depending on this movement, the rays originating at different altitudes will present a different degree of vertical alignment [6]. As shown in the figure, it is discernible that the extensive spatial coverage results in a limited vertical alignment among these rays, which can also be appreciated in Figure 1b.

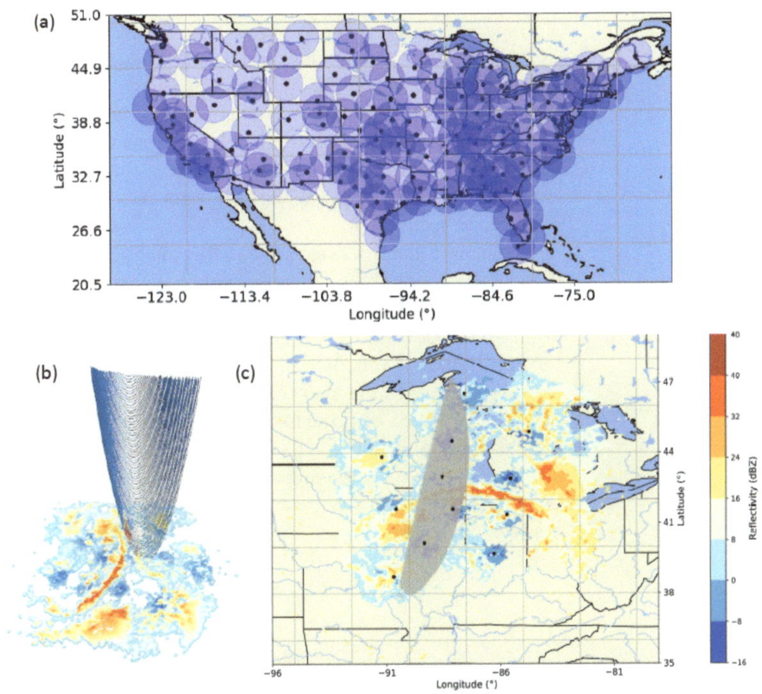

Figure 1. In panel (**a**), the distribution of NEXRAD radar across the continental United States is shown. Black points indicate radar locations, with the blue areas illustrating the approximate range of the radars. The panel (**b**) showcases a particular colocated observation of PAZ and NEXRAD in 3D, while panel (**c**) shows the same observation in a 2D image. Only the portion of the rays below 20 km is shown. The gray region in panel (**c**) represents the 2D projection of the PRO rays, and the coloured map is a Plan Position Indicator (PPI) of the reflectivity measured by the selected radars for that observation.

2.4. Calculation of K_{dp} and $\Delta\Phi$ from NEXRAD

To obtain vertical profiles of the observable $\Delta\Phi$ using NEXRAD data, it is essential to calculate the variable K_{dp} since it is not directly provided in the NEXRAD Level II files. To accomplish this task and process the radar files, we use the Py-Art python module [16]. Once the radars for each observation are selected, the variable K_{dp} is calculated for each one of them. After evaluating various algorithms, we have found that the method described in [17,18] suits our purposes best, providing appropriate values of K_{dp}.

The algorithm in question, as well as the majority of algorithms subjected to testing, have been explained and compared in a prior publication, as documented in [19]. This specific algorithm that we use consists on a four-step process for retrieving the K_{dp}, and also allows us to adjust certain input parameters. These parameters are the number of iterations of the four step process, the dimensions of the sliding window employed in the smoothing, and a pre-filter procedure applied to the variable representing the total differential phase shift (Ψ_{dp}), defined as:

$$\Psi_{dp} = \Phi_{dp} + \delta_{hv}, \tag{5}$$

being Φ_{dp} the differential phase shift measured by the radar and δ_{hv} the differential backscatter phase shift.

Regarding the number of iterations, it has been empirically ascertained that variations therein do not yield a discernible impact on the final value of K_{dp}; consequently, the default value of 10 iterations has been retained. The window size is defined as the extent of the sliding window employed for smoothing the Ψ_{dp} before the K_{dp} calculation. As outlined in [19], smaller window sizes are better suited for scenarios featuring substantial values and steep gradients in K_{dp}, whereas larger window sizes are more appropriate for those characterized by gradual gradients in such variables. In the present study, it has been observed that the final $\Delta\Phi$ profile undergoes significant variations in response to alterations in this parameter. To establish an optimal fit for the outcomes, a comprehensive analysis of the window size has been undertaken, with detailed findings presented in the subsequent section.

The pre-filter process applied to the raw Ψ_{dp} values encompasses several steps, namely: the exclusion of values characterized by a correlation coefficient (ρ_{hv}) below a specified threshold, herein set at 0.65; the identification and removal of portions marked by pronounced discontinuities; the elimination of exceedingly brief sequences of valid data; and the application of a median filter to each profile, as documented in [16]. The K_{dp} obtained from this algorithm is also corrected in terms of the elevation angle. This correction is essentially carried out due to the differences in the geometry of both techniques that we aim to compare. While in NEXRAD the signals exhibit a difference in the angle of incidence as the elevation emission angle from the radar changes, in PRO, the signals are practically tangential to the surface, meaning that the corresponding elevation angle is very close to 0 degrees. The correction is expressed by the following expression [20]:

$$K_{dp}(0) \approx \frac{K_{dp}(\theta)}{cos^2(\theta)}, \tag{6}$$

where θ is the elevation angle, $K_{dp}(0)$ is the specific differential phase shift at an elevation angle of $\theta = 0^\circ$, and $K_{dp}(\theta)$ is the specific differential phase shift at an elevation angle θ.

The next step is to map the K_{dp} values obtained from NEXRAD to the PRO rays. To do so, we perform the interpolation with each of the selected radars for each PRO event, and calculate the mean of total overlapping interpolations. Through this process, we obtain the K_{dp} values corresponding to the radars for each PRO ray, measured in units of degrees per kilometer. Integrating the K_{dp} from NEXRAD along each PRO ray directly yields the $\Delta\Phi$ at the S-band for each of the rays, which are associated with a specific tangential height h_t. Finally, two conversions are performed to obtain the final profile: one to convert the units from degrees to millimeters and the other to transition from S-band to L-band. Considering that both bands are within the Rayleigh regime, the approximation made is a conversion factor [21]:

$$\Delta\Phi_L(mm) = \Delta\Phi_S(degrees) \cdot \frac{\lambda_S}{\lambda_L} \cdot \frac{\lambda_L}{360}, \tag{7}$$

where λ represents the wavelength, and the subscripts S and L refer to the frequency bands. The first conversion term accounts for the conversion from S to L band, while the second term refers to the conversion from degrees to millimeters.

A smoothing process is also applied to the resulting vertical profile of $\Delta\Phi$, which uses a sliding window of five adjacent elements, with the aim of mitigating noise and accentuating the general trends within the profiles. Figure 2 shows an scheme of the steps followed in order to obtain the $\Delta\Phi_{NEX}$.

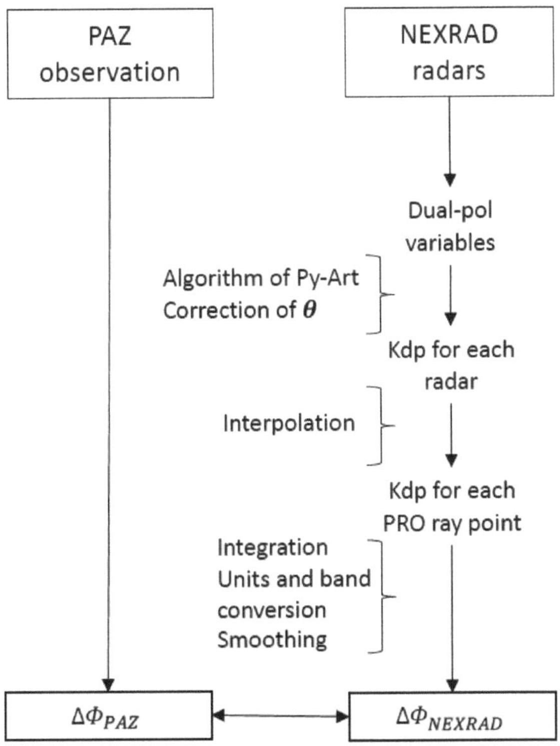

Figure 2. Diagram showing the steps followed to obtain the vertical profiles of $\Delta\Phi$ from NEXRAD.

3. Results and Discussions

3.1. Sensitivity of $\Delta\Phi_{NEX}$ to Window Size

To ascertain the optimal window size that suited our specific cases, we systematically assessed three distinct window dimensions, specifically, 4, 6, and 8. This parameter is expressed in units of number of gates, thus meaning that 4, 6, and 8 represent the number of gates we are considering for a specific window. Here, gate refers to a longitudinal element within the radar beam, representing a discrete volume of space. The length of one gate depends on the configuration at which the radar is operating. However, typically, the value of one gate is around 0.25 km. The window size has to be an even number according to the algorithm specifications that we are using for retrieving K_{dp}. The three different window sizes were employed to perform the analysis in which we compute the vertical profiles of $\Delta\Phi_{NEX}$ for each PAZ observation within our subset of coincidences. We have started by evaluating the correlation coefficient between the PAZ observations and the profiles derived from NEXRAD for these three window sizes, thereby facilitating the identification of the most appropriate window size for a particular observation.

The computation of the correlation coefficient was restricted to the altitude range spanning from 3.5 km to 12 km to avoid low-altitude ambiguities in the PAZ retrievals [6]. On the other hand, an upper boundary was established as a safeguard measurement, prompted by the fact that when we reach the altitude at which no targets are detected the values of the NEXRAD profiles asymptotically approach zero, whereas the PAZ data continues to exhibit small fluctuations due to noise. This upper threshold was imposed to ensure that this disparity does not influence the correlation coefficient.

In an analogue way, observations where we do not have precipitation are characterized by a vertical profile of $\Delta\Phi_{NEX}$ that is completely or almost completely zero, and a vertical profile of $\Delta\Phi_{PAZ}$ that contains some noise. For this reason, the correlation coefficient that we will have associated with non-rain events will be practically zero. Therefore, in Figure 3 we present the correlation coefficient values computed exclusively for cases that involve rain events. This selection process encompasses isolating cases from the dataset of PAZ NEXRAD coincidences based on specific criteria: we considered cases where the mean $\Delta\Phi$ value from PAZ, between 0 and 10km, exceeded 1.5 mm and where radar coverage of the PRO area exceeded 60%. We have also discarded four cases where the $\Delta\Phi_{NEX}$ had unrealistic values.

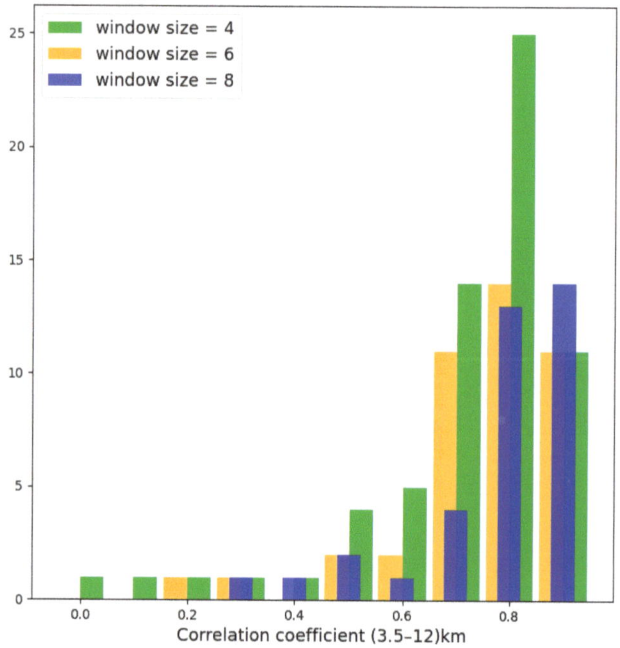

Figure 3. Histogram showing the correlation coefficient for the cases considered as rainy events (see text), for the different window sizes represented in different colors (as indicated in the legend).

The analysis reveals that, except for two (three) cases when employing a window size of 4 (6) gates, most PAZ observations demonstrate a notable concordance with the corresponding profiles of $\Delta\Phi_{NEX}$. These three cases that exhibit a lack of correlation in the $\Delta\Phi$ profiles appear to manifest anomalous data retrieval issues within the NEXRAD radar system. We also have a larger amount of cases that fit the previous restrictions for the window size of four gates.

While the correlation coefficient provides an indication on the agreement about the shape of the vertical profiles, computing the mean values of the $\Delta\Phi$ between two heights provide further indication on the agreement between the magnitudes. The mean $\Delta\Phi$ between 2 km and 8 km for PAZ and NEXRAD has been computed for the profiles obtained with the three window sizes, and the results of the comparison between the two are shown in Figure 4. When no or little precipitation is observed (small values of $\Delta\Phi_{PAZ}$), it can be seen how there is larger dispersion in the $\Delta\Phi_{NEX}$ when the window size is smaller. This was the same result as obtained in [19], where they have seen that smaller window sizes performed better for larger values of K_{dp}, while larger window sizes did for smaller values of K_{dp}. This is mainly because smaller values of K_{dp} are typically associated to the presence of small raindrops or drizzle, and therefore, employing a larger window size could help reduce excessive noise and average the small-scale variations and provide a smoother representation. Whereas when measuring higher values of K_{dp}, there may be more pronounced variations over shorter distances so a smaller window size might be preferred in order to capture these variations at a finer spatial scale.

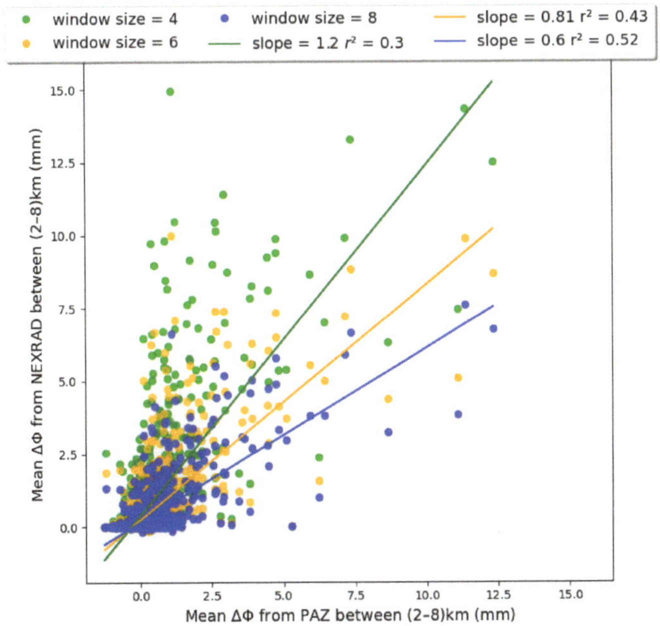

Figure 4. Mean ΔΦ (mm) between 2 km and 8 km for both NEXRAD and PAZ and for three window sizes. The slope and the coefficient of determination are also displayed in the legend. The colocated observations presented here are the ones where the radar coverage exceeds 60%.

Similarly, it is observed that for larger values of $\Delta\Phi_{PAZ}$, there exists an inverse proportionality with respect to the window size on the $\Delta\Phi_{NEX}$, where smaller window sizes are associated with larger values of $\Delta\Phi_{NEX}$. This is expected due to the fact that smaller window sizes in the smoothing of the Ψ_{dp} observations allow for rapidly varying regions to be inverted to K_{dp}, which could correspond to noisy observations or areas of precipitation.

All in all, the results in Figure 4 indicate that, for the majority of cases, the $\Delta\Phi_{PAZ}$ and $\Delta\Phi_{NEX}$ agree better for small window sizes in large ΔΦ regimes, while the agreement is better using a larger window size under low ΔΦ conditions. This suggests that an adaptative window size to specific precipitation regimes would be a good idea, as has been previously suggested in [22]. This, however, is not implemented in the Py-Art K_{dp} retrievals and is out of the scope of this work to perform such modifications.

Continuing with the analysis, we have also examined situations in which there is no precipitation. Since the correlation coefficient would not be a representative value of the agreement between both profiles, based on what we have already commented, we show the histograms present in Figure 5a,b. In Figure 5a, we have selected, for various window sizes, those cases in which the mean of $\Delta\Phi_{NEX}$ below 10 km is less than 1 mm. For these observations, we have plotted the histogram of corresponding values of the mean $\Delta\Phi_{PAZ}$ below 10 km and calculated the percentage of these observations with a $\Delta\Phi_{PAZ} \leq 1$ mm. The analogue has been represented in Figure 5b where the values of the mean $\Delta\Phi_{NEX}$ are represented for those observations with a mean $\Delta\Phi_{PAZ} \leq 1$ mm. From the results depicted in Figure 5a,b, it is observed that for both figures and various window sizes, over 85% of the cases exhibit $\Delta\Phi$ values below 1 mm. This agrees with the fact that the noise values considered for this observable are less than 1.5 mm as stated in [5,6]. This implies that for most observations where PAZ does not detect rainfall, the NEXRAD radars do not detect precipitation either.

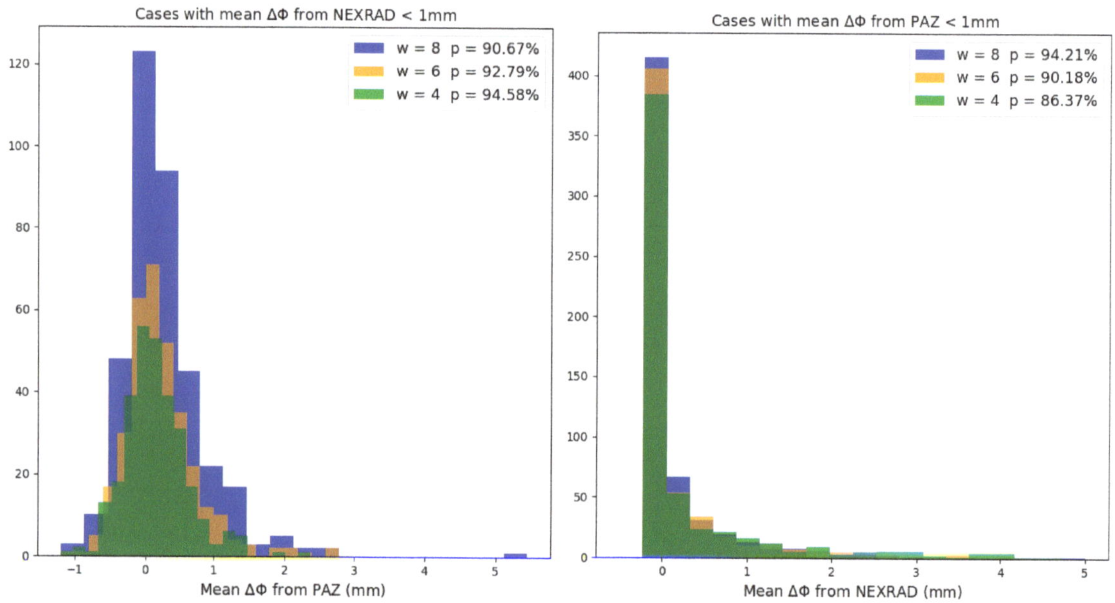

(a) (b)

Figure 5. Mean $\Delta\Phi_{PAZ}$ between 0–10 km (**a**) for those observations where the mean $\Delta\Phi_{NEX}$ between 0–10 km is below 1 mm, and $\Delta\Phi_{NEX}$ between 0–10 km (**b**) for those observations where the mean of $\Delta\Phi_{PAZ}$ between 0–10 km is below 1 mm. For each of the window sizes we have calculated the percentage of the cases that have the mean $\Delta\Phi_{PAZ}$ between 0–10 km below 1 mm (**a**) and $\Delta\Phi_{NEX}$ between 0–10 km below 1 mm (**b**).

Table 1 shows the number of colocated observations that we have depending on the restrictions that we apply.

Table 1. Number of cases employed in the analysis depending on the restrictions applied. The $\langle\Delta\Phi\rangle$ represents the mean $\Delta\Phi$ between 0–10 km and cc is the correlation coefficient between the profiles. In some cases the number of observations will depend on the window size, such as $\langle\Delta\Phi_{NEX}\rangle$, so for this tables the number of cases correspond to the window size of four gates.

	Number of Cases
Total	3208
≥60% area covered	1076
$\langle\Delta\Phi_{PAZ}\rangle \geq 1$ mm	221
$\langle\Delta\Phi_{PAZ}\rangle \geq 1.5$ mm	117
≥60% area covered and $\langle\Delta\Phi_{PAZ}\rangle \geq 1.5$ mm	51
cc ≥ 0.6	290
cc ≥ 0.8	127
$\langle\Delta\Phi_{PAZ}\rangle \leq 1$ mm	2987
$\langle\Delta\Phi_{NEX}\rangle \leq 1$ mm	2898
$\langle\Delta\Phi_{PAZ}\rangle \leq 1$ mm and $\langle\Delta\Phi_{NEX}\rangle \leq 1$ mm	2818

We should also take into consideration that we are using the radar data that is temporally closest to the observations. Radar data are generated approximately every 8 min, and the time we consider as the PAZ observation time is the start of the occultation, which lasts for approximately 2 min. Hence, in situations involving rapidly evolving precipitation events, there may be an associated error that warrants careful consideration. Furthermore, it is possible that there are situations in which a significant portion of the precipitation is not covered by the radar observations. The selection of the 60% area coverage threshold was made with the aim of ensuring an adequate number of cases for statistical analysis while also ensuring that a substantial portion of the occultation area is covered.

3.2. Illustrative Examples of Vertical Profiles of $\Delta\Phi$

The main aim of this investigation is to assess the agreement between vertical profiles of $\Delta\Phi$ obtained from PAZ and from NEXRAD. As demonstrated in Figures 3 and 4, the agreement for well-covered, precipitating cases is good. Furthermore, in this section we scrutinize three specific cases of coincident observations between PAZ and NEXRAD, where PRO rays intersect precipitation regions (see Figures 6–8). Through a meticulous side-by-side comparison, our goal is to identify significant similarities or differences in the $\Delta\Phi$ measurements obtained by both instruments during precipitation events. In addition, we present a statistical analysis of the difference between $\Delta\Phi_{PAZ}$ and $\Delta\Phi_{NEXRAD}$ to provide a broader and more comprehensive perspective on the results.

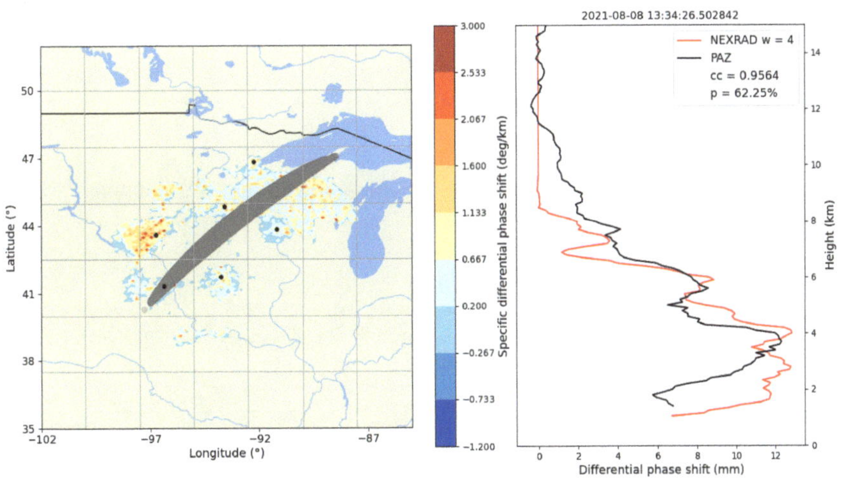

Figure 6. Same as in Figure 7, but corresponding to PAZ profile ID PAZ1.2021.220.13.34.G01.

Upon examination of these profiles, a noteworthy degree of similarity becomes evident. The absence of values at lower altitudes is intentional to mitigate ambiguities in the PAZ profiles at those heights. Nevertheless, in all cases, both the shape and magnitude of the profiles exhibit a high degree of concordance. While most $\Delta\Phi$ peaks are consistently represented in both profiles, they may not align precisely in terms of altitude. Discrepancies in the profiles may be attributed, in part, to factors such as instrumental characteristics, variations in measurement methodology, time differences in the case of rapidly evolving precipitation, and retrieval errors.

Moreover, a substantial presence of hydrometeors at lower altitudes (around 2–4 km, as observed in Figures 7 and 6) corresponds to a higher level of agreement between PAZ and NEXRAD profiles, in terms of the shape of the profile. Conversely, for peaks situated at higher altitudes (approximately 5–7 km, as is seen in the case shown in Figure 8), the agreement diminishes. This reduction may be attributed to the potential influence of mixed-phase hydrometeors around these altitudes and also to the presence of smaller particles that are more difficult to be sensed by PAZ because of the lower frequency employed. A window size of 4 gates was employed for the three cases presented here. However, while our analysis indicates that a window size of 4 gates yields a superior correlation coefficient for the majority of cases, observations exist where an alternative window size demonstrates a better fit. This variation emphasizes the importance of customizing the analysis to specific cases in order to ensure the most accurate results when calculating K_{dp}.

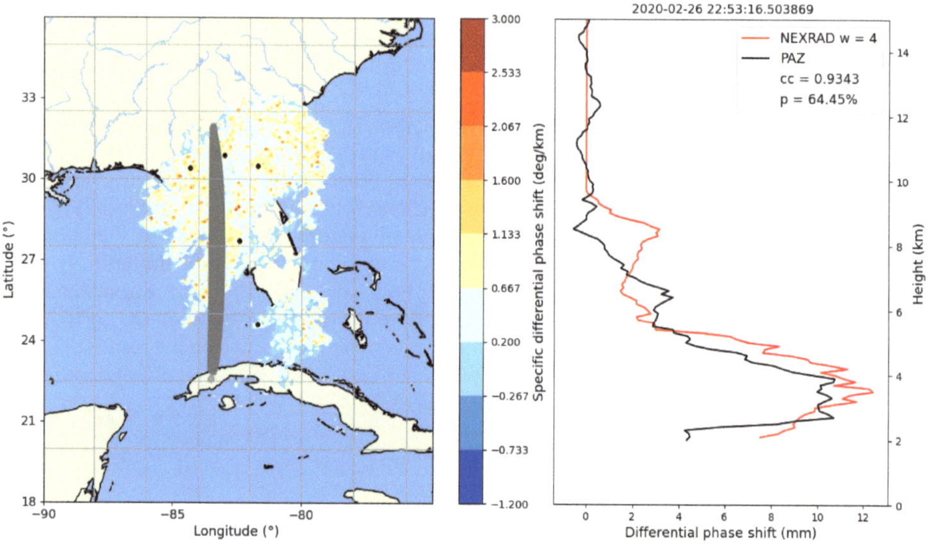

Figure 7. PAZ observation (ID: PAZ1.2020.057.22.53.G09) colocated with NEXRAD radars and the associated vertical profiles for both instruments. The left panel shows the K_{dp} composite field captured by the radars (black points indicate radar locations) with the area of the projection on the surface of the portion of PRO rays below 20 km, in grey. Right panel shows the corresponding vertical profiles of $\Delta\Phi$ as obtained using NEXRAD data (red) and PAZ (black). In the legend we also present the corresponding values of window size (w), the correlation coefficient (cc) and the percentage of the area covered by the radars (p).

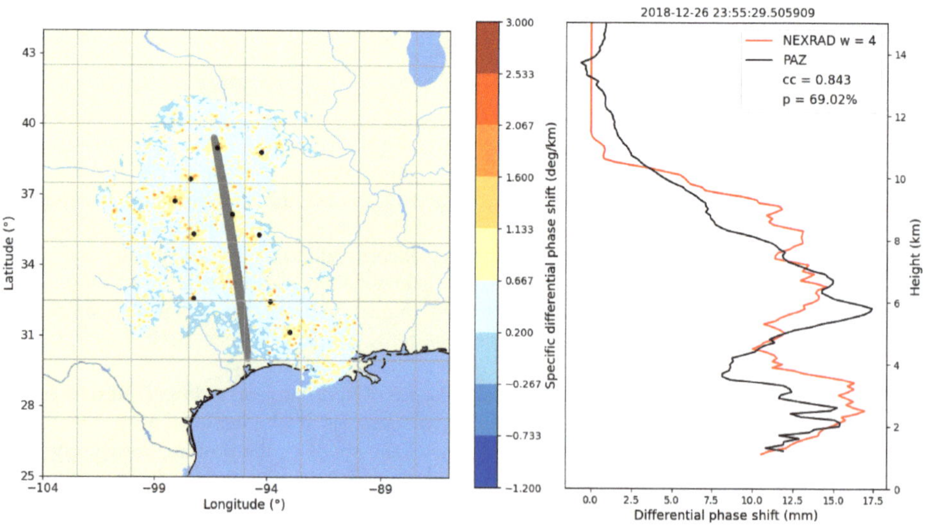

Figure 8. Same as in Figure 7, but corresponding to PAZ profile ID PAZ1.2018.360.23.55.G11.

3.3. Statistical Analysis of $\Delta\Phi$ Differences

From a broader perspective, statistics on the difference between $\Delta\Phi_{PAZ}$ and $\Delta\Phi_{NEX}$ for all cases fulfilling the coverage condition (i.e., coverage > 60%) have been computed. The outcomes of this analysis are depicted in Figure 9. Each column in the figure corresponds to a different window size, while each row imposes a distinct constraint on the mean $\Delta\Phi_{PAZ}$ within the altitude range of 0–10 km. The first row illustrates scenarios where $\langle\Delta\Phi_{PAZ}\rangle \geq 4$ mm (associated to heavy precipitation regime), the second row comprises observations with $\langle\Delta\Phi_{PAZ}\rangle \leq 4$ mm (linked to lower precipitation regimes), and the third row includes colocated cases in which PAZ has not detected precipitation, denoted by -0.5 mm $\leq \langle\Delta\Phi_{PAZ}\rangle \leq 0.5$ mm.

In Figure 9a–c, corresponding to the heavier precipitation regime, the discernible trend reveals an increasing positivity in the median difference between $\Delta\Phi$ from PAZ and NEXRAD with the increase of the window size. This observation aligns with expectations, as commented earlier, where the values of $\Delta\Phi_{NEX}$ exhibit a diminishing trend with an increasing window size. Notably, the optimal window size appears to be for 6 gates, a choice supported to some extent by favorable outcomes in Figures 3 and 4 when employing this size for the moving window. Nevertheless, at altitudes spanning 6–8 km, the window size of 4 gates seems to be a potentially better choice, again revealing discrepancies in regions characterized by a higher concentration of mixed-phase hydrometeors and possibly by smaller hydrometeors.

For Figure 9d–f, corresponding to lighter precipitation than the previous case, a similar pattern is observed, with the difference between PAZ and NEXRAD exhibiting an increasing positivity correlated with the window size. However, here the optimal window size is inferred to be 8 gates, a fact consistent with the previous comments that a larger window size is preferable for scenarios with lower values of K_{dp}.

Figure 9g–i depict the statistics for cases with no or very little precipitation, where the difference remains practically invariant across the different window sizes. For this cases, where K_{dp} values from NEXRAD are expected to be near 0, the result of the difference between $\Delta\Phi_{PAZ}$ and $\Delta\Phi_{NEX}$ is expected to yield values consistent with the PAZ $\Delta\Phi$ noise. The results presented here resemble those obtained in [6], with small standard deviations of $\Delta\Phi$ increasing with decreasing altitude.

In summary, for instances where precipitation has been identified, the importance of the selected window size becomes relevant in ensuring accurate determinations of K_{dp}. Throughout this analysis, the same window size has been applied uniformly across all radars surrounding an observation. Nevertheless, it warrants consideration to investigate whether better results could be achieved by adapting the window size based on the specific precipitation characteristics encountered along each radar ray. Such an approach may offer valuable insights into optimizing the accuracy and relevance of K_{dp} estimations.

3.4. Echotop Height Comparison

A final study of the concordance between $\Delta\Phi_{NEX}$ and $\Delta\Phi_{PAZ}$ has been performed, consisting on comparing the echotop height values extracted from both datasets, as depicted in Figure 10. The determination of NEXRAD's echotop consists on the interpolation of reflectivity data from the radars onto the PRO rays, analogous to the approach applied to K_{dp}. The following thresholds were applied to the reflectivity values: $Z \geq 20$ dBZ and $Z \geq 10$ dBZ. By identifying the maximum height associated with these thresholds, we then select the associated tangential height of the ray that actually crosses that point, instead of the actual height. By doing this, we are not obtaining the "real" echotop height but the "projected" one to the tangent point, and this is done to ensure a fair comparison between the two observational techniques.

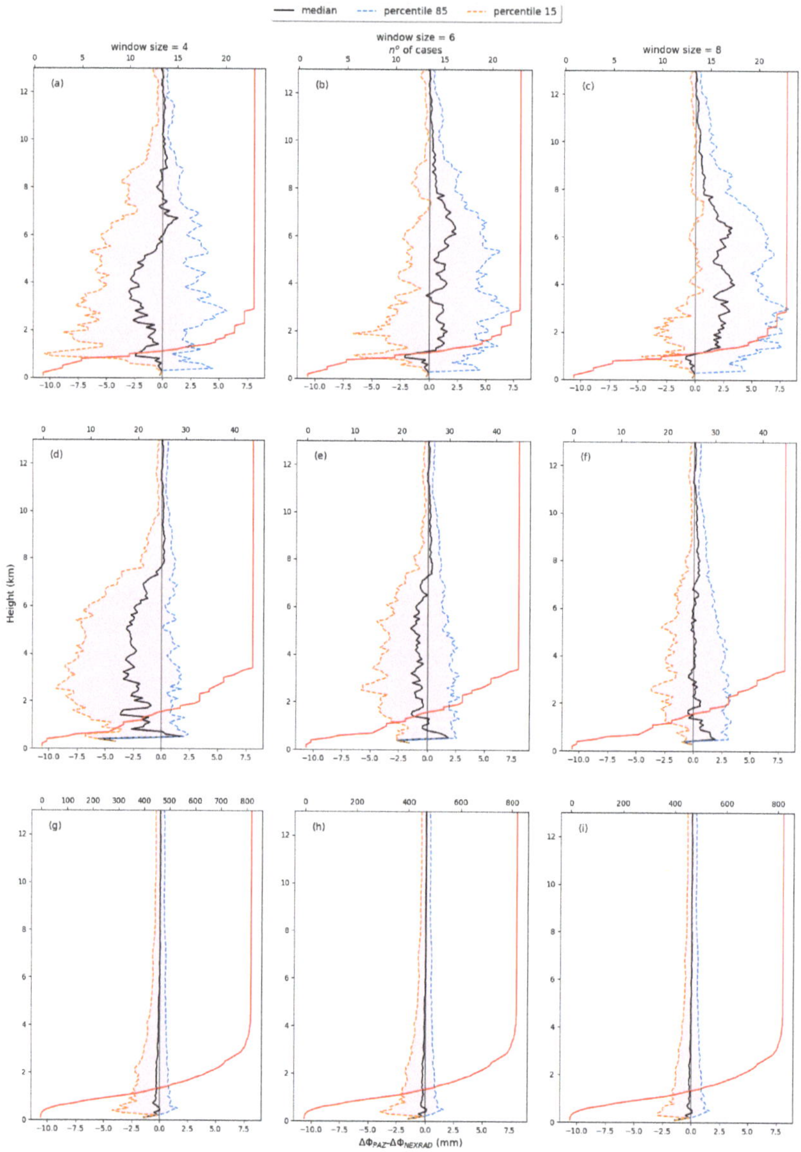

Figure 9. Difference between $\Delta\Phi_{PAZ}$ and $\Delta\Phi_{NEX}$ for those cases that are covered more than 60% by the radars. Each column represents a different window size ((**a**,**d**,**g**) represent a window size = 4, etc.), while each row represents a different condition for the mean $\Delta\Phi_{PAZ}$ between 0–10 km. For the first row, the cases represented are the ones were PAZ has detected rain and $\langle\Delta\Phi_{PAZ}\rangle$ is larger than 4 mm. In second row, we represent as well cases where PAZ has detected precipitation but $\langle\Delta\Phi_{PAZ}\rangle$ is lower than 4mm. The third row represents those cases where PAZ has not detected precipitation, this means that $\langle\Delta\Phi_{PAZ}\rangle$ is between $\pm$0.5mm. For each figure the number of valid points for each altitude is represented by a red line (top x-axis).

The echotop equivalent for PAZ observations was defined as the highest point where $\Delta\Phi_{PAZ}$ surpasses a threshold for five consecutive measurements. This threshold was calculated by taking the mean of $\Delta\Phi_{PAZ}$ values above 20 km (where neither rain nor clouds are anticipated) and adding four times the standard deviation. This rigorous criterion ensures that if surpassed, the $\Delta\Phi_{PAZ}$ signature unequivocally originates from precipitation or cloud-related effects.

Figure 10 shows roughly a linear relationship between the two, with exceptions attributable to cases where precipitation was not detected. This can be appreciated by the values of the slope for the linear regression, being the one corresponding to $Z \geq 10$ dBZ the closest to one. For the case where $Z \geq 10$ dBZ, it is evident that "echotop" heights recorded by PAZ are lower than those of NEXRAD. This suggests that PAZ underestimates the echotop height compared to NEXRAD, registering lower upper boundaries of the precipitation layers. These differences could be attributed to factors such as respective frequencies employed by each instrument for detection, being NEXRAD more sensitive to smaller particles than PAZ. This may indicate that the specific hydrometeors associated with such thresholds are smaller than what PAZ can effectively detect. Besides, the instruments could be more sensitive to different atmospheric layers. NEXRAD is more sensitive to smaller particles, so they might be detecting features in a higher atmospheric layer than PAZ.

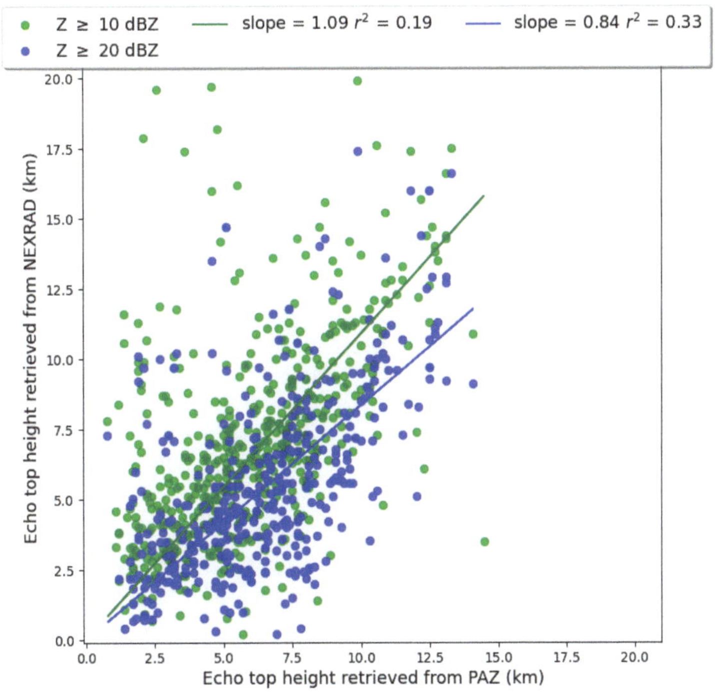

Figure 10. Values of the echotop height obtained from NEXRAD and PAZ datasets for two different thresholds, $Z \geq 20$ dBZ and $Z \geq 10$ dBZ.

4. Conclusions

In this study, we have selected colocated observations from the PAZ satellite and NEXRAD ground-based radars in order to validate the vertical structure of the PRO $\Delta\Phi$ observable. Such validation is achieved by using NEXRAD to obtain profiles of differential phase shift. These profiles were then subjected to a comparative analysis against their analogues retrieved from PAZ. Furthermore, an investigation focused on optimizing the smoothing window parameter for calculating the K_{dp} variable was conducted with the aim of achieving the best fit between the profiles. Subsequently, both the distinctions and similarities within these profiles were discussed, and some properties were subject to individual examination. It has been shown that the agreement holds for both the shape and magnitude of the observable. Moreover, statistical comparison using selected profiles grouped by different precipitation regimes exhibit mean differences and dispersion consistent with the used window size; that is, agreement increases with lower window size when considering heavier precipitation, whereas the agreement is better using larger window size when precipitation is lighter. Also, for the non-rainy cases, the are no significant biases, and the dispersion agrees with that reported in previous studies.

The good agreement observed between the vertical profiles obtained from both platforms show the potential of PAZ PRO observations in characterizing, to some extent the vertical structure of heavy precipitation events. This underscores the potential for multi-platform validation of precipitation measurements, particularly in regions with limited ground-based radar coverage. Besides, it makes clear the PAZ mission's capabilities to contribute to heavy precipitation research, emphasizing the method's importance as a powerful tool for further enhancing the understanding and characterization of precipitation events and their associated thermodynamics.

Notably, we have identified remarkable consistency in the detection altitudes of hydrometeors between PAZ and NEXRAD, with numerous observations showing that both instruments identify hydrometeors at similar altitudes. However, some differences were observed under specific meteorological conditions, particularly around altitudes where the presence of small frozen particles is more pronounced. Due to the higher frequency employed by NEXRAD compared to PAZ, the radars are more sensitive to such small particles. This sensitivity leads to variations in detection altitudes for specific hydrometeor types. Nevertheless, PAZ observations demonstrate increased sensitivity to particles such as snow. This is thought to be due to the geometry employed by the technique. Therefore, the combination of data from PAZ and NEXRAD could enhance our ability to interpret radar data in regions experiencing mixed-phase precipitation.

It is imperative to acknowledge that certain observed disparities may be attributed to factors such as instrumental characteristics and variations in measurement methodologies, as well as retrieval errors. Consequently, a prudent consideration of these limitations is essential when drawing conclusions from this study. Besides, radar-based data are subject to some uncertainties such as anomalous propagation, partial beam filling, beam overshooting and spatio-temporal resolution, among others.

Also, it is worth mentioning that we encountered certain challenges in processing NEXRAD Level II data. Two of the algorithms mentioned in [19] were impractical to apply due to their computational cost, and another provided K_{dp} values lacking physical significance. Regarding the window size, existing literature on the optimal choices based on radar-derived variable values was limited, as was the guidance for selecting a specific window size for a particular case study. Since the K_{dp} is a polarimetric variable independent of attenuation and miscalibration, it is a valuable variable for many analyses, so in regard to the different processes to calculate K_{dp}, the data obtained from PAZ could prove useful in refining this procedure.

In summary, this research solidifies the PRO technique as an instrument for quantifying heavy precipitation events. It emphasizes the role played by satellite-based systems, such as PAZ, in advancing our comprehension of hydrometeor detection and characterizing the microphysical properties of the atmosphere, especially in remote regions experiencing complex weather phenomena.

As for future work, there is potential to adapt the K_{dp} algorithm to consider variations in window size based on the polarimetric and non-polarimetric variables measured by the radar. We have seen in this study that some precipitation characteristics could be taken into consideration when selecting the proper window size. Additionally, exploring algorithms that classify hydrometeors could be valuable for a more in-depth study of areas with mixed-phase hydrometeors, providing insights into the sensitivity of $\Delta\Phi$ to these conditions.

Author Contributions: Conceptualization, A.P., R.P. and E.C.; methodology, A.P.; software, A.P.; validation, A.P., R.P. and E.C.; formal analysis, A.P.; investigation, A.P, R.P. and E.C.; resources, E.C.;

data curation, A.P.; writing—original draft preparation, A.P.; writing—review and editing, A.P., R.P. and E.C.; visualization, A.P.; supervision, R.P. and E.C.; project administration, E.C.; funding acquisition, E.C. All authors have read and agreed to the published version of the manuscript.

Funding: The ROHP-PAZ project is part of the Grant PID2021-126436OB-C22 funded by the Spanish Ministry of Science and Innovation MCIN/AEI/10.13039/501100011033 and by "ERDF A way of making Europe" of the "European Union". R.P is funded by the grant RYC2021-033309-I, included in the «NextGenerationEU»/PRTR. This work was also partially supported by the program Unidad de Excelencia María de Maeztu CEX2020-001058-M. Part of the investigations are done under the EUMETSAT ROM SAF CDOP4.

Data Availability Statement: Data from PAZ is available at: https://paz.ice.csic.es/ [12] (accessed on 18 March 24). Level II data from NEXRAD is available at: https://s3.amazonaws.com/noaa-nexrad-level2/index.html [15] (accessed on 18 March 24).

Acknowledgments: The authors would like to thank the developers of Py-Art for creating and making available this Python module to easily analyze data obtained from weather radars. We express our gratitude to three anonymous reviewers whose comments contributed to the improvement of the manuscript.

Conflicts of Interest: The authors declare no competing interests.

Abbreviations

The following abbreviations are used in this manuscript:

GNSS	Global Navigation Satellite Systems
PRO	Polarimetric Radio Occultation
NEXRAD	Next Generation Weather Radars
ROHP	Radio Occultation and Heavy Precipitation
ICE-CSIC	Institut de Ciéncies de l'Espai-Consejo Superior de Investigaciones Científicas
IEEC	Institut d'Estudis Espacials de Catalunya
NOAA	National Oceanic and Atmospheric Administration
UCAR	University Corporation for Atmospheric Research
NASA	National Aeronautics and Space Administration
GPS	Global Positioning System
LEO	Low Earth Orbit
IGOR+	Integrated GPS Occultation Receiver
NWP	Numerical Weather Prediction
GPM	Global Precipitation Mission
GMI	GPM Microwave Imager
PPI	Plan Position Indicator
PSD	Particle Size Distribution
IMERG	Integrates Multi-satellitE Retrievals for GPM
DPR	Dual frequency Precipitation Radar
NWS	National Weather Service
QC	Quality Control

References

1. Fjeldbo, G.; Eshleman, V.R. The Bistatic Radar-Occultation Method for the Study of Planetary Atmospheres. *J. Geophys. Res.* **1965**, *70*, 3217–3225. [CrossRef]
2. Kursinski, E.R.; Hajj, G.A.; Schofield, J.T.; Linfield, R.P.; Hardy, K.R. Observing Earth's atmopshere with radio occultation measurements using the Global Positioning System. *J. Geophys. Res. Atmos.* **1997**, *102*, 23429–23465. [CrossRef]
3. Healy, S.B.; Jupp, A.M.; Marquardt, C. Forecast impact experiment with GPS radio occultation measurements. *Geophys. Res. Lett.* **2005**, *32*, L3804. [CrossRef]
4. Cardellach, E.; Tomás, S.; Oliveras, S.; Padullés, R.; Rius, A.; de la Torre Juárez, M.; Turk, F.; Ao, C.; Kursinski, E.R.; Schreiner, B.; et al. Sensitivity of PAZ LEO Polarimetric GNSS Radio-Occultation Experiment to Precipitation Events. *IEEE Trans. Geosci. Remote Sens.* **2014**, *53*, 190–206. [CrossRef]
5. Cardellach, E.; Oliveras, S.; Rius, A.; Tomás, S.; Ao, C.; Franklin, G.W.; Iijima, B.A.; Kuang, D.; Meehan, T.K.; Padullés, R.; et al. Sensing Heavy Precipitation With GNSS Polarimetric Radio Occultations. *Geophys. Res. Lett.* **2019**, *46*, 1024–1031. [CrossRef]

6. Padullés, R.; Ao, C.; Turk, F.; de la Torre Juárez, M.; Iijima, B.; Wang, K.N.; Cardellach, E. Calibration and validation of the Polarimetric Radio Occultation and Heavy Precipitation experiment aboard the PAZ satellite. *Atmos. Meas. Tech.* **2020**, *13*, 1299–1313. [CrossRef]
7. Turk, F.; Padullés, R.; Cardellach, E.; Ao, C.; Wang, K.N.; Morabito, D.D.; de la Torre Juárez, M.; Oyola, M.; Hristova-Veleva, S.; Neelin, J.D. Interpretation of the Precipitation Structure Contained in Polarimetric Radio Occultation Profiles Using Passive Microwave Satellite Observations. *J. Atmos. Ocean. Technol.* **2021**, *38*, 1727–1745. [CrossRef]
8. Padullés, R.; Cardellach, E.; Turk, F.; Ao, C.; de la Torre Juárez, M.; Gong, J.; Wu, D.L. Sensing Horizontally Oriented Frozen Particles With Polarimetric Radio Occultations Aboard PAZ: Validation Using GMI Coincident Observations and Cloudsat a Priori Information. *IEEE Trans. Geosci. Remote Sens.* **2021**, *60*, 1–13. [CrossRef]
9. Padullés, R.; Cardellach, E.; Turk, F. On the global relationship between polarimetric radio occultation differential phase shift and ice water conten. *Atmos. Chem. Phys.* **2023**, *23*, 2199–2214. [CrossRef]
10. Wulfmeyer, V.; Hardesty, R.M.; Turner, D.D.; Behrendt, A.; Cadeddu, M.P.; Di Girolamo, P.; Schlüssel, P.; Baelen, J.V.; Zus, F. A review of the remote sensing of lower tropospheric thermodynamic profiles and its indispensable role for the understanding and the simulation of water and energy cycles. *Rev. Geophys.* **2015**, *53*, 819–895. [CrossRef]
11. Tomás, S.; Padullés, R.; Cardellach, E. Separability of systematic effects in polarimetric GNSS radio occultations for precipitation sensing. *IEEE Trans. Geosci. Remote Sens.* **2018**, *56*, 4633–4649. [CrossRef]
12. Padullés, R.; Cardellach, E.; Oliveras, S. resPrf [Dataset]. 2024. Available online: https://digital.csic.es/handle/10261/348253 (accessed on 18 March 24). [CrossRef]
13. Klazura, G.E.; Imy, D.A. A Description of the Initial Set of Analysis Products Aivailable from the NEXRAD WSR-88D System. *Bull. Am. Meteorol. Soc.* **1993**, *74*, 1293–1312. [CrossRef]
14. Crum, T.D.; Saffle, R.E.; Wilson, J.W. An Update on the NEXRAD Program and Future WSR-88D Support to Operations. *Weather Forecast.* **1998**, *13*, 253–262. [CrossRef]
15. NOAA National Weather Service (NWS) Radar Operations Center. *NOAA Next Generation Radar (NEXRAD) Level 2 Base Data*; National Centers for Environmental Information: Asheville, NC, USA, 1991. [CrossRef]
16. Helmus, J.J.; Collis, S.M. The Python ARM Radar Toolkit (Py-Art), a library for working with weather radar data in the Python programmin language. *J. Open Res. Softw.* **2016**, *4*, e25. [CrossRef]
17. Vulpiani, G.; Montopoli, M.; Passeri, L.D.; Gioia, A.G.; Giordano, P.; Marzano, F.S. On the use of dual-polarized C-band radar for operational rainfall retrieval in mountainous areas. *J. Appl. Meteorol. Climatol.* **2012**, *51*, 405–425. [CrossRef]
18. Vulpiani, G.; Baldini, L.; Roberto, N. Characterization of Meditarranean hail-bearing storms using an operational polarimetric X-band radar. *Atmos. Meas. Tech.* **2015**, *8*, 4681–4698. [CrossRef]
19. Reimel, K.J.; Kumjian, M. Evaluation of Kdp Estimation Algorithm Performance in Rain Using a Known-Truth Framework. *J. Atmos. Ocean. Technol.* **2021**, *38*, 587–605. [CrossRef]
20. Griffin, E.M.; Schuur, T.J.; Ryzhkov, A.V. A Polarimetric Analysis of Ice Microphysical Processes in Snow, Using Quasi-Vertical Profiles. *J. Appl. Meteorol. Climatol.* **2018**, *57*, 31–50. [CrossRef]
21. Bringi, V.N.; Chandrasekar, V. *Polarimetric Doppler Weather Radar: Principles and Applications*; Cambridge University Press: Cambridge, UK, 2001.
22. Wang, Y.; Chandrasekar, V. Algorithm for Estimation of the Specific Differential Phase. *J. Atmos. Ocean. Technol.* **2009**, *26*, 2565–2578. [CrossRef]

remote sensing

MDPI

Article

An Ambient Adaptive Global Navigation Satellite System Total Electron Content Predictive Model for Short-Term Rapid Geomagnetic Storm Events

Renato Filjar [1,2,3,*], Ivan Heđi [4], Jasna Prpić-Oršić [5] and Teodor Iliev [3,6]

1 Department of Computer Engineering, Faculty of Engineering, University of Rijeka, 51000 Rijeka, Croatia
2 Laboratory for Data Mining, Open and Big Data, Centre for Artificial Intelligence and Cybersecurity, University of Rijeka, 51000 Rijeka, Croatia
3 Laboratory for Spatial Intelligence, Krapina University of Applied Sciences, 49000 Krapina, Croatia; tiliev@uni-ruse.bg
4 Department of ICT, Virovitica University of Applied Sciences, 33000 Virovitica, Croatia; ivan.hedi@vuv.hr
5 Department of Naval Architecture and Ocean Engineering, Faculty of Engineering, University of Rijeka, 51000 Rijeka, Croatia; jasna.prpic.orsic@riteh.uniri.hr
6 Department of Telecommunications, University of Ruse, 7004 Ruse, Bulgaria
* Correspondence: renato.filjar@gmail.com or renato.filjar@vhzk.hr

Abstract: Satellite navigation is an essential component of the national infrastructure. Space weather and ionospheric conditions are the prime sources of GNSS (global navigation satellite system) positioning, navigation, and timing (PNT) service disruptions and degradations. Protection, toughening, and augmentation (PTA) of GNSS PNT services require novel approaches in ionospheric effects mitigation. Standard global ionospheric correction models fail in the mitigation of high-dynamics and local ionospheric disturbances. Here, we demonstrate that in the case of the short-term fast-developing geomagnetic storm, a machine learning-based environment-aware GNSS ionospheric correction model for sub-equatorial regions may provide a substantial improvement over the existing global Klobuchar model, considered a benchmark. The proposed machine learning-based model utilises just the geomagnetic field density component observations as a predictor to estimate TEC/GNSS ionospheric delay as the prediction model target. Further research is needed to refine the methodology of machine learning model development selection and validation and to establish an architecture-agnostic framework for GNSS PTA development.

Keywords: satellite navigation; GNSS positioning, navigation and timing; machine learning-based environments; TEC/GNSS ionospheric delay

Citation: Filjar, R.; Heđi, I.; Prpić-Oršić, J.; Iliev, T. An Ambient Adaptive Global Navigation Satellite System Total Electron Content Predictive Model for Short-Term Rapid Geomagnetic Storm Events. *Remote Sens.* **2024**, *16*, 3051. https://doi.org/10.3390/rs16163051

Academic Editor: Haobo Li

Received: 11 July 2024
Revised: 4 August 2024
Accepted: 15 August 2024
Published: 19 August 2024

1. Introduction

Standard GNSS (global navigation satellite system) ionospheric delay correction models suffer from shortcomings caused by their global nature and coverage, and constraints of the broadcast model parameters update on a daily basis. In such a manner, the standard GNSS ionospheric correction models, such as the Klobuchar model [1], used for GPS positioning, fail to account for local and sudden ionospheric events. Failure in the characterisation of the actual TEC propagates into GNSS pseudorange measurement errors, resulting in increased GNSS position estimation errors, and the delay affects the growing number of GNSS-based technology and socio-economic applications, as modern civilisation becomes reliant on GNSS positioning, navigation and timing (PNT) services and their guaranteed performance levels [2,3].

The ionospheric delay results from the conditions the satellite radio wave encounters during its propagation through the Earth's ionosphere [4]. The impact propagation process that leads to the formation of the ionospheric delay, and, consequently, the GNSS pseudorange measurement errors and GNSS position estimation errors, was described with the Space weather–GNSS positioning performance coupling model [5].

The analytical expression of the ionospheric delay may be derived from the Appleton-Hartree equation [4]. Derivation yields the relation between the ionospheric delay Δt_{iono} [s] and the vertical ionospheric profile $N(h)$ [electrons/m^3], an analytical model of the free electron density at a given height h above the Earth's mean sea level, as given in (1). Physical constants used in (1) denote, as follows, e, unit electron charge (1.6×10^{-19} C); m_e, unit electron mass (9.1×10^{-31} kg); c, velocity of light in vacuum (2.99792458×10^8 m/s); ε_0 permittivity of vacuum (8.854×10^{-12} F/m); and ω, angular wave frequency in [rad/s] Integration bounds in (1) are determined with the lower h_{lower} and upper h_{upper} boundary of the ionosphere.

$$\Delta t_{iono} = \frac{e^2}{2\epsilon_0 m_e \omega^2 c} \int_{h_{lower}}^{h_{upper}} N(h)dh \quad (1)$$

The introduction of numerical values of physical constants yields a relationship between the ionospheric delay of a radio signal and the vertical ionospheric profile, as expressed with (2) [4], with Δt_{iono} [s] denoting the ionospheric time delay, $N(h)$ [electrons/m^3] denoting the vertical ionospheric profile, h [m] denoting height above the mean sea level, c denoting the velocity of light in vacuum (2.99792458×10^8 m/s) and f denoting radio carrier frequency.

$$\Delta t_{iono} = \frac{40.3}{cf^2} \int_{h_{lower}}^{h_{upper}} N(h)dh \quad (2)$$

Satellite navigation systems operate under the presumption of satellite signal propagation at the velocity of light in vacuum along its path from a satellite aerial to a receiver aerial, a condition that is not met during the passage through the ionosphere and troposphere [6].

Multiplication of both sides of (2) with the velocity of light in vacuum c will yield an equivalent expression, describing the relationship between the error of the measured distance between a satellite and a receiver aerial $\Delta\rho_{iono}$ [m], the so-called pseudorange and the vertical ionospheric profile $N(h)$, as given in (3) [2,7,8].

$$\Delta\rho_{iono} = \frac{40.3}{f^2} \int_{h_{lower}}^{h_{upper}} N(h)dh = \frac{40.3}{f^2} TEC \quad (3)$$

The integral factor in Equations (1) and (3) is known as Total Electron Content (TEC). *TEC*, expressed in [electrons/m^2], denotes the surface density of free electrons encountered by a satellite radio signal traveling along its path. *TEC* takes large values and is commonly expressed in TECU units (1 TECU = 1×10^{16} electrons/m^2). *TEC* results from the ionospheric conditions described with the vertical ionospheric profile *N(h)*, which renders *TEC* the outcome of the ionospheric conditions, rather than the ionospheric descriptor.

The unmet presumption of the satellite radio signal propagation at the constant velocity of light in vacuum is the prime single cause of the satellite positioning error [6]. Ionospheric conditions cause a complex behaviour of the GNSS ionospheric delay, described with a bias and random error components [5]. Quiet space weather, geomagnetic and ionospheric conditions render the bias component of the ionospheric delay dominant, while in disturbed ionospheric conditions, the influence of the random component dominates. Standard ionospheric delay correction models, such as the Klobuchar model for GPS, Beidou and Glonass (CDMA) systems, address the bias component of the GNSS ionospheric delay. This causes minor to considerable problems for the GNSS ionospheric delay prediction, and the resulting GNSS PNT degradation, in times of space weather, geomagnetic and ionospheric disturbances, as shown in Figure 1.

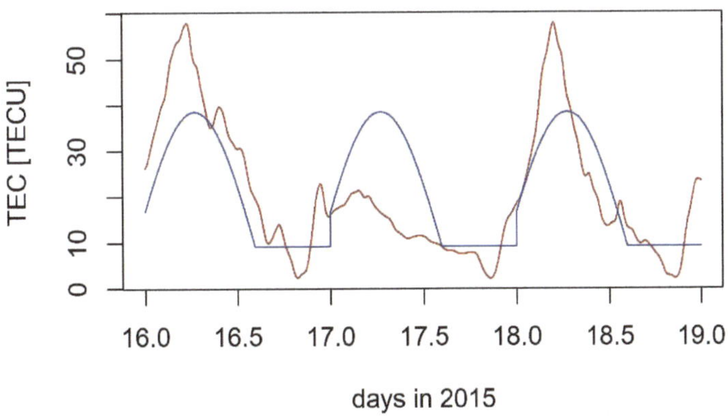

Figure 1. Klobuchar model (blue line) failure in prediction of the *TEC*/GNSS (red line) pseudorange error (St Patrick's Day 2015 geomagnetic storm, at Darwin, NT).

Sudden, localised and short-term geomagnetic and ionospheric disturbances are of particular concern, as such conditions are not described correctly with the standard correction models, which have a global nature and extent; do not consider local disturbances; and have the correction model parameters updated rarely (once a day) [5,6].

Ionospheric delay may be mitigated successfully using simultaneous pseudorange measurements at two different carrier frequencies [6]. The dual-frequency method is commonly applied in specially authorised GNSS positioning processes [2,7,8]. However, the vast majority of GNSS receivers on the market utilise a single-frequency approach. In an application of the reverse-engineering process, the dual-frequency method may be utilised for the determination of *TEC*.

Thus, a GNSS receiver becomes a *TEC* sensor [3,9]. It may be shown that the actual *TEC* encountered on the satellite signal path seen from the receiver perspective as coming at the elevation angle *E* [rad] may be determined using (4) [6], where the related symbols denote the following: *STEC* denotes slant (actually observed) *TEC*, at elevation angle *E*; $\rho(f_1)$ and $\rho(f_2)$ denote simultaneously observed (measured) pseudoranges in [m] at frequencies f_1, and f_2, respectively; b_s denotes satellite bias in [m]; and b_r denotes receiver bias in [m]. Various implementations of the *TEC* estimation procedure deploy different approaches in the estimation of satellite and receiver bias.

$$STEC = \frac{\rho(f_2) - \rho(f_1) - b_s - b_r}{40.31 \cdot \left[\frac{1}{f_2^2} - \frac{1}{f_1^2}\right]} \tag{4}$$

TEC observations should be normalised for satellite signals traveling different paths and distances, passing different segments of the Earth's ionosphere. A mapping function *m*(*E*) was introduced to determine the normalised vertical *TEC* (*VTEC*) [2,6,8], as given in Equations (5) and (7), with R_{Earth} denoting the Earth's radius and *h* denoting height above the mean sea level.

$$STEC = m(E) \cdot VTEC \tag{5}$$

$$m(E) = \frac{1}{\sqrt{1 - \left(\frac{R_{Earth}}{R_{Earth}+h} \cdot cos(E)\right)^2}} \tag{6}$$

$$TEC = \frac{\sqrt{1 - \left(\frac{R_{Earth}}{R_{Earth}+h} \cdot cos(E)\right)^2}}{40.31} \cdot \frac{f_1^2 f_2^2}{f_1^2 - f_2^2} \cdot [\rho(f_2) - \rho(f_1) - b_s - b_r] \tag{7}$$

Recent *TEC* prediction model developments for the purpose of GNSS position estimation improvement were focused on the traditional time-series techniques, with the utilisation of spherical harmonics for GNSS position estimation improvement [10,11]. Ref. [3] proposed the adaptive GNSS-based positioning process, which respects the actual state of the local environment for satellite positioning. Dubbed the ambient adaptive PNT, it exploits the abundance of precise sensors accompanying GNSS receivers, such as those in smartphones, which are capable of the GNSS PNT environment observation, as well as trusted detailed third-party data on the same subject. The adaptiveness to the GNSS PNT environment is based on the situation awareness obtained either using trusted third-party data for the region in question and/or direct measurements of descriptors of the GNSS PNT environment performed at the position of a GNSS receiver. Development of the adaptive GNSS-based positioning process involves the introduction of advanced position estimation methods [3], as well as observations-based and statistical learning-founded [12] prediction correction models. While statistical learning methods have been utilised in space weather research [13], their utilisation in satellite navigation for mitigation of the ionospheric effects is still novel [5,9,14].

Here, we contribute to the subject with a proposal for and a demonstration of a method for an ambient-aware tailored personalised GNSS ionospheric delay correction model development based on observations of the local geomagnetic environment (geomagnetic field density). The research aims at the provision of a reliable and robust GNSS TEC prediction model based on the current observations of the immediate ambient (positioning environment) conditions and utilisation of machine learning methods for the GNSS TEC predictive model development and operation. The proposal targets single-frequency commercial-grade GNSS receivers, a class of GNSS receivers prevailing on the market. Considering its intended cross-disciplinary adoption and self-sustainable personalised deployment, the method and the correction model are anticipated to extend model development and deployment characteristics, such as (i) accuracy and precision in terms of both the bias and the variance, (ii) conceptual simplicity, (iii) fast model development and (iv) high efficiency and low energy consumption for model development and deployment. The model development and deployment methods are to serve the increasing number of GNSS PNT processes implemented in mobile and stationary GNSS PNT applications, including smartphones, autonomous road vehicles, aircraft, vessels and Internet-of-Things (IoT) devices, with a wide range of computational capacity levels and available energy constraints. The proposed GNSS TEC prediction model aims at the provision of an alternative to the standard *TEC* correction models, such as the Klobuchar model, thus becoming an integral component of the GNSS PNT process and algorithm.

Integrated into the GNSS PNT process and algorithm [3], the GNSS pseudorange measurement error/TEC model aims at the provision of adaptiveness to the GNSS ambient (positioning environment) conditions and improved mitigation of the GNSS ionospheric delay, compared with the Klobuchar model set up as the reference (benchmark) model.

2. Methods and Materials

The GNSS ambient conditions in the immediate vicinity of a GNSS signal-collecting mobile unit determine the degradation level of the GNSS PNT performance [6,8,15]. The statement holds for both a traditional GNSS receiver and a mobile unit of a positioning-as-a-service system [3]. The research presented hypotheses that the near-real-time situation awareness of positioning environment conditions may significantly reduce positioning performance degradation due to both natural and artificial adversarial effects. Furthermore, it is argued here that a bespoke GNSS correction model based on the situation awareness of the positioning environment conditions may be developed, maintained and operated by the reception side of the GNSS system. The concept relies on the assumptions of (i) internet-based connectivity; (ii) a mobile unit equipped with appropriate sensing devices, such as magnetometers, to be utilised for the positioning environment condition assessment; and (iii) the computational capacity of mobile units. All three presumptions are fulfilled in mass-market devices, such as smartphones, automobiles or personal computers, and will be in a vast range of Internet-of-Things devices. The proposed method may be considered a valuable contribution to the protection, toughening and augmentation efforts of the core GNSS without the need for expensive and complicated infrastructure development.

The proposed GNSS TEC predictive model is aimed to serve the GNSS community, and those utilising single-frequency GNSS receivers in particular, through harvesting ambient condition awareness. Its purpose is to provide a valuable alternative to the standard GNSS ionospheric correction models by exploiting the sensing, computational and information resources available to a mobile unit (a GNSS receiver) during its operation.

The complexity of space weather, geomagnetic and ionospheric disturbances creates a range of effects on the GNSS PNT performance and its degradation. Statistical properties of variables describing both the ionospheric conditions and the GNSS PNT performance differ significantly in different scenarios of the ionospheric disturbances. Separate assessments of various scenarios of ionospheric disturbances and the GNSS PNT performance degradations are, therefore, required. This research focuses on short-term rapidly developing ionospheric disturbances, one of the extreme scenarios of ionospheric disturbances that causes unexpected, fast and significant GNSS PNT performance degradation.

This section details the proposal of the concept, method and model, as well as material (data) used in practical implementation for a proof-of-principle demonstration.

2.1. TEC/GNSS Ionospheric Delay Prediction Model Development

Statistical learning methods for prediction model development and real-time observations of geomagnetic conditions and GNSS pseudorange measurements are used in the candidate sub-equatorial short-term rapidly developing ionospheric storm *TEC* prediction model. The B_x, B_y and B_z components of the geomagnetic field density vector in [T] are considered predictors of the *TEC* prediction model. The *TEC* experimental values are derived from the raw GPS pseudorange observations, using the common methodology described in Section 1, Equation (7). *TEC* derivation using model (7) in Section 1 is selected in consideration of the computational capacity of the targeted market of single-frequency commercial-grade GNSS receivers, mobile devices containing them and positioning-as-a-service systems. The experimental *TEC* values are considered true values for the purpose of the GNSS TEC predictive model development. *TEC* is considered the outcome of the sub-equatorial short-term rapidly developing ionospheric storm TEC prediction model. The TEC prediction model development procedure is outlined in Figure 2.

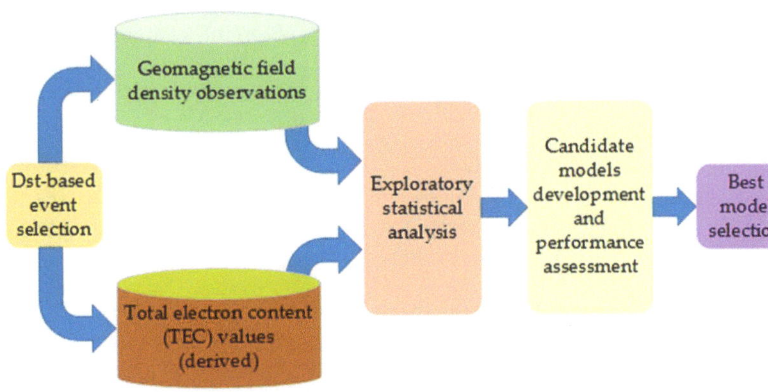

Figure 2. A methodology for TEC prediction model development.

The Disturbance Storm-Time (D_{st}) index, a geomagnetic condition descriptor, is considered a selector of short-term rapidly developing geomagnetic storm scenarios [4,16]. Geomagnetic field density component observations and raw dual-frequency GNSS pseudorange observations collected during the selected short-term and rapidly developing geomagnetic and ionospheric storms are aggregated into a single set of original observations. Raw dual-frequency GNSS pseudoranges are used for the derivation of experimental *TEC* values. The exploratory statistical analysis is performed on components of geomagnetic field density (predictors) and derived *TEC* (outcome) to determine their statistical models. Results of the exploratory statistical analysis are used in the selection of statistical learning methods for candidate TEC prediction model developments. Models developed are validated on the independent testing set of *TEC* and geomagnetic field density component observations. The performance of candidate models is compared mutually and with the performance of the standard Klobuchar model to identify the best performer to be pronounced the sub-equatorial short-term rapidly developing ionospheric storm TEC prediction model.

2.2. Statistical Learning-Based Model Development Methods

This research embraces the concept of statistical learning on experimental observations of related statistical variables [17,18] for the development of candidates of the sub-equatorial short-term rapidly developing ionospheric storm TEC prediction model. The results of exploratory statistical analysis of the aggregated set of predictors and outcome observations lead to the selection of two statistical learning methods for the development of candidates for the sub-equatorial short-term rapidly developing ionospheric storm TEC prediction model.

2.2.1. Boosted Generalised Additive Model (GAMB) Development Method

The boosted generalised additive model (GAMB) development method is a machine learning method based on the generalised additive model introduced by [17] and its boosting enhancement [18–20]. The method is aimed at modelling the non-linear and non-parametric relations between the target variable and predictors. The generalised additive model (GAM) method allows for modelling the non-linear and non-parametric relations between the expectation of the target variable y and predictors $\{x_1, x_2, \ldots, x_n\}$ by extending the concept of linear regression through the development of the smoothing function $g(E(y))$ of one or more predictors (8), based on the penalised regression approach [17], with $E(y)$ denoting the expectation of y, β_0 denoting a constant and $f()$ denoting a function.

$$g(E(y)) = g(y) = \beta_o + \sum_{i=1}^{n} f_i(x_i) \tag{8}$$

The boosting principle contributes to model development in a sense similar to the random forest approach. Through the boosting process, the predictions of multiple additive models, trained on subsets of the original observations, are combined in the optimisation sense to yield the response of the GAMB model. The GAMB model benefits from the deployment of boosting in terms of a reduction in the bias and variance of individual/simple models, thus achieving improved accuracy and robustness.

The boosting process is of iterative nature and involves the following repeating tasks: (i) development of a weak learner (a simple GAM) based on the observation subset, (ii) calculation of residuals from the weak learner, (iii) calculation of the gradient of the loss function with respect to residuals, (iv) update of the weak learner and (v) repetition of (i) to (iv) until the optimisation criterion is reached [20].

The GAMB model development method has been implemented in various machine learning (ML) programming environments, including the R programming environment for statistical computing [19–21].

2.2.2. Stochastic Gradient Boosting (SGB) Model Development Method

The stochastic gradient boosting (SGB) model development method was introduced by [22]. Given a system of outcome variable y and a set of explanatory variables (predictors) $x = \{x_1, x_2, \ldots x_n\}$ with related values arranged in a training set $\{y_i, x_i\}_1^N$, the method is to yield a function $F'(x)$ that maps x to y for all of their values, so the expected value E of a specified loss function $\Psi(y, F(x))$ is minimised, creating an optimisation problem, as described with (9).

$$F'(x) = \underset{F(x)}{argmin} E_{y,x}\Psi(y, F(x)) \tag{9}$$

The boosting procedure is implemented through $F'(x)$ approximation with a polynomial expansion of $F(x)$ in the form as given by (10).

$$F(x) = \sum_{m=0}^{M} \beta_m h(x; a_m) \tag{10}$$

Function $h(x; a_m)$ is called the 'base learner' and is usually selected as a simple function with parameters $a = \{a_1, a_2, a, \ldots a_M\}$. In the gradient tree boosting method deployment, the '*base learner*' is defined as an L-terminal node regression tree.

An iterative method may be established to solve for $F(x)$, starting with an initial guess of $F_0(x)$ and continuing with the procedure depicted in (11).

$$(\beta_m, a_m) = \underset{\beta,a}{argmin} \sum_{i=1}^{M} \Psi(y_i, F_{m-1} + \beta h(x_i; a))\ F_m(x) = F_{m-1}(x) + \beta_m h(x; a_m) \tag{11}$$

At every iteration, a regression tree partitions the x-space into L non-overlapping sub-spaces $\{R_{lm}\}_{l=1}{}^{L}$ and determines a separate constant value of h for each sub-space. The approach reduces the problem to a 'location' estimate γ_{lm} based on the Ψ criterion, as given by (12).

$$\gamma_{lm} = \underset{\gamma}{argmin} \sum_{x_i} \Psi(y_i, F_{m-1}(x_i) + \gamma) \tag{12}$$

The iterative procedure for $F_m(x)$ determination may be expressed with Equation (13), where the parameter ν, $0 < \nu < 1$, controls the learning rate.

$$F_m(x) = F_{m-1}(x) + \nu\gamma_{lm} 1(x \in R_{lm}) \tag{13}$$

Randomness was introduced in the gradient boosting method with the introduction of a sub-sample of the training data drawn from the original training set without replacement using a random permutation $\{\pi i\}_1{}^N$ of the integers {1, 2, …, N} to extract a random training sub-sample $\{y_{\pi(i)}, x_{\pi(i)}\}_1{}^{\tilde{N}}$ of the size $\tilde{N} < N$. The enhancement completes the definition of the SGB method, as outlined by [22,23]. The SGB method is summarised in Algorithm 1 below.

Algorithm 1 Stochastic Gradient Boosting (SGB) Methodology

1: $F_0(x) = \underset{\gamma}{argmin} \sum_{i=1}^{N} \Psi(y_i, \gamma)$
2: **for** m = 1 **to** M **do**
3: $\{\pi(i)\}_1^N = rand\_perm\{i\}_1^N$
4: $\tilde{y}_{\pi(i)m} = -\left[\frac{\partial\Psi(y_{\pi(i)}, F(x_{\pi(i)}))}{\partial F(x_{\pi(i)})}\right]_{F(x)=F_{m-1}(x)}, i = 1, \tilde{N}$
5: $R_{lm1}^{L} = L - terminalnodetree\left(\tilde{y}_{\pi(i)m}, x_{\pi(i)1}^{\tilde{N}}\right)$
6: $\gamma_{lm} = \underset{\gamma}{argmin} \sum_{x_{\pi(i)} \in R_{lm}} \Psi\left(y_{\pi(i)}, F_{m-1}\left(x_{\pi(i)}\right) + \gamma\right)$
7: $F_m(x) = F_{m-1}(x) + \nu \cdot \gamma_{lm} 1(x \in R_{lm})$
8: **end**

The presented research utilised the stochastic gradient boosting method implementation in the caret package [12] of the open-source R environment for statistical computing [21].

2.2.3. Bagged CART (BCART) Model Development Method

The bagged classification and regression tree (CART) model is an ensemble of decision trees developed on the subsets of the original set of observations [24]. The bagged CART decision is made as an average of decisions of individual decision trees in the BCART model [23], as depicted in Figure 3.

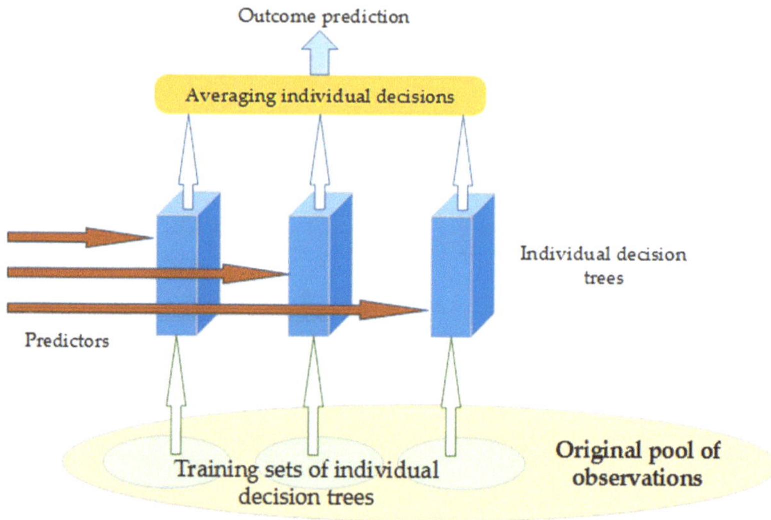

Figure 3. The bagged CART model development method.

The BCART method is implemented in the caret package [12] of the R environment for statistical computing [21].

2.2.4. Model Performance Assessment

The residual analysis-based model performance assessment procedure [12,23,25] is utilised here to examine the properties and success of developed candidates for the TEC prediction model and to allow for comparison between the candidate models and the standard Klobuchar model.

A residual r is defined as a difference between the predicted y_i and observed y outcome values for the same set of predictor values, as given in (14).

$$r_i = y_i - y \tag{14}$$

Performance indicators are selected to describe the quality of a model assessed as follows. The predicted vs. observed (P-O) diagram, a graphical representation of the prediction–observation outcome pairs, extends the goodness of fit and indicates the range of outcome values in which the model performs well. The root-mean-square error (*RMSE*) value of a set of residuals extends the ability of the model to describe bias (systematics of a phenomenon considered). *RMSE* is determined using (15).

$$RMSE = \sqrt{\frac{1}{N} \cdot \left(r_1^2 + r_2^2 + \ldots + r_N^2\right)} = \sqrt{\frac{1}{N} \cdot \sum_{i=1}^{i=N} r_i^2} \tag{15}$$

The coefficient of determination, defined using (16), and commonly known as the R^2 coefficient, extends the ability of the model to describe the original variance contained in the original data set.

$$R^2 = 1 - \frac{\sum_{i=1}^{i=N} r_i^2}{\sum_{i=1}^{i=N} (y_i - \overline{y})^2} \tag{16}$$

The R^2 coefficient of determination extends the percentage of the variance of the original data set (sample) explained with the regression model. The performance indicator defined by (16) is related to the number of predictors p used in the model and the number of observations in the original set of observations, n. The more objective performance indicator, called the adjusted coefficient of determination ($adjR^2$) and derived from the R^2 coefficient, is defined in (17), with n denoting number of observations in the sample and p denoting number of predictors.

$$adjR^2 = 1 - \left(1 - R^2\right)\frac{n-1}{n-p-1} \tag{17}$$

The $adjR^2$ coefficient allows for comparison between models with training sets of different sizes and of different numbers of predictors.

The three aforementioned indicators are used in the performance assessment of the candidates for the sub-equatorial short-term rapidly developing ionospheric storm TEC prediction model. A tailored model performance assessment software is developed in the R environment for statistical analysis.

2.3. Overview of the Four Rapid Short-Term Geomagnetic Storms Scenarios and Data

Ionospheric conditions are the prime individual source of GNSS positioning performance degradation [1,26]. Ref. [27] proposed the space weather–GNSS positioning performance coupling model that is utilised as a framework for this research. We hypothesise that *TEC*, as the result of the ionospheric conditions and the model outcome, may be modelled based on the local geomagnetic conditions, represented and described solely by the near-real-time observations of the local geomagnetic field density. In that sense, *TEC* would serve as the outcome and components of the geomagnetic field density as predictors of the proposed TEC prediction model. With a reference to the space weather–GNSS positioning performance coupling model [27], the geomagnetic conditions result from space weather conditions, and *TEC* further affects the quality of satellite-based positioning. This research contributes to the description of the geomagnetic conditions–*TEC* development–GNSS pseudorange measurement coupling and allows for the prediction of GNSS positioning performance deterioration due to the ionospheric delay of a GNSS signal.

A short-term rapidly developing ionospheric disturbance has the potential for a sudden GNSS positioning performance deterioration of a dominantly random nature. Prospects for the correction of such a source of GNSS positioning error using traditional global standard models are rather dire. Furthermore, the extent of the ionospheric disturbance effects is more pronounced in sub-equatorial regions by a specific pattern of free electron transfer in the upper atmospheric layers [4]. This research aims at a statistical description of the class of short-term rapidly developing ionospheric disturbances to support the tailored personalised ambient-aware GNSS TEC prediction model for improved PNT performance.

The development of a geomagnetic storm takes a common three-phase pattern, which is described in morphological terms using the Disturbance Storm-Time (D_{st}) index [4], although the ability of the D_{st} index to serve as a predictor of GNSS performance degradation events was challenged [16]. The D_{st} index points out the geomagnetic events of global significance, although it is based on processed observations in sub-equatorial regions. A geomagnetic storm starts with a short-duration positive phase, when the D_{st} index increases compared with a common condition. The positive phase of a geomagnetic storm is then followed by a rapid negative through phase, when the D_{st} index suddenly drops significantly towards the extreme negative values. The rapid negative through phase transforms into a prolonged recovery phase, during which D_{st} index values gradually rise towards the pre-storm conditions.

A D_{st}-based geomagnetic storm description is used here for the selection of the short-term and rapidly developing geomagnetic events used as scenarios of the research presented. Scenarios are selected additionally based on the additional criterion of the absence of any considerable geomagnetic disturbance at least a week prior to the geomagnetic storm outbreak to avoid a possible memory effect. The time series of the D_{st} index values taken from the internet archive [28], for the four geomagnetic storms selected are depicted in Figure 4.

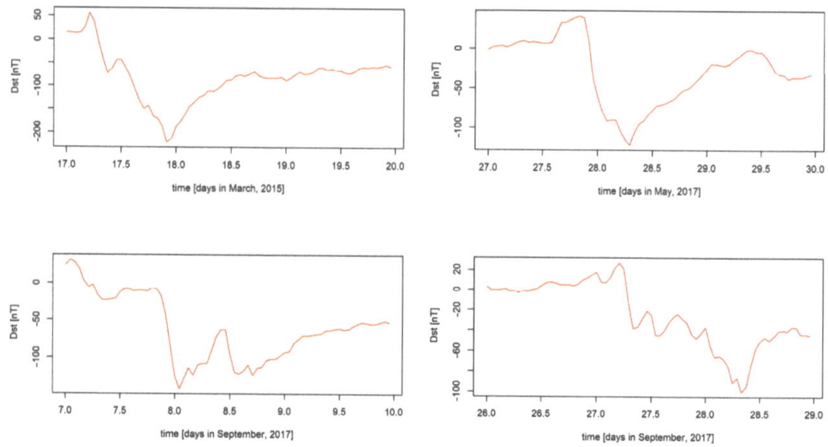

Figure 4. A D_{st}-based selection of rapidly developing short-term geomagnetic storms in mid-March, 2015; late May, 2017; early September, 2017; and late September, 2017.

The short-term rapidly developing geomagnetic storms of global outreach were identified in mid-March 2015, May 2017 and early and late September, 2017. All four storms lasted for three days each, extending a three-phase development pattern of a significant geomagnetic field disruption, with the potential to affect *TEC* development and, consequently, the GNSS positioning performance.

The selected class of geomagnetic storms establishes the four scenarios for the research presented. The March 2015 storm, known also as the St Patrick's Day storm, occurred between 17 March 2015 (DOY76 in 2015) and 19 March 2015 (DOY78 in 2015). The May 2017 storm occurred between 27 May 2017 (DOY147 in 2017) and 29 May 2017 (DOY149 in 2017). The early-September 2017 storm occurred between 7 September 2017 (DOY250 in 2017) and 9 September 2017 (DOY252 in 2017). The late-September, 2017 storm occurred between 26 September 2017 (DOY269 in 2017) and 28 September 2017 (DOY271 in 2017).

The original experimental observations of *TEC* and geomagnetic field density, aimed for utilisation in the TEC prediction model development, should be collected in the close vicinity and provided by trusted sources. Two internet-based trusted sources are identified that provide the required data collected in the sub-equatorial region of the Northern Territories, Australia, as detailed in subsequent sections.

2.4. True TEC Derivation from Dual-Frequency GPS Pseudoranges at IGS Reference Station Darwin, NT

The International GNSS Service [29] operates a global network of stationary GNSS reference stations that systematically collect the raw GNSS pseudoranges uncorrected for ionospheric effects every 30 s on a daily basis. Structured in the RINEX format, the internet-based IGS observation archive serves as an invaluable source of experimental GNSS-related observations.

Single-frequency commercial-grade GNSS receivers on the market utilise different combinations of GNSS signals, with the GPS ones being common with all of them. For that reason, this research utilises the GPS pseudorange observations for the derivation of experimental (true) *TEC*. The GPS pseudorange observations taken at the IGS reference station in Darwin, NT, Australia (Figure 5), for four scenarios of geomagnetic storms identified in Section 2.3 are used in this research. The selection of the IGS Darwin reference stations was driven by its position in the sub-equatorial region, with pronounced ionospheric disturbance effects and with its proximity to the INTERMAGNET [30] Kakadu, NT, Australia, reference station. The true *TEC* is estimated from dual-frequency GNSS pseudorange observations using the procedure outlined in (7) (Section 1), with the GPS-TEC Programme software, revision 3.0, developed by Dr Gopi Seemala [31]. The GPS-TEC Programme deploys estimates of the satellite bias b_s as provided by the University of Bern. The receiver bias b_r is estimated by using the re-scaling standardisation procedure applied to the raw GPS *TEC* estimates [31].

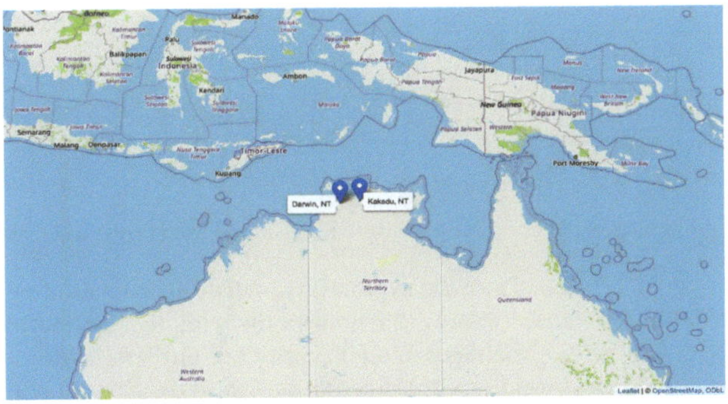

Figure 5. Positions of the IGS reference station Darwin, NT, and the INTERMAGNET reference station Kakadu, NT. Figure designed with bespoke software in the R environment for statistical computing using the leaflet R package and icons and the Open Street Map background layer.

2.5. Geomagnetic Field Density Observations at INTERMAGNET Reference Station Kakadu, NT

The INTERMAGNET operates the world network of stationary reference sites that systematically collect the observations of the geomagnetic field density vector components B_x, B_y and B_z [30]. The observation procedure requires the measurements to be taken on a daily basis, every minute. Collected observations are stored in structured text files openly available to interested parties. Observations taken at the INTERMAGNET reference station Kakadu, NT, Australia (Figure 5), for four scenarios of geomagnetic storms identified in Section 2.3 are used in the presented research.

The selection of the INTERMAGNET Kakadu reference stations as the source of geomagnetic field density observations was driven by its proximity to the IGS Darwin reference stations. The research assumes similar geomagnetic and ionospheric conditions, resulting in similar GNSS pseudorange measurement degradations, in the locations of two reference stations separated by a distance of 178.5 km.

2.6. Material Summary Per Geomagnetic Storm Scenario

As described in Section 2.2, this research utilises four sets of data (time series) per scenario: *TEC* values and three components of the geomagnetic field density vector. Data sets of geomagnetic field density components and the associated experimental *TEC* are statistically analysed to assist the development of the ambient-aware GNSS TEC prediction model for PNT in the case of short-term rapidly developing ionospheric storms. The results of the statistical analysis are presented in a box-plot form. The exploratory statistical analysis results are summarised in the rest of this section for the four scenarios defined in Section 2.3.

2.6.1. The Mid-March 2015 Geomagnetic Storm Scenario (The St Patrick's Day 2015 Storm, Storm 1)

Box plots of predictors B_x, B_y and B_z and the experimentally derived *TEC* target are presented in Figure 6.

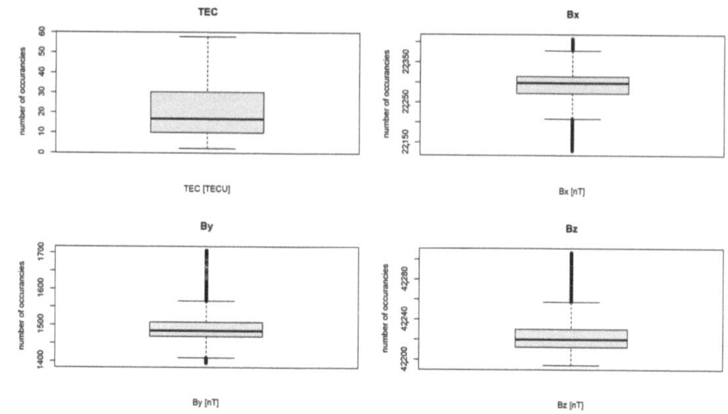

Figure 6. Box plots of predictors B_x, B_y and B_z and target *TEC* data during Storm 1.

The results of the exploratory statistical analysis of related time series of *TEC*, B_x, B_y and B_z variables show that none of them follow a normal statistical distribution. The *TEC*, B_y and B_z variables extend a number of outliers, with the respective long right tails of the corresponding experimental statistical distributions. The B_x variable extends several outliers at the left tail of its experimental statistical distribution.

2.6.2. The Late-May 2017 Geomagnetic Storm Scenario (Storm 2)

Box plots of predictors B_x, B_y and B_z and the experimentally derived *TEC* target are presented in Figure 7.

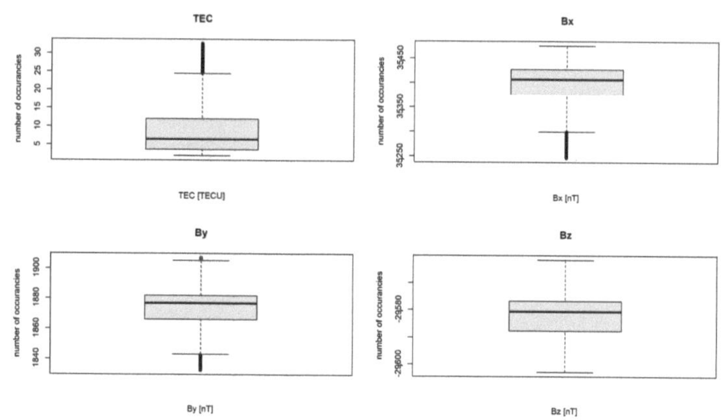

Figure 7. Box plots of predictors B_x, B_y and B_z and target *TEC* data during Storm 2.

The results of the exploratory statistical analysis of related time series of *TEC*, B_x, B_y and B_z variables show that none of them follow a normal statistical distribution. The *TEC* variable yields numerous outliers at the right tail, while the B_x and B_y variables extend outliers at the left tails of their corresponding experimental statistical distributions. Additionally, the B_y variable yields a few outliers at the right tail.

2.6.3. The Early-September 2017 Geomagnetic Storm Scenario (Storm 3)

Box plots of predictors B_x, B_y and B_z and the experimentally derived *TEC* target are presented in Figure 8.

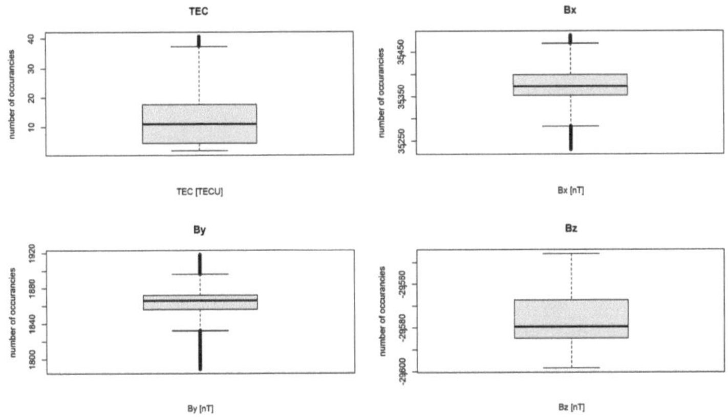

Figure 8. Box plots of predictors B_x, B_y and B_z and target *TEC* data during Storm 3.

The results of the exploratory statistical analysis of related time series of *TEC*, B_x, B_y and B_z variables show that none of them follow a normal statistical distribution. While *TEC* values extend a few outliers on the right tail of the statistical distribution, the B_x and B_y variables yield numerous outliers at both tails of their corresponding experimental statistical distributions.

2.6.4. The Late-September 2017 Geomagnetic Storm Scenario (Storm 4)

Box plots of predictors B_x, B_y and B_z and the experimentally derived *TEC* target are presented in Figure 9.

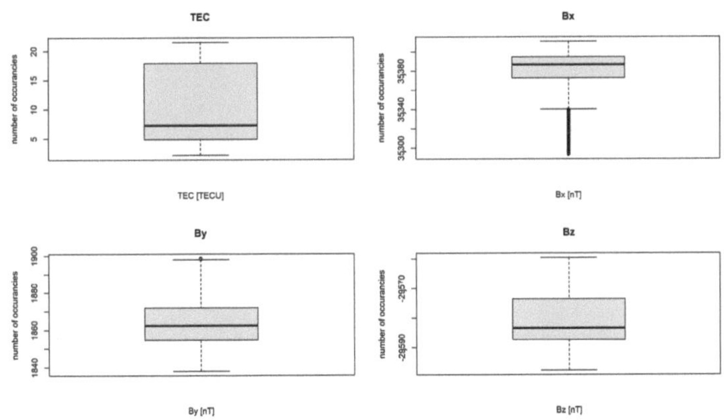

Figure 9. Box plots of predictors B_x, B_y and B_z and target *TEC* data during Storm 4.

The results of the exploratory statistical analysis of related time series of variables show *TEC* and B_z as following a normal statistical distribution. The B_x and B_y variables experienced a number of outliers, with slight tails, left and right.

2.6.5. Analysis and Discussion

Overall, the exploratory analysis of *TEC* and geomagnetic field density component observations leads to the conclusion of short-term rapidly developing storms as a well-described class of space weather events affecting the GNSS positioning performance. Additional analysis is conducted to obtain a deeper insight into the nature of *TEC* dynamics during four geomagnetic storms under consideration. The Cullen–Frey method [32] is applied to estimate the theoretical statistical distribution that fits data in all four *TEC* sets concerned. The Cullen–Frey method examines the relationship between kurtosis and the square of skewness of bootstrapped samples (subsets) of the original data.

The Cullen and Frey graph analysis reveals the beta statistical distribution as the most promising fit to the experimental data of all four cases considered. Three of them extend a high similarity of the theoretical statistical distribution fit, while the May 2017 storm extends a somewhat larger square of skewness. The findings confirm the case of short-term rapidly developing geomagnetic storms as a well-defined class of GNSS-related space weather events.

Additional exploratory statistical analysis is performed to identify the processes behind *TEC* dynamics for all four cases of rapidly developing short-term geomagnetic storms including the following statistical tests [33]: (i) the two-sample *t*-test to determine whether the means of two sets of *TEC* observations of different geomagnetic storms are equal, (ii) the two-sample *F*-test to determine whether the variances of two sets of *TEC* observations of different geomagnetic storms are equal and (iii) the two-sample Kolmogorov–Smirnoff test to determine whether two sets of *TEC* observations of different geomagnetic storms follow the same statistical distribution. The exploratory statistical analysis finds that no pairs of *TEC* sets share either the same mean, variance or result from the same statistical distribution. Given the complexity of the *TEC* generation processes, the results of the exploratory analysis confirm the expectations. Additionally, the results of statistical tests indicate the need for an advanced method for TEC correction model development. The inference leads to the selection of machine learning-based methods as a suitable approach in the solution of TEC prediction model development.

The resulting Cullen and Frey diagrams are depicted in Figure 10.

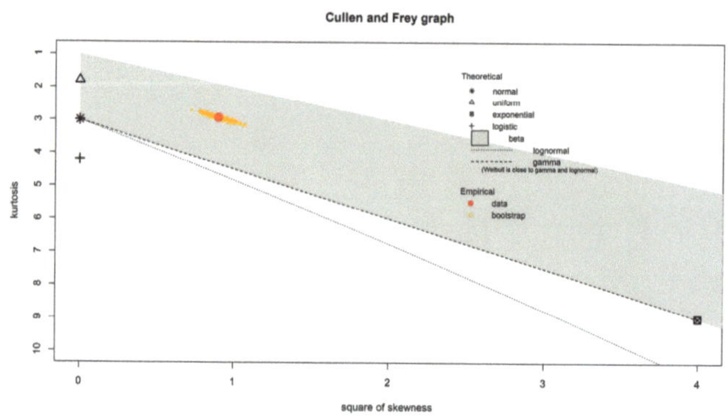

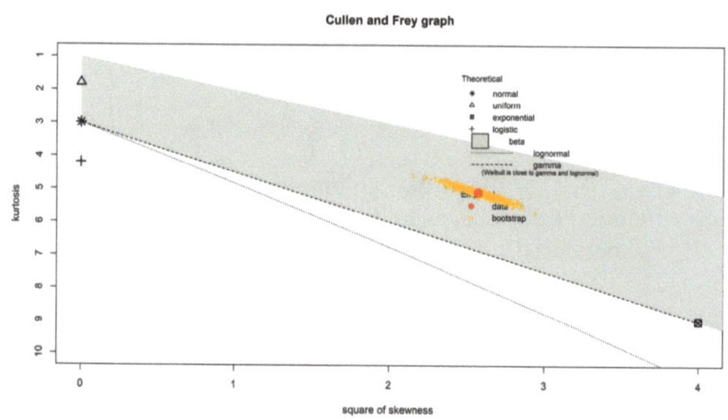

Figure 10. *Cont.*

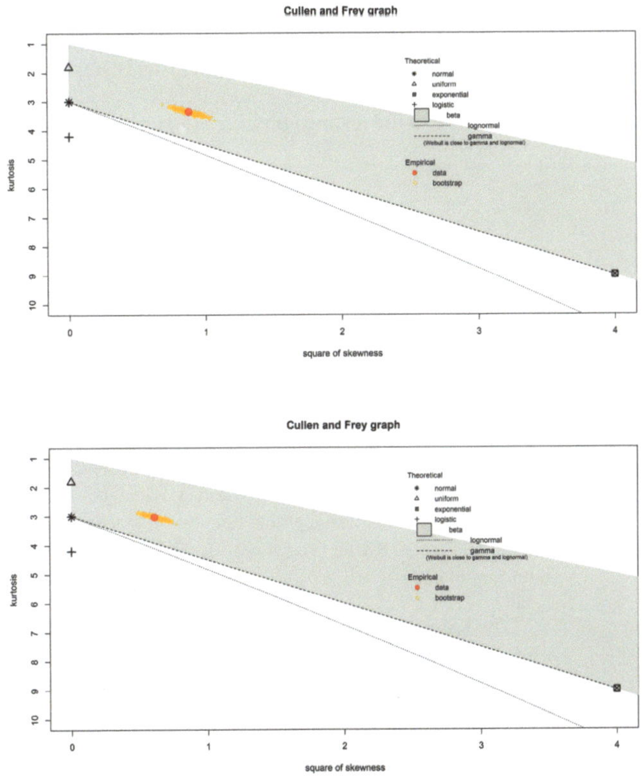

Figure 10. Cullen and Fray diagrams of four *TEC* sets under consideration, starting from top to bottom: St Patrick's Day storm, (DOY76–DOY78 in 2015); May 2017 storm (DOY147–DOY149 in 2017); early-September 2017 storm (DOY250–DOY252 in 2017); and late-September 2017 storm (DOY269–DOY271 in 2017).

The Cullen and Fray analysis, the exploratory data analysis and statistical tests [33] are performed in the R environment for statistical computing [21] using the R package fitdistrplus [32] for the former and the standard packages for the latter analyses.

3. Research Results

We aggregate the time series of all four scenarios into a single pool of observations while keeping the variable-related structure, thus composing a set of observations as a representative sample comprising descriptions of different variances of short-term rapidly developing geomagnetic storms. The aggregated original pool consists of 13,817 observations of *TEC* (outcome) and B_x, B_y and B_z (predictors) variables from the four selected scenarios (Section 2.3). We split the pool of observations into training (model development) and testing (model evaluation) subsets of the original pool of observations using the 80–20 Pareto principle [34,35]. The cross-validation procedure is involved in the development of both the SGB-based and BC-based TEC prediction model candidates to mitigate the effects of a non-normal experimental distribution and randomisation involved in observation selection for training and testing subsets of the original data. The testing subset is used for the assessment of Klobuchar model performance to provide a benchmark (reference) model for additional comparisons of the quality of developed TEC prediction model candidates. Section 2.2.4 outlines the method performance assessment criteria, including root-mean-square error (*RMSE*) for bias modelling performance assessment, the adjusted coefficient of determination ($adjR^2$) for variance modelling performance assessment and the P-O diagram for graphical assessment of the model agility. Model development and model performance validation tasks are performed using the tailored software our team developed in the R environment for statistical computing. Assessment results of the ability of candidate PPR-based, SGB-based, BCART-based and Klobuchar models to describe bias and variance in the testing subset are depicted in Figure 11 and outlined in Table 1.

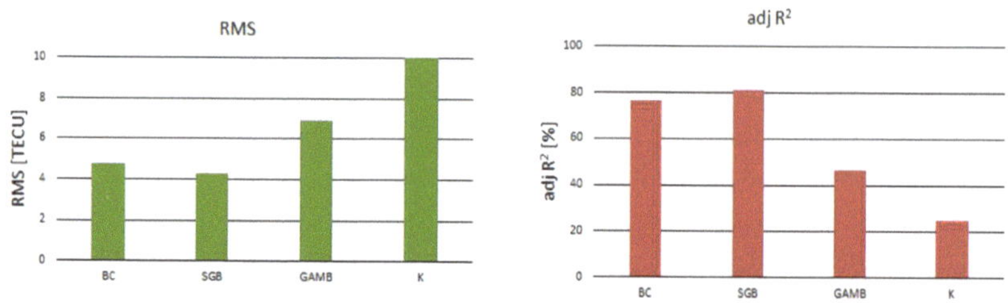

(a) (b)

Figure 11. Performance assessment results of TEC prediction model candidates (BC denotes bagged CART-based model, SGB denotes stochastic gradient boosting-based model, GAMB denotes boosted generalized additive model, and K denotes Klobuchar (standard) GPS ionospheric delay correction model), where *RMSE* (**a**), denotes the root-mean-square error (*RMSE*) value and (**b**) adjusted coefficient of determination ($adjR^2$) denotes the adjusted coefficient of determination value.

The Klobuchar model, the standard GPS error correction model considered a reference model in this research, performs poorly during short-term rapidly developing geomagnetic storms in sub-equatorial regions. It extends a large *RMSE* and describes only 25% of the original variance. Contenders for the TEC prediction model perform far better than Klobuchar, in support of the hypothesis of improved GNSS ionospheric correction estimation based solely on the near-real-time local geomagnetic field density vector observations. The PPR model reduces by nearly 30% the Klobuchar model *RMSE*, and doubles the original variance coverage, compared with the Klobuchar model. The BCART model halves the Klobuchar model *RMSE* and covers more than 76% of the original variance. The SGB-based TEC prediction model extends an even better *RMSE* value than the BCART model and is capable of modelling more than 81% of the original variance.

Statistical learning models develop as a result of experience. They may be designed to improve their predictive capacity and performance. The time required to complete model development may indicate the computational effort needed to develop the model as related information for GPS positioning process developers and operators. Model development times for the TEC predictive model contenders are examined, with the results presented in Table 1.

Table 1. TEC prediction model development time.

	Bagged CART (BCART)	**Stochastic Gradient Boosting (SGB)**	**Boosted Generalized Additive Model (GAMB)**
Model development time [s]	17.89	32.15	13.43

The SGB-based model requires the most time to develop, almost twice as much as needed for BCART model development. Considering the performance accomplished, the selection of the BCART model may be a good trade-off for applications where computing resources are critical. The PPR model requires just about one-fifth of the SGB model development time, which trades with a significantly reduced performance in comparison with the SGB model.

The P-O diagrams reveal the agility of the TEC model candidates, as shown in Figure 12.

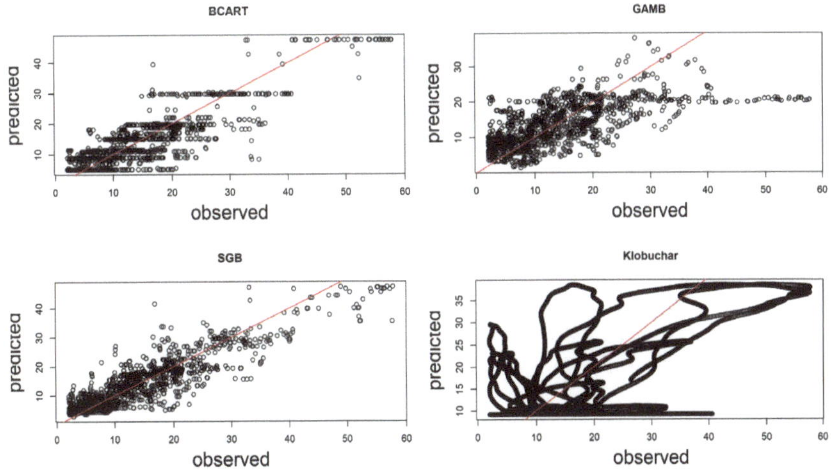

Figure 12. Prediction vs. observed (P-O) diagrams of the projection pursuit regression (PPR), bagged CART (BCART), stochastic gradient boosting (SGB), boosted generalized additive model (GAMB) and Klobuchar (K) TEC prediction models, with reference line (red).

Considering the performance assessment indices defined in Section 2.2.4, the stochastic gradient boosting (SGB) TEC prediction model extends the best performance during short-term rapidly developing geomagnetic storms in the sub-equatorial region of all three models assessed.

4. Discussion

This research addresses the development of the ambient-aware GNSS TEC prediction model suitable for integration within the ambient-aware GNSS PNT framework as an alternative to standard GNSS ionospheric correction models, such as the Klobuchar model. The proposed ambient-aware GNSS TEC prediction model development methodology is demonstrated in the scenario of short-term rapidly developing ionospheric storms, one of the extreme cases of ionospheric conditions that may cause significant degradation of the GNSS PNT performance. The proposed ambient-aware GNSS TEC prediction model returns the *TEC* estimate for the particular case of the ionospheric conditions, determined by the values of predictors (B_x, B_y and B_z) at the time of prediction.

Based on the statistical properties of four selected cases of short-term rapidly developing ionospheric storms, three ambient-aware GNSS TEC prediction models are developed and their performance is assessed and compared mutually and in relation to the Klobuchar model's performance in the same cases. As a result, the stochastic gradient boosting (SGB) TEC prediction model is found to be the best performer in the group. The SGB GNS TEC prediction model covers bias with a root-mean-square error (*RMSE*) of 4.28 TECU, a 60% improvement compared with the Klobuchar model. Further to this, the stochastic gradient boosting (SGB) TEC prediction model describes 82% of the original variance in derived experimental *TEC* observations, compared with just 25% as extended by the Klobuchar model. The stochastic gradient boosting (SGB) TEC prediction model requires more time and effort to be developed. However, once developed, it provides the best performance, with reasonable execution time concerning deployment in modern computationally improved devices, such as smartphones, IoT devices, cars, drones and others.

The proposed GNSS TEC prediction model aims at deployment within the ambient-aware GNSS PNT framework, either on mobile devices or within the positioning-as-a-service framework. Particular concern is given to implementation on devices utilising single-frequency GNSS PNT, with the aim to provide an alternative to standardised global ionospheric correction models.

The implementation of the proposed method and the model are rather simple and straightforward in modern software-defined radio (SDR)-based GNSS receivers and even more elegant and efficient in the positioning-as-a-service distributed GNSS processes. Utilisation of the SDR concept renders the GNSS PNT process and algorithm transparent and flexible in terms of improvement of the existing PNT algorithm and for the introduction of new services by exploitation of methods and techniques of statistics, computer science and mobile communications. We demonstrated the deployment of the proposed ambient aware GNSS TEC prediction model within a laboratory ambient-aware PNT framework which includes the open-source RTKLIB SDR, in both real-time and post-processing simulations. In the post-processing scenario, the ionospheric corrections were calculated using the proposed ambient-aware GNSS TEC prediction model, with data structured in the IONEX format.

Sources of data may be either the mobile unit's own measurements of the positioning environment conditions (components of the geomagnetic field in the vicinity of a GPS/GNSS receiver) using the unit's own sensors, trusted third-party data (NOAA, NASA EU Copernicus, INTERMAGNET, etc.) delivered through a dedicated and encrypted communications protocol via the mobile internet or both. The actual benefit achieved depends on the mobile unit's ability to measure the geomagnetic field components accurately and correctly and on the third party's ability to provide near-real-time data of high accuracy Furthermore, thorough and systematic consideration should be given to communications safety and to means of deployment and operation of machine learning methods to safeguard them from adversarial cyber-attacks [36,37]. A case of geomagnetic data-based spoofing may be overcome with authentication, sensor information fusion and additional analysis of time series of data.

This research provides the proposal for the method, and its proof-of-principle justification, thus establishing a solid framework for further refinements and developments planned to be accomplished by this group. Future research will focus on model development and validation for different levels of ionospheric disturbances and ambients of PNT (geographic latitudes, urban/rural environments, inclusion of information from other ambient sensors, etc.).

5. Conclusions

Satellite navigation has become one of the pillars of modern civilisation and an essential component of the national infrastructure. Space weather and ionospheric conditions render the prime source of single-frequency GNSS PNT service disruptions and degradations. The PTA of GNSS PNT services requires novel approaches in tackling the ionospheric effects on GNSS PNT. Standard global ionospheric correction models cannot mitigate the local ionospheric disturbances, as well as those of high dynamics. A self-adaptive positioning environment-aware GNSS position estimation algorithm, which engages a bespoke machine learning-based GNSS ionospheric correction model, offers huge promises in the PTA of GNSS. Here, we show that even in the case of demanding ionospheric conditions, such as during a short-term fast-developing geomagnetic storm in a sub-equatorial region, a machine learning-based environment-aware GNSS ionospheric correction model developed and operated by a position estimation entity, either a traditional GNSS receiver or a positioning-as-a-service system, may provide a substantial improvement over the existing global Klobuchar model, which is considered as a benchmark.

This research evaluates three candidates for the ambient-aware GNSS PNT ionospheric correction models based on machine learning methods and large sets of experimental observations of geomagnetic field density components as predictors and *TEC*/single-frequency GPS ionospheric delay as the target. Machine learning development methods for three models are selected based on the results of the exploratory statistical analysis of predictors and target observations. The performance of the three GPS ionospheric model candidates, (i) the bagged CART model, (ii) the boosted generalized additive model

(GAMB) and (iii) the stochastic gradient boosting (SGB) model, are assessed and compared with the Klobuchar model as the benchmark.

The ambient-aware SGB *TEC*/GNSS pseudorange measurement error predictive model is proposed as the result of the comparison, based on experimental observations and a statistical/machine learning model development technique, with the component of the geomagnetic field density vector as the sole predictor and *TEC* as the target. The TEC prediction model is developed and validated for GNSS ionospheric delay corrections during short-term rapidly developing geomagnetic storms in a sub-equatorial region, which significantly reduces (60%) bias error compared with standard Klobuchar model and describes 82% of the original *TEC* variance. The research finds D_{st} to be a good classifier for the ionospheric condition scenarios.

Further research is needed to refine the methodology for machine learning-based model development method selection and validation to be deployed for various classes and scenarios of ionospheric conditions and geographic latitudes, enhance the robustness of the machine learning-based model to safe-guard it against malicious attacks and establish an architecture-agnostic framework for operational deployment of the resulting optimal machine learning-based and positioning environment-aware bespoke ionospheric correction model that contributes to GNSS resilience development through advanced PTA deployment.

Author Contributions: R.F. conceived the study. R.F., I.H. and J.P.-O. prepared the problem statement and assessed the state-of-the-art research. R.F. and J.P.-O. developed the methodology and the tailored software in the R environment for statistical computing. I.H. and T.I. aggregated and pre-processed sets of observations using performed exploratory statistical analysis, analysed statistical learning-based model development methods and performed research in accordance with the methodology set and utilising the R-based tailored software. All authors contributed to the inference of the results and the formulation of the conclusion. R.F. and I.H. contributed equally to the presented work. All authors have read and agreed to the published version of the manuscript.

Funding: J.P.-O. acknowledges support of her research contribution by the Croatian Science Foundation under the project IP-2022-10-2821.

Data Availability Statement: Restrictions apply to the availability of these data. Data were obtained from [28–30] and are available on [28–30] with the permission of [28–30].

Acknowledgments: J.P.-O. acknowledges support of her research contribution by the Croatian Science Foundation under the project IP-2022-10-2821. The results presented in this paper rely on the geomagnetic data collected at Kakadu National Park, NT, Australia. The authors thank Geoscience Australia, Symonston, ACT, Australia, for supporting its operation and INTERMAGNET for promoting the high standards of magnetic observatory practice (www.intermagnet.org (accessed on 24 July 2024)). All authors consent the published acknowledgments.

Conflicts of Interest: The authors declare no conflicts of interest.

References

1. Klobuchar, J.A. Ionospheric Time-Delay Algorithm for Single-Frequency GPS Users. *IEEE Trans. Aerosp. Electron. Syst.* **1987**, *23*, 325–331. [CrossRef]
2. Betz, J.W. *Engineering Satellite-Based Navigation and Timing: Global Navigation Satellite Systems, Signals, and Receivers*; Wile-IEEE Press: Piscataway, NJ, USA, 2015; ISBN 978-1118615973.
3. Filjar, R. An application-centred resilient GNSS position estimation algorithm based on positioning environment conditions awareness. In Proceedings of the 2022 International Technical Meeting of The Institute of Navigation, Long Beach, CA, USA, 25–27 January 2022; pp. 1123–1136. [CrossRef]
4. Hunsucker, R.D.; Hargreaves, J.K. *The High-Latitude Ionosphere and Its Effects on Radio Propagation*; Cambridge University Press: Cambridge, UK, 2003.
5. Filić, M.; Filjar, R. On correlation between SID monitor and GPS-derived TEC observations during a massive ionospheric storm development. In Proceedings of the URSI AP-RASC 2019 Meeting, New Delhi, India, 9–15 March 2019. [CrossRef]
6. US Department of Defense. *Global Positioning System Standard Positioning Service Performance Standard*, 5th ed.; Office of the Department of Defense: Washington, DC, USA, 2020. Available online: https://www.gps.gov/technical/ps/2020-SPS-performance-standard.pdf (accessed on 30 June 2024).

7. Davies, K. *Ionospheric Radio*; Peter Peregrinus Ltd.: London, UK, 1990; ISBN 978-0863411861.
8. Groves, P.D. *Principles of GNSS, Inertial, and Multisensor Integrated Navigation Systems*, 2nd ed.; Artech House: London, UK; Boston, MA, USA, 2015; ISBN 978-1-60807-005-3.
9. Filić, M. On Development of the Forecasting Model of GNSS Positioning Performance Degradation due to Space Weather and Ionospheric Conditions. In Proceedings of the 2nd URSI Atlantic Radio Science Meeting (AT-RASC), Gran Canaria, Spain, 27 September 2018. [CrossRef]
10. Abhigna, M.S.R.; Sridhar, M.; Harsha, P.B.S.; Krishna, K.S.; Ratnam, D.V. Broadcast ionospheric delay correction algorithm using reduced order adjusted spherical harmonics function for single-frequency GNSS receivers. *Acta. Geophys.* **2021**, *69*, 335–351. [CrossRef]
11. Mehmood, M.; Filjar, R.; Saleem, S.; Hah, M.; Ahmed, A. TEC derived from local GPS network in Pakistan and comparison with IRI-2016 and IRI-PLAS 2017. *Acta. Geophys.* **2021**, *69*, 381–389. [CrossRef]
12. Kuhn, M. The Caret Package. Available online: https://topepo.github.io/caret/index.html (accessed on 4 July 2024).
13. Hemant Chandrokar, M. Machine Learning in Space Weather: Forecasting, Identification & Uncertainty Quantification. Ph.D. Thesis, Eindhoven University of Technology, Eindhoven, The Netherlands, 2019.
14. Ulukavak, M. Deep learning for ionospheric TEC forecasting at mid-latitude stations in Turkey. *Acta. Geophys.* **2021**, *69*, 589–606. [CrossRef]
15. Morton, Y.J.; van Diggelen, F.; Spilker, J.J.; Parkinson, B.W. *Position, Navigation, and Timing Technologies in the 21st Century: Integrated Satellite Navigation, Sensor Systems, and Civil Applications*; Wiley–IEEE Press: Piscataway, NJ, USA, 2020; Volume I, ISBN 978-1119458418.
16. Filjar, R.; Kos, S.; Krajnovic, S. Dst index as a potential indicator of approaching GNSS performance deterioration. *J. Navig.* **2013**, *66*, 149–160. [CrossRef]
17. Hastie, T.; Tibshirani, R. Generalized Additive Models. *Statist. Sci.* **1986**, *1*, 297–310. [CrossRef]
18. Efron, B.; Hastie, T. *Computer Age Statistical Inference: Algorithms, Evidence, and Data Science (Student Edition)*; Cambridge University Press: Cambridge, UK, 2021; ISBN 978-1108823418.
19. Hofner, B.; Mayr, A.; Robinzonov, N.; Schmid, M. Model-based boosting in R: A hands-on tutorial using the R package mboost. *Comput. Stat.* **2014**, *29*, 3–35. [CrossRef]
20. Hofner, B.; Mayr, A.; Robinzonov, N.; Schmid, M. Model-Based Boosting in R: A Hands-On Tutorial Using the R Package mboost (the R mboost Package Vignette). Available online: https://cran.r-project.org/web/packages/mboost/vignettes/mboost_tutorial.pdf (accessed on 30 June 2024).
21. R Core Team. *R: A Language and Environment for Statistical Computing*; R Foundation for Statistical Computing: Vienna, Austria, 2024. Available online: http://www.R-project.org (accessed on 30 June 2024).
22. Friedman, J.H. Stochastic gradient boosting. *Comput. Stat. Data Anal.* **2002**, *38*, 367–378. [CrossRef]
23. James, G.; Witten, D.; Hastie, T.; Tibshirani, R. *An Introduction to Statistical Learning with Applications in R*, 2nd ed.; Springer Science+Business Media: New York, NY, USA, 2021; ISBN 978-1071614174.
24. Mason, L.; Baxter, J.; Bartlett, P.; Frean, M. Boosting algorithms as gradient descent. In In Proceedings of the NIPS'99: Proc 12th International Conference on Neural Information Processing Systems, Denver, CO, USA, 29 November–4 December 1999; pp. 512–518.
25. Biecek, P.; Burzykowski, T. With examples in R and Python. In *Explanatory Model Analysis: Explore, Explain, and Examine Predictive Models*; CRC Press: Boca Raton, FL, USA, 2020; ISBN 978-0-367-13559-1.
26. Strang, G.; Borre, K. *Linear Algebra, Geodesy, and GPS*; Wellesley-Cambridge Press: Wellesley, MA, USA, 1997; ISBN 0-9614088-6-3.
27. Filić, M.; Filjar, R. Modelling the Relation between GNSS Positioning Performance Degradation, and Space Weather and Ionospheric Conditions using RReliefF Features Selection. In Proceedings of the 31st International Technical Meeting ION GNSS+, 1999–2006, Miami, FL, USA, 24-28 September 2018. [CrossRef]
28. ISGI. *International Service of Geomagnetic Indices Kyoto Dst Data Archive*; International Service of Geomagnetic Indices: Strasbourg, France, 2024. Available online: http://isgi.unistra.fr/data_download.php (accessed on 30 June 2024).
29. IGS. International GNSS Service GNSS RINEX Data Archive (Repository Hosted by NASA). Available online: https://cddis.nasa.gov/archive/gnss/data/daily/ (accessed on 30 June 2024).
30. INTERMAGNET. The International Real-Time Magnetic Observatory Network Data Archive. Available online: https://intermagnet.github.io (accessed on 30 June 2024).
31. Seemala, G. GPS-TEC Programme, Version 3.0. Available online: https://seemala.blogspot.com/ (accessed on 30 June 2024).
32. Delignette-Muller, M.L.; Dutang, C. Fitdistrplus: An R Package for Fitting Distributions. *J. Stat. Softw.* **2020**, *64*, 1–34, (revised May 2020). [CrossRef]
33. Maindonald, J.; Brown, W.J. *Data Analysis and Graphics Using R–An Example-Based Approach*; Cambridge University Press: Cambridge, UK, 2010; ISBN 978-0-521-76293-9.
34. Pareto, V. *Cours d'Économie Politique (Tome Second)*; Edition par F. Rouge, Laussane et F. Pichon; Rouge: Lausanne, Switzerland, 1897.
35. Newman, M.E.J. Power laws, Pareto distributions, and Zipf's Law. *Contemp. Phys.* **2005**, *46*, 323–351. [CrossRef]

36. Joseph, A.D.; Nelson, B.; Rubinstein, B.I.P.; Tygar, J.D. *Adversarial Machine Learning*; Cambridge University Press: Cambridge, UK, 2019; ISBN 978-1-107-04346-6.
37. Comiter, M. *Attacking Artificial Intelligence AI's Security Vulnerability and What Policymakers Can Do about It*; Belfer Center for Science and International Affairs, Harvard Kennedy School: Cambridge, MA, USA, 2019. Available online: https://www.belfercenter.org/publication/AttackingAI (accessed on 15 August 2024).

remote sensing

MDPI

Article

Diurnal Cycles of Cloud Properties and Precipitation Patterns over the Northeastern Tibetan Plateau During Summer

Bangjun Cao [1], Xianyu Yang [2,*], Yaqiong Lu [3], Jun Wen [2] and Shixin Wang [1]

1 School of Ecology and Environmental Science, Qinghai Institute of Technology, Xining 810016, China; caobj1989@163.com (B.C.); wangshx12@lzu.edu.cn (S.W.)
2 School of Atmospheric Science, Chengdu University of Information Technology, Chengdu 610225, China; jwen@cuit.edu.cn
3 Institute of Mountain Hazards and Environment, Chinese Academy of Sciences, Chengdu 610041, China; yaqiong@imde.ac.cn
* Correspondence: xyang@cuit.edu.cn; Tel.: +86-199-4082-6090

Abstract: In the context of rising temperatures and increasing humidity in Northwest China, substantial gaps remain in understanding the mechanisms of land–atmosphere cloud–precipitation coupling across the northeastern Tibetan Plateau (TP), Loess Plateau (LP), and Huangshui Valley (HV). This study addresses these gaps by investigating cloud properties and precipitation patterns utilizing the Fengyun-4 Satellite Quantitative Precipitation Estimation Product (FY4A-QPE) and ERA5 datasets. We specifically focus on Lanzhou, a pivotal city within the LP, and Xining, which epitomizes the HV. Our findings reveal that diurnal variations in precipitation are significantly less pronounced in the eastern regions compared to northeastern TP. This discrepancy is attributed to marked diurnal fluctuations in convective available potential energy (CAPE) and wind shear between 200 and 500 hPa. While both cities share similar wind shear patterns and moisture transport directions, Xining benefits from enhanced snowmelt and effective water retention in surrounding mountains, resulting in higher precipitation levels. Conversely, Lanzhou suffers from moisture deficits, with dry, hot winds exacerbating the situation. Notably, precipitation in Xining is strongly correlated with CAPE, influenced by diurnal variability, and intensified by valley and lake–land breezes, which drive afternoon convection. In contrast, Lanzhou's precipitation exhibits a weak relationship with CAPE, as even elevated values fail to generate significant cloud formation due to insufficient moisture. The ongoing trends of warming and humidification may lead to improved precipitation patterns, especially in the HV, with potential ecological benefits. However, concentrated rainfall during summer afternoons and midnights raises concerns regarding extreme weather events, highlighting the susceptibility of the HV to geological hazards. This research underscores the need to further explore the uncertainties inherent in precipitation dynamics in these regions.

Keywords: northeastern Tibetan Plateau; diurnal cycle; cloud; precipitation

Citation: Cao, B.; Yang, X.; Lu, Y.; Wen, J.; Wang, S. Diurnal Cycles of Cloud Properties and Precipitation Patterns over the Northeastern Tibetan Plateau During Summer. *Remote Sens.* **2024**, *16*, 4059. https://doi.org/10.3390/rs16214059

Academic Editor: Filomena Romano

Received: 21 September 2024
Revised: 22 October 2024
Accepted: 28 October 2024
Published: 31 October 2024

1. Introduction

Renowned as "the roof of the world" for its towering average elevation exceeding 4000 m, the Tibetan Plateau (TP), located at 26°00′–39°47′N, 73°19′–104°47′E, is also hailed as "the Asian water tower" because it nurtures the origins of major Asian rivers like the Yangtze, Yellow, and Lancang [1]. Precipitation across the TP is pivotal in sustaining these vital waterways, while its thermal output significantly contributes to atmospheric energy dynamics [1–3]. Furthermore, the TP profoundly influences precipitation patterns in adjacent and downstream regions [4–7]. Since the 1980s, Northwest China has witnessed a remarkable climatic shift from a "warm and dry" to a "warm and wet" era. This shift is evidenced by increased precipitation, runoff, and glacier melt, a significant rise in inland lake levels, and an escalation in flood events. Concurrently, there have been improvements in vegetation and a reduction in the frequency of dust storms [8].

Accurately capturing the diurnal cycle of clouds and precipitation patterns over the TP through the integration of remote sensing technologies, ground-based observations, and modeling efforts is pivotal for unraveling the intricate weather and climate dynamics within this region, as exemplified by notable atmospheric experiments such as the Second (TIPEX2) and Third Tibetan Plateau Atmospheric Experiments (TIPEX3) [9]. Previous research, based on observations and simulation outcomes, has delineated that China experiences three primary rainy seasons: the pre-summer rainy season in southern China (preceding the mei-yu period), the mei-yu season itself, and the midsummer rainy season [10–14]. Within the TP, the primary rainy seasons manifest from June to August, encompassing both the mei-yu and midsummer seasons, with total summer precipitation averaging below 400 mm, lower than in neighboring regions [15]. The monthly mean precipitation rate approximates 0.3 mm h^{-1}, while event-specific precipitation rates range from 1 to 20 mm h^{-1}, yielding an average daily precipitation of less than 15 mm [16,17]. Notably, cloud tops frequently soar above 12 km above ground level (AGL), occasionally exceeding 16 km AGL [18], highlighting the unique meteorological characteristics of this region.

Summer precipitation over TP exhibits a distinctive diurnal cycle marked by regional variations. In central TP, precipitation peaks in the evenings, displaying the most significant fluctuations [19–22]. The eastern TP foothills experience nocturnal rainfall before midsummer [23–26], while Himalayan valleys predominantly receive precipitation from midnight to sunrise hours, accompanied by a transition of cloud cover from ridges to valleys [27]. The diurnal patterns of precipitation within the East Asian Summer Monsoon (EASM) region are intricately shaped by a myriad of factors, including monsoonal flow, sea–land breeze, boundary-layer dynamics, low-level jets, and inertial oscillations within the mid-troposphere (~500 hPa) of the horizontal wind field [28–30]. Cao et al. (2022) examined the diurnal variation and influencing factors of summer precipitation and cloud parameters over TP and Sichuan Basin (SB). They found that, during the mei-yu season, the daily maximum precipitation and cloud parameters over TP occur in the evening, while the minimums occur in the morning. Over TP, CAPE significantly impacts precipitation, whereas low-level winds and cloud liquid water content in SB are the primary influences [31].

Satellite precipitation observations provide global-scale data that surpass the capabilities of conventional rain and snow gauges and surface-based radar measurements. Numerous advanced satellite algorithms [32–39], such as Fengyun-4 Satellite Quantitative Precipitation Estimation Product (FY4A-QPE) [39], leverage infrared and passive microwave data for improved accuracy. However, mountainous regions pose significant challenges for these satellite products. Furthermore, the spatial distribution of cloud optical thickness and the cloud water path over TP, as derived from satellite retrievals, are closely linked to increases in water-vapor transport flux divergence.

As temperatures rise and humidity increases across Northwest China, northeastern TP exhibits a notable transition in topography from the Kunlun Mountains in the west to the Qilian Mountains in the east. This topographical gradient sharply contrasts with the SB in southwestern TP, suggesting potential disparities in the diurnal characteristics of clouds and precipitation between these regions. Significant differences in precipitation patterns have already been observed between the Loess Plateau (LP) and the Huangshui Valley (HV), highlighting the necessity for a comprehensive investigation [31]. However, the mechanisms of land–atmosphere cloud–precipitation coupling over these areas remain poorly understood. We used Fengyun-4 satellite's FY4A-QPE product with the ERA5 reanalysis data to delineate northeastern TP into two distinct subregions. We specifically focus on Lanzhou, a pivotal city within the LP, and Xining, which epitomizes the HV. Notably, these two cities are situated less than 200 km apart. We then analyzed the diurnal variations in cloud parameters and precipitation during the meiyu and midsummer periods to elucidate the underlying mechanisms that drive these regional disparities. This study contributes to a deeper understanding of this vital region's complex interactions between topography, atmospheric conditions, and precipitation dynamics.

2. Study Area and Data Description

2.1. Fengyun-4 Satellite Quantitative Precipitation Estimation Product (FY4A-QPE)

The precipitation data are sourced from the FY-4A QPE. The FY-4A satellite has the Advanced Geosynchronous Radiation Imager (AGRI). The AGRI sensor provides Level 1 to Level 4 meteorological products, with Level 2 products offering the most comprehensive data. The FY4A-QPE is a Level 2 operational precipitation product for the China region, derived from the FY-4A satellite's AGRI. FY4A-QPE monitors precipitation intensity, range, and trend in China, supporting weather analysis, forecasting, flood monitoring, and warning services. It has a temporal resolution of 4.5 min and a spatial resolution of 4 km [39]. The data can be accessed at http://satellite.nsmc.org.cn/PortalSite/Data/Satellite.aspx (accessed on 1 March 2024). The data files contain quantitatively estimated pixel instantaneous precipitation rates derived from AGRI's precipitation retrieval algorithm by converting instantaneous brightness temperature data observed in the infrared channel into hourly precipitation amounts. The selected data period is for the summer months of June, July, and August from 2019 to 2023.

2.2. ERA5 Data

The fifth iteration of ECMWF's atmospheric reanalysis for the global climate, ERA5, encompasses a comprehensive timeline stretching from January 1940 to the present. This cutting-edge dataset is the product of the esteemed Copernicus Climate Change Service (C3S) housed within ECMWF. ERA5 offers a granular view of the Earth's climate, furnishing hourly estimates for many atmospheric, terrestrial, and oceanic variables. Utilizing a 31 km grid, the data comprehensively cover the globe, enabling detailed insights into regional and global climate patterns. The atmosphere is meticulously resolved with 137 vertical levels, spanning from the surface to an altitude of 80 km, revealing intricate vertical structures and dynamics. Moreover, ERA5 recognizes the importance of uncertainty quantification, integrating information on uncertainties for all variables, albeit at a compressed spatial and temporal scale. Cloud LWC, IWC, CBH, CAPE, u, and v are sourced from ERA5. ERA5 combines vast historical observations into global estimates using advanced modeling and data assimilation systems. ERA5 has been widely used in various studies. Using sounding observations, the boundary layer height from ERA5 is evaluated [40].

2.3. Research Area

For this study, two specific periods were selected: (1) the mei-yu period, from June 1 to June 25, and (2) the midsummer period, from July 1 to August 10. The study area encompasses the eastern TP and its downstream region (32°–40°N, 90°–108°E), further divided into two subregions based on elevation: 32°–36°E (Section A) and 36°–40°E (Section B) (see Figure 1). Data from ERA5 and observational sources were categorized into the mei-yu and midsummer seasons to analyze the diurnal cycle of precipitation.

Figure 1 shows that the southern part of the study area has higher terrain, with elevation gradually decreasing from west to east. The elevation generally decreases from west to east in the northern part (Section A). We specifically focus on Lanzhou, a pivotal city within the LP, and Xining, which epitomizes the HV. Notable valleys are present between 95°E and 100°E, corresponding to the area around Xining and Qinghai Lake, and between 103°E and 106°E, corresponding to Lanzhou and its surroundings. Mountain ranges lie between these valleys.

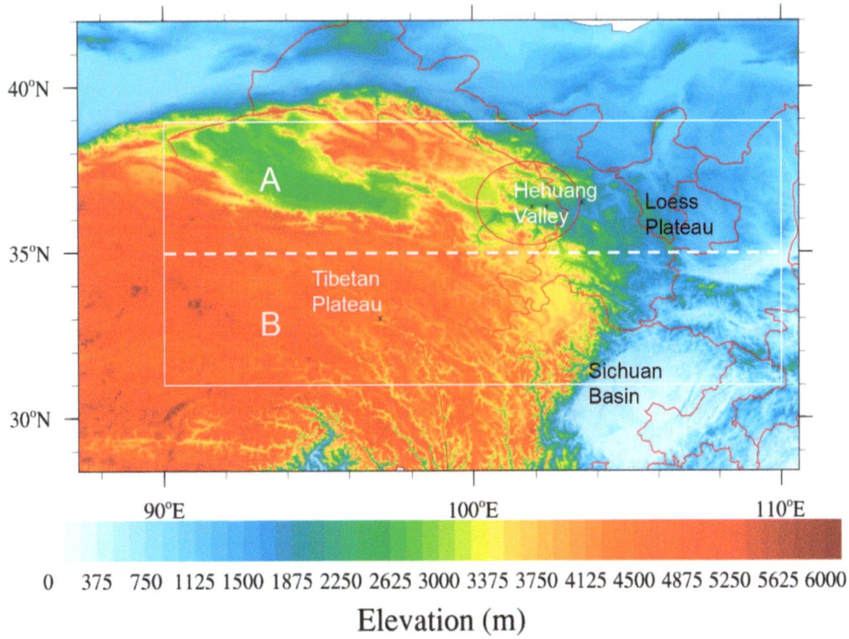

Figure 1. Topographic features of the eastern Tibetan Plateau, highlighting Section A as the northern region and Section B as the southern region. The stars indicate the locations of Yushu, Xining, Haidong, and Lanzhou city from west to east. Annotations include the Loess Plateau, Huangshui Valley, and northeastern Tibetan Plateau (TP).

3. Methods

Hovmöller diagrams [24] were used to show the diurnal cycle of CBH, precipitation, dewpoint spread, IWC, and LWC, as well as their variations with latitude. The analysis employs a local time (LT) framework, defined as Co-ordinated Universal Time (UTC) plus 7 h, aligning with the regional time zone under investigation. The construction of Hovmöller diagrams follows a standardized approach, where longitude is systematically plotted along the horizontal (x-) axis, while time, typically in hourly increments, is recorded on the vertical (ordinate) axis. This configuration facilitates the visualization of how a selected physical field evolves spatially across longitude and temporally throughout the day. The contour values of each variable are represented using a color scheme or shading, ensuring that the spatial and temporal patterns are readily apparent. Additionally, the height of the LCL is derived using the relationship $Z_{lcl} = 123(T - T_d)$, where T is the air temperature at 2 m and T_d is the dewpoint temperature, with the LCL determined by the dewpoint spread [41,42].

4. Results

4.1. Patterns of Precipitation and Cloud

We first analyzed the diurnal variation in precipitation from the FY4A-QPE dataset during the mei-yu period. In Section A (i.e., the northern part), spanning 95°E to 102°E (around Xining and Haidong), precipitation shows pronounced diurnal variation, with a dispersion pattern extending from afternoon to night from west to east. However, it lacks a distinct nocturnal rainfall signature. In contrast, the region between 103°E and 106°E (around Lanzhou) and further eastward (i.e., a general location toward the east between 106°E and 110°E) displays smaller diurnal variations and reduced total precipitation, with significant regional variability (Figure 2a). During midsummer, Section A exhibits marked diurnal variation, mainly concentrated between 1500 and 2100 LT in the 98°E to 103°E range (Figure 2b), without west-to-east dispersion seen earlier. In contrast, the region from 105°E to 110°E shows significant dispersion from 1200 to 2400 LT.

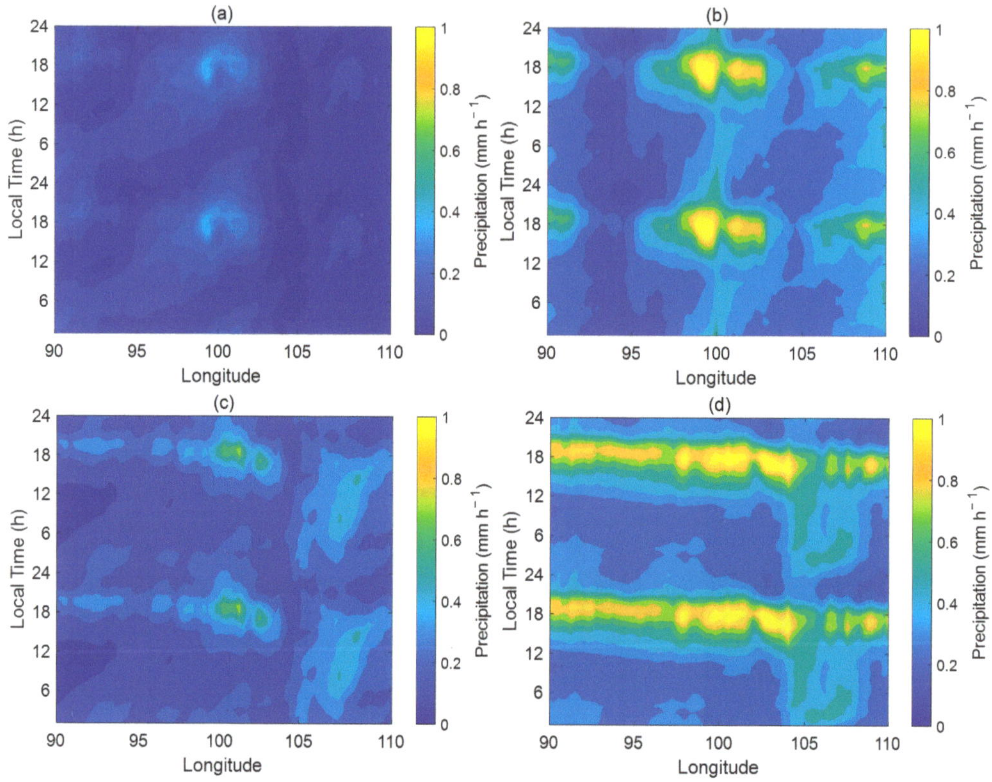

Figure 2. Diurnal variation in precipitation during the mei-yu ((**a**) for Section A, (**c**) for Section B) and midsummer periods ((**b**) for Section A, (**d**) for Section B), based on quantitative precipitation estimates from the Fengyun 4A Satellite (FY4A QPE).

In Section B (the southern part), precipitation generally shows substantial diurnal variation, characterized predominantly by nocturnal rainfall commencing after 1800 LT. This pattern shows minimal dispersion between 90°E and 103°E, but a notable dispersion is observed in the area eastward (i.e., a general location toward the east between 103°E and 110°E), with overall smaller regional differences (Figure 2c). During midsummer, the diurnal variation remains pronounced in Section B, with negligible dispersion from 90°E to 103°E, but significant dispersion is seen from 105°E to 110°E, spanning from 0000 to 1900 LT (Figure 2d).

When precipitation data from ERA5 (Figure S1) and FY4A-QPE are compared, the results indicate that, regardless of the region (A or B) or the season (mei-yu or midsummer), the precipitation patterns derived from ERA5 closely align with the spatial and temporal distribution observed in FY4A-QPE. While the two datasets have minor numerical discrepancies, the overall congruence in their precipitation patterns is evident.

When FY4A-QPE data are compared to ground measurements (Figure S2), significant discrepancies between Lanzhou and Xining emerge. Specifically, observational data for Lanzhou indicate markedly lower precipitation totals, with an absence of notable nocturnal rainfall. Conversely, Xining displays significant nocturnal precipitation, with FY4A-QPE reporting an hourly averaged intensity peak at 2100 LT, which occurs three hours earlier than observed in ground measurements. Additionally, the precipitation estimates derived from FY4A-QPE consistently surpass those recorded by observations.

We subsequently investigated the diurnal variation of cloud base height (CBH) over the northeastern TP. This analysis diverges from the work of Cao et al. (2022) [31], which focused on the southern TP and its relationship with the surface boundary layer, as well as Zhao et al. (2023) [28], which explored cloud amount and vertical distribution across the TP. In Section A, during the mei-yu period, the CBH remains relatively low, with minimal diurnal variation spanning 95°E to 102°E (around Xining and Haidong) (Figure 3a). In contrast, regions located between 103°E and 106°E (around Lanzhou) and the 91–93°E area exhibit higher CBH, with more pronounced diurnal fluctuations. The CBH patterns observed during midsummer are consistent with those recorded during the mei-yu period (Figure 3b). In Section B, the CBH generally shows smaller diurnal variation, with less pronounced changes occurring in the western part and larger variations between 90°E and 104°E (Figure 3c)—however, regions located between 103°E and 106°E display higher CBH with significant diurnal variability. Like Section A, the CBH characteristics during midsummer closely resemble those observed during the mei-yu period (Figure 3d).

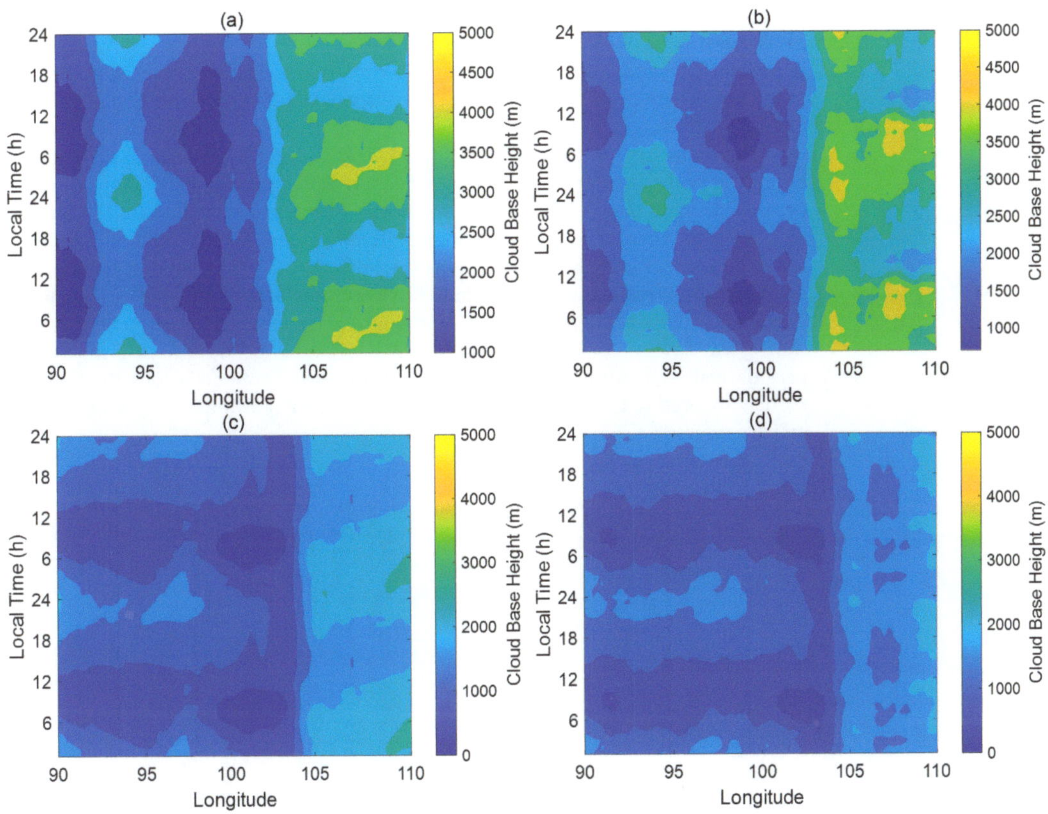

Figure 3. Diurnal variations in cloud base height (CBH) during the mei-yu ((**a**) for Section A, (**c**) for Section B) and midsummer periods ((**b**) for Section A, (**d**) for Section B) derived from ERA5 data.

We further assessed the liquid water content within clouds, distinguishing between daytime and nighttime across the mei-yu and midsummer periods and within Sections A and B. This analysis diverges from the work of Cao et al. (2022) [31], which focused on the southern TP and its relationship with the surface boundary layer, as well as Zhao et al. (2023) [28]. During both mei-yu and midsummer periods in A section, the liquid water content is notably high around 100°E to 102°E (near Xining), extending from near the surface to higher altitudes, with a peak around 500 hPa (Figure 4a,b). This elevated content is attributed to Xining's location in a moisture-rich valley. In contrast, regions around 103°E to 104°E (near Lanzhou) exhibit significantly lower daytime liquid water content (Figure 4a,b). At night, the liquid water content over Xining diminishes compared to daytime levels. However, it remains higher than over Lanzhou (Figure 4c,d), highlighting the contrasting moisture conditions where Xining has abundant moisture and dry, hot winds influence Lanzhou. Furthermore, regardless of whether it is daytime or nighttime, the cloud liquid content within the boundary layer of the area to the east of Lanzhou is significantly lower than that in the Hehuang Valley.

Section B, characterized by a different terrain with no valley between 95°E and 100°E (Figure 4e,g), exhibits a distinct cloud liquid water content distribution. During the daytime in both mei-yu and midsummer periods, a liquid water center emerges near 95°E. As one moves eastward, the cloud water content progressively decreases. Along steep slopes between 103°E and 105°E, the liquid water content near the surface is larger than in the A section. There is another liquid water center near 108°E (around Qinling Mountain). Overall, nighttime liquid water content is consistently lower than during the day.

The investigation of ice water content, segregated by daytime and nighttime conditions, reveals distinct patterns across Section A during the mei-yu and midsummer periods. During the day, ice water content exhibits peaks at 90°E to 92°E, 97°E to 99°E, and 102°E to 110°E, with concentrations centered around 300 hPa (Figure 5a,b). At night, ice water content increases notably between 98°E and 103°E, encompassing both Xining and Lanzhou. This increase is attributed to lower nocturnal temperatures enhancing ice formation (Figure 5c,d). In Section B, a pronounced center of ice water content is observed during the daytime between 92°E and 103°E during both the mei-yu and midsummer periods, centered around 300 hPa. Overall, the ice water content is consistently greater at night compared to the daytime, reflecting the influence of diurnal temperature variations on ice processes.

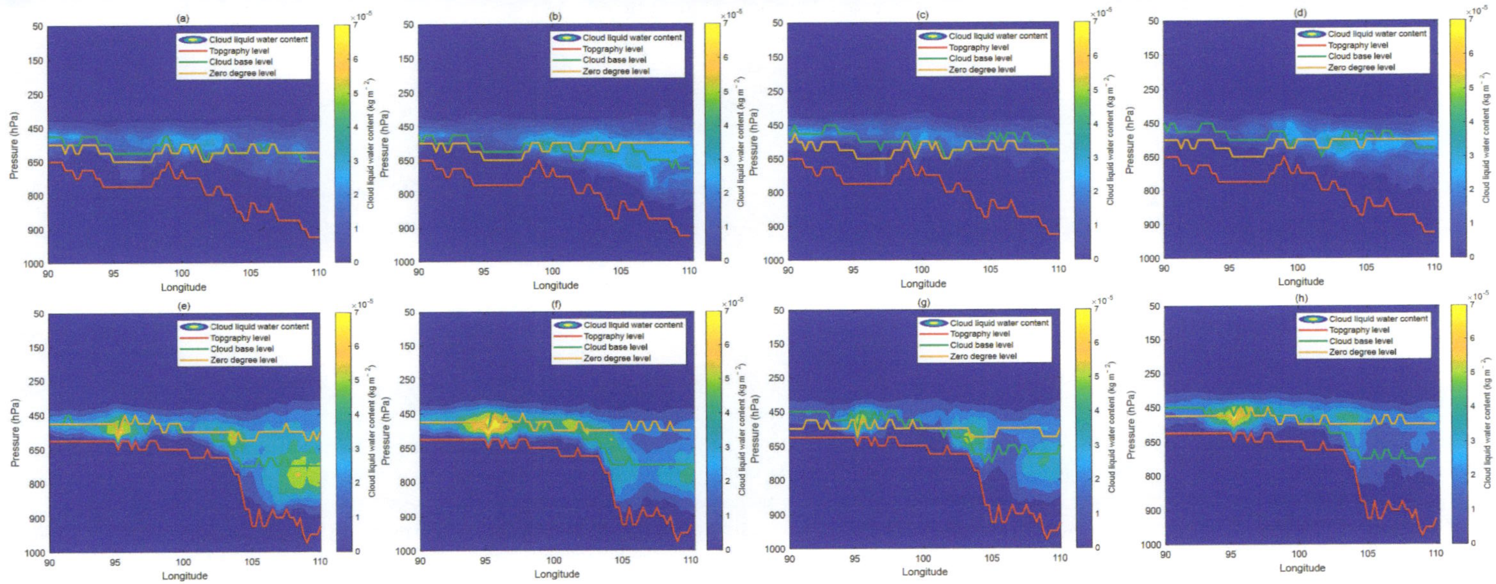

Figure 4. Pressure-level profiles showing cloud liquid water content (shaded color), cloud base (green line), zero-degree level (orange line), and topography (red line) along longitude for (**a**,**b**) daytime ((**a**) during the mei-yu, (**b**) during midsummer periods) and nighttime ((**c**) during the mei-yu, (**d**) during midsummer periods) in Section A, and daytime ((**e**) during the mei-yu, (**f**) during midsummer periods) and nighttime ((**g**) during the mei-yu, (**h**) during midsummer periods) in Section B.

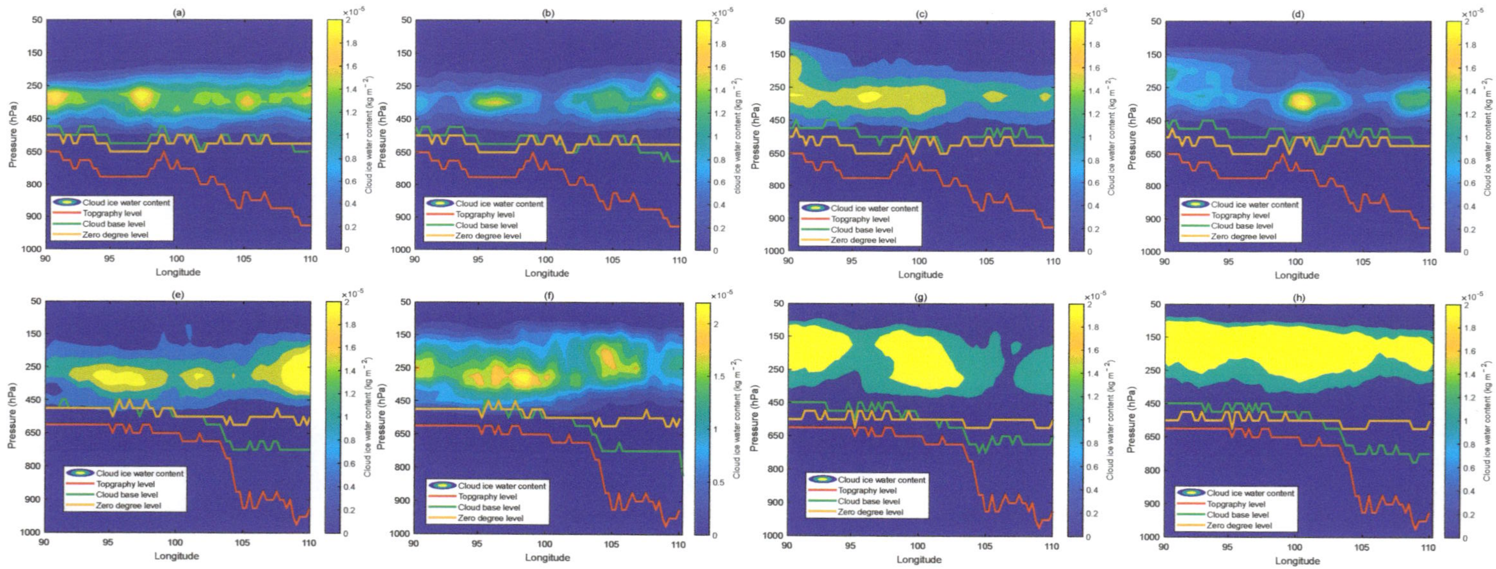

Figure 5. Similar with Figure 4, but for cloud ice water content. Pressure-level profiles showing cloud ice water content (shaded color), cloud base (green line), zero-degree level (orange line), and topography (red line) along longitude for (**a**,**b**) daytime ((**a**) during the mei-yu, (**b**) during midsummer periods) and nighttime ((**c**) during the mei-yu, (**d**) during midsummer periods) in Section A, and daytime ((**e**) during the mei-yu, (**f**) during midsummer periods) and nighttime ((**g**) during the mei-yu, (**h**) during midsummer periods) in Section B.

The analysis of precipitation types during the mei-yu season reveals distinct regional contributions from liquid and solid precipitation. In Section A, liquid precipitation is the dominant form between 96°E and 102°E, whereas solid precipitation has a more significant presence around 95°E and 101°E (Figure 6a). Liquid precipitation generally prevails across 90°E to 103°E but diminishes notably between 103°E and 104°E. Conversely, solid precipitation shows minimal contribution from 90°E to 104°E, with a gradual increase observed from 104°E to 110°E. During midsummer, liquid precipitation is most pronounced at 96°E, 102°E, and 106°E in the A region, whereas solid precipitation contributions are higher between 93°E and 95°E, at 97°E, and near 102°E and 104°E (Figure 6b). In Section B, liquid precipitation is particularly prominent at 92°E, 94°E to 96°E, and 102°E to 104°E. Solid precipitation is significant between 90°E and 100°E, but its presence diminishes at 102°E to 104°E before becoming prominent again from 104°E to 110°E (Figure 6c,d). The topographic map indicates that both regions A and B exhibit a steep elevation gradient between 102°E and 105°E, with altitudes decreasing from approximately 3000 m to nearly several hundred meters. This abrupt drop in elevation likely contributes to the observed consistency of the curve at 105° longitude, as the significant change in altitude affects various correlations. However, the uncertainties associated with these findings necessitate further investigation using additional data and methodologies.

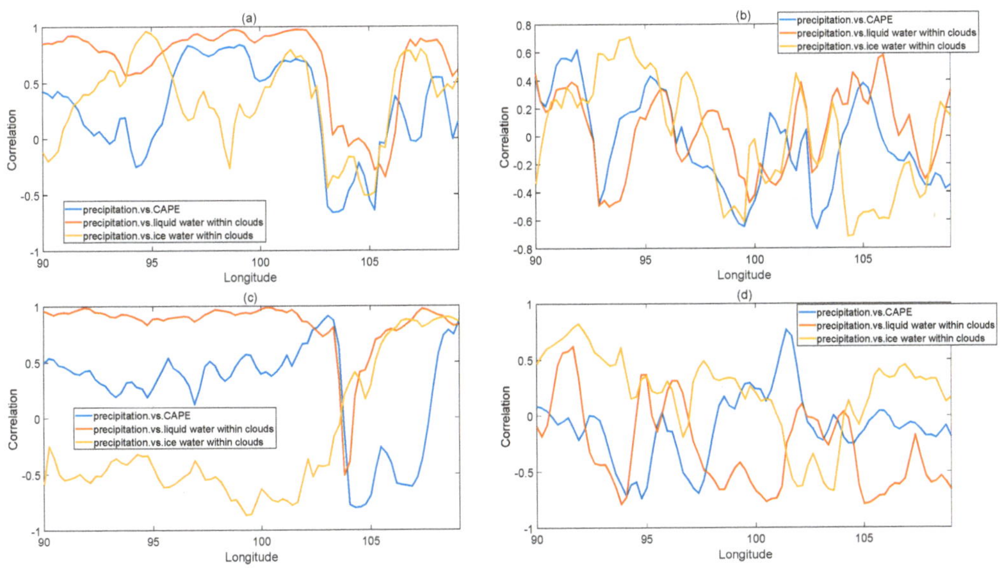

Figure 6. Correlation of CAPE, cloud LWC, and IWC with precipitation rate during the mei-yu ((**a**) for Section A, (**c**) for Section B) and midsummer periods ((**b**) for Section A, (**d**) for Section B), derived from ERA5.

Examining water vapor transport patterns unveils significant large-scale disparities between northeastern TP and eastern areas. Water vapor in this region is mainly transported from the west or southwest (Figure 7). Furthermore, distinctions are observed within the northeastern TP, particularly between Xining and Lanzhou. The diurnal precipitation variation over the eastward area is more subtle than that of eastern TP. Furthermore, the intensity of water vapor transport over Xining peaks significantly at 1400 LT during the daytime, contrasting sharply with the weaker transport observed over Lanzhou (Figure 7a,c). At night, the water vapor transport intensity is comparable between Xining and Lanzhou city, remaining consistently low (Figure 7b,d). These findings suggest that precipitation is predominantly concentrated during the daytime in Xining, where water vapor in the atmosphere is effectively intercepted. Conversely, Lanzhou, positioned downstream at the base of a slope, experiences considerably less precipitation. To elucidate this pattern's mechanisms, subsequent analyses will explore thermodynamic and dynamic factors.

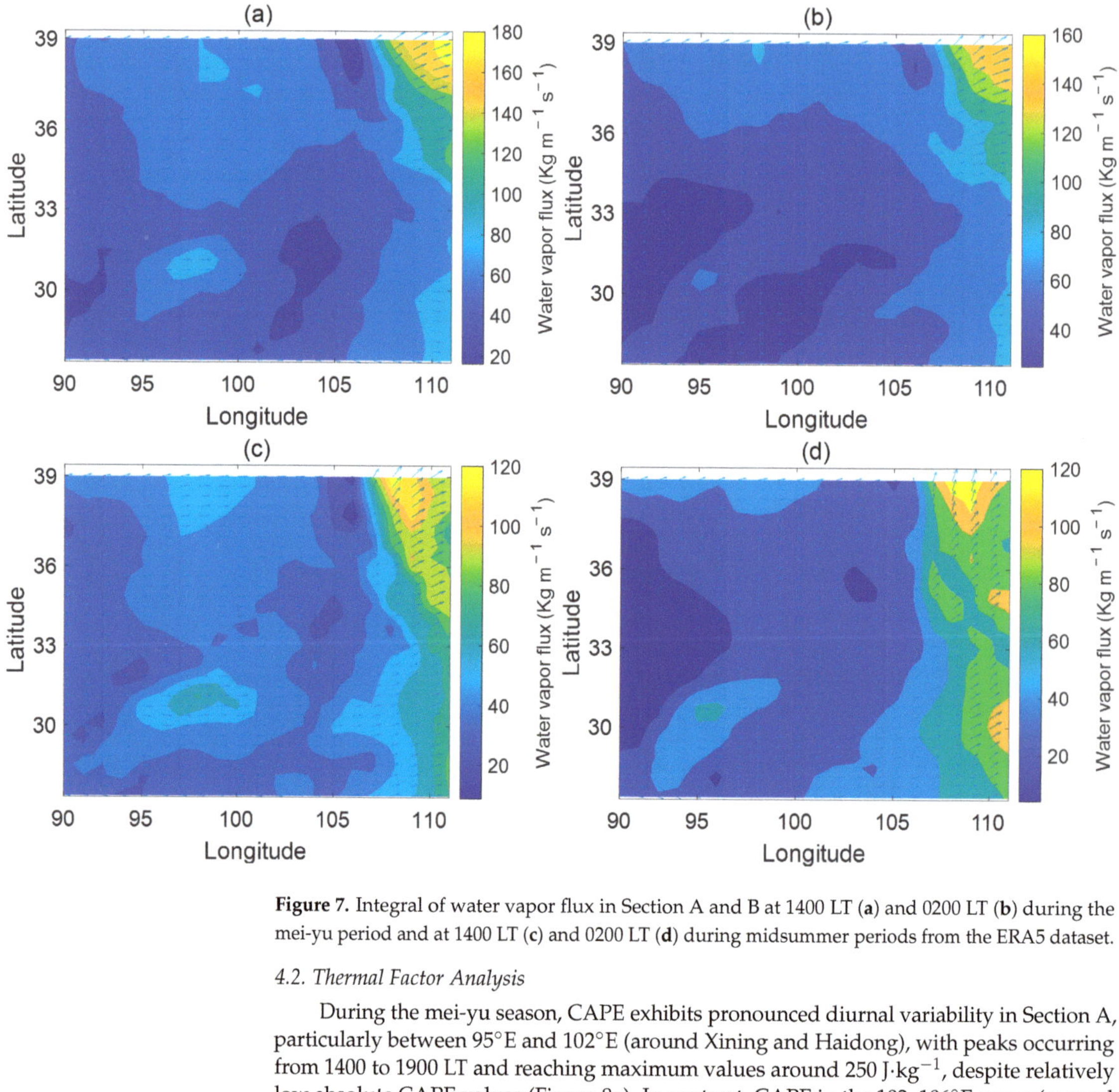

Figure 7. Integral of water vapor flux in Section A and B at 1400 LT (**a**) and 0200 LT (**b**) during the mei-yu period and at 1400 LT (**c**) and 0200 LT (**d**) during midsummer periods from the ERA5 dataset.

4.2. Thermal Factor Analysis

During the mei-yu season, CAPE exhibits pronounced diurnal variability in Section A, particularly between 95°E and 102°E (around Xining and Haidong), with peaks occurring from 1400 to 1900 LT and reaching maximum values around 250 J·kg^{-1}, despite relatively low absolute CAPE values (Figure 8a). In contrast, CAPE in the 103–106°E range (around Lanzhou) displays minimal diurnal variation, maintaining values near 200 J·kg^{-1}. In Section B, CAPE shows more substantial diurnal fluctuations, with significantly higher values between 1400 and 1900 LT in the 95–102°E range, peaking at approximately 700 J·kg^{-1}, while values between 103°E and 106°E remain lower, around 250 J·kg^{-1} (Figure 8c). During midsummer, CAPE in the A region (90–102°E) around Xining and Haidong exhibits peaks between 1200 and 1900 LT, ranging from about 50 to 250 J·kg^{-1} (Figure 8b). Similar patterns are observed in Section B between 90°E and 105°E, where CAPE also reaches its maximum between 1200 and 1900 LT, with values around 250 J·kg^{-1} (Figure 8d). Notably, the overall CAPE values between 103° and 110° E during midsummer do not show significant differences.

Correlation analysis reveals a strong association between CAPE and precipitation in the A region, particularly between 96°E and 102°E (around Xining and Haidong), with correlation coefficients exceeding 0.6 (Figure 6a). In contrast, this relationship between 103°E and 106°E (around Lanzhou) is insignificant. In Section B, only the 100–102°E range shows a relatively strong correlation, while other areas exhibit weaker associations than the A region. These findings suggest that elevated CAPE levels are generally associated with increased precipitation near Xining, driven by favorable thermal conditions that enhance valley and lake–land breezes. Conversely, the link between CAPE and precipitation around Lanzhou and the eastward area is less clear; due to limited moisture availability, even high CAPE values around Lanzhou seldom result in cloud formation and precipitation.

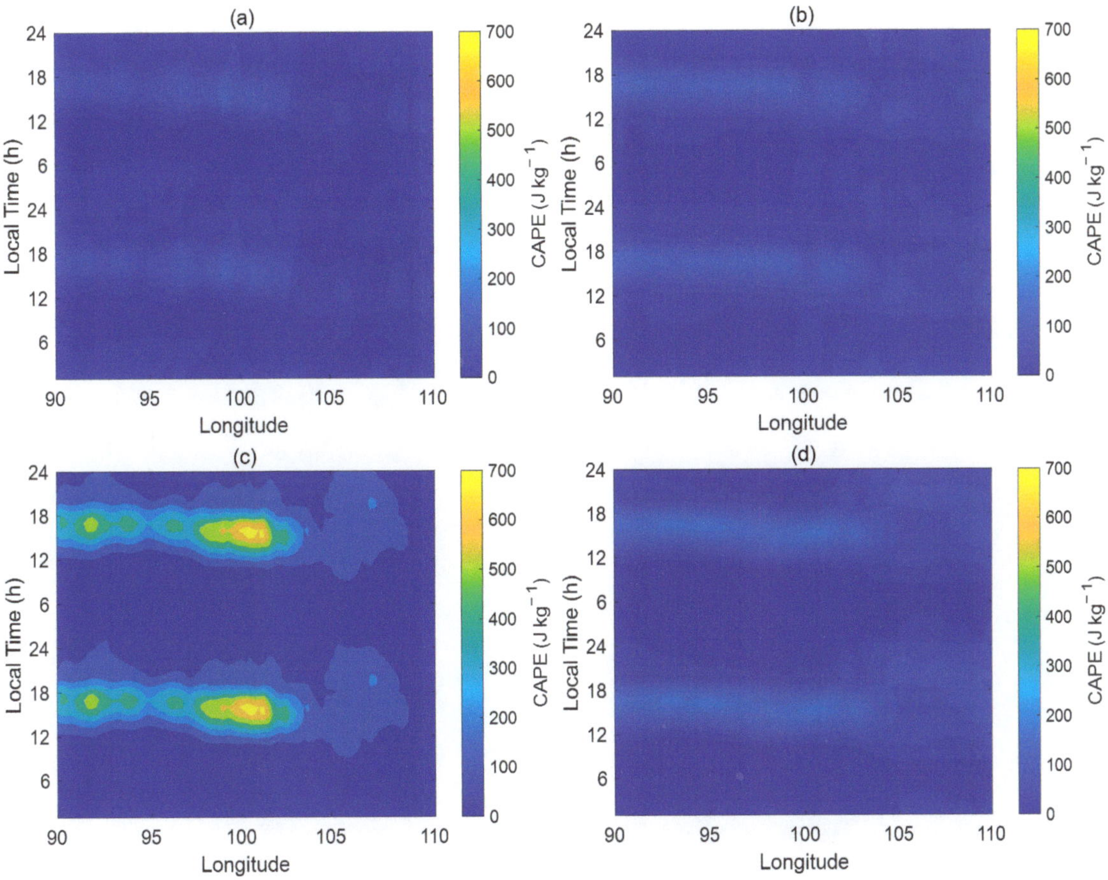

Figure 8. Diurnal variation in convective available potential energy (CAPE) during the mei-yu ((**a**) for Section A, (**c**) for Section B) and midsummer periods ((**b**) for Section A, (**d**) for Section B) from ERA5.

During the mei-yu season, notable diurnal variation in the temperature–dew point difference is observed in the A region, with pronounced centers located between 91°E and 95°E and between 103°E and 105°E. The temperature–dew point difference peaks at 1800 LT, up to 20 °C, and reaches a minimum of approximately 2 °C at 0600 LT (Figure 9a). In contrast, the smallest diurnal variation is found between 96°E and 102°E (around Xining and Haidong) and between 107°E and 110°E, where the difference peaks at just 8 °C at 1800 LT and drops to 1 °C at 0600 LT. In the B region, spanning from 90°E to 106°E, there is significant diurnal variation from west to east, with the temperature–dew point difference also peaking at 1800 LT and reaching its minimum at 0600 LT (Figure 9c).

During midsummer, similar patterns emerge in the A region, with significant diurnal variation between 90°E and 96°E and between 103°E and 106°E. The temperature–dew point difference peaks at 1800 LT, with a maximum of 20 °C, and reaches a minimum at 0600 LT. Conversely, the diurnal variation remains minimal between 96°E and 102°E (around Xining and Haidong) and from 107°E to 110°E, with the difference peaking at 1800 LT and minimizing at 0600 LT (Figure 9b,d). Overall, the temperature–dew point difference in the A region is consistently larger compared to the B region. This difference aligns with the distribution characteristics of CBH, indicating a close relationship between the temperature–dew point difference and CBH.

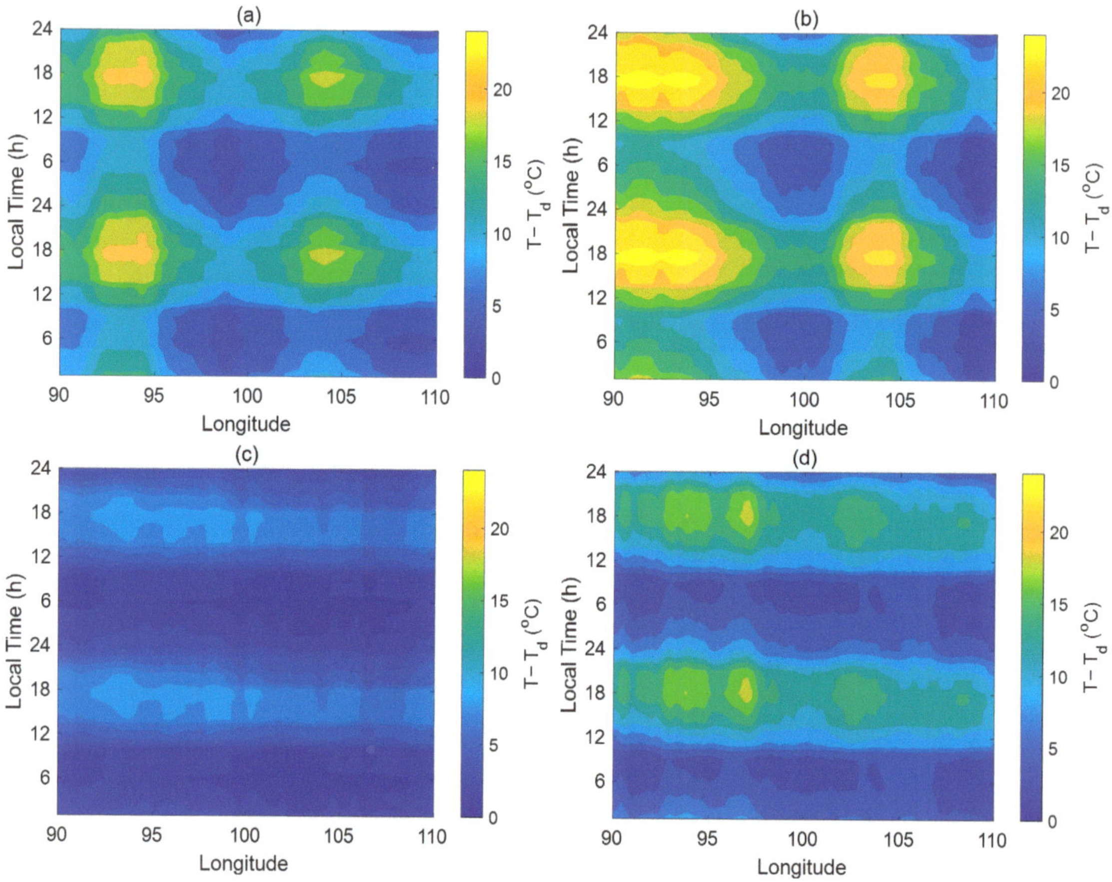

Figure 9. Diurnal variation in dewpoint spread during the mei-yu ((**a**) for Section A, (**c**) for Section B) and midsummer periods ((**b**) for Section A, (**d**) for Section B).

4.3. Dynamic Factors Analysis

During the mei-yu season, westerly winds dominate the wind field at 200 hPa across both the A and B regions, with speeds reaching up to 40 m·s^{-1} (Figure S3). Moreover, the A region continues to experience dominant westerlies, though wind speeds decrease in the B region by midsummer. During the mei-yu season in the A region, the wind field is characterized by westerly and southwesterly winds at 500 hPa height, while the B region shows pronounced diurnal variations due to inertial oscillations across most areas

(Figure S4). The difference in wind fields between 500 and 200 hPa reveals that the wind shear between these levels is relatively minor in the A region, exerting limited influence on cloud formation and precipitation. Conversely, the B region experiences more substantial wind shear between 500 hPa and 200 hPa, which has a more pronounced impact on clouds and precipitation. The TP exhibits notable and rapid downward momentum transfer. Significant diurnal variations in low-level wind speed over the central and western parts of the TP are associated with an increased likelihood of precipitation the following night. The northeastern TP is influenced by both inertial oscillations and local lake–land breezes and valley wind circulations, leading to longitudinal dispersion in precipitation patterns.

Calculations based on the index from Cao et al. (2022) show that, during the mei-yu season [29], the A region exhibits a gradual decrease in values from west to east, peaking between 90°E and 92°E at 1900 LT to 0500 LT, with a maximum of 28 $m{\cdot}s^{-1}$, and reaching a minimum of 22 $m{\cdot}s^{-1}$ at 1200 LT (Figure 10). The lowest values are observed between 100°E and 106°E, with maximum and minimum values around 21 $m{\cdot}s^{-1}$. In the B region, values are higher between 95°E and 100°E from 2000 LT to 0600 LT, reaching up to 22 $m{\cdot}s^{-1}$, while the lowest values are around 16 $m{\cdot}s^{-1}$ between 90°E and 92°E.

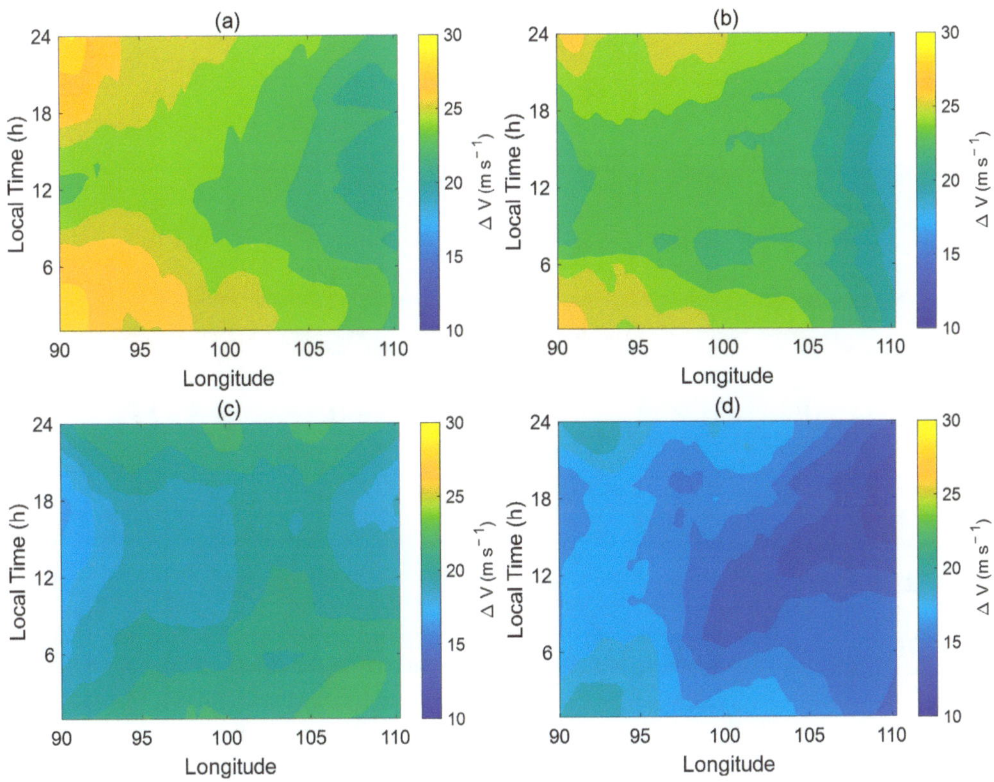

Figure 10. Diurnal variation in ΔV during the mei-yu ((**a**) for Section A, (**c**) for Section B) and midsummer periods ((**b**) for Section A, (**d**) for Section B).

During midsummer, a similar west-to-east decrease is observed in the A region, with higher values between 90°E and 100°E reaching 26 $m{\cdot}s^{-1}$ from 2000 LT to 0400 LT and the lowest values around 19 $m{\cdot}s^{-1}$ at 110°E. In the B region, values also decrease from west to east, peaking at 19 $m{\cdot}s^{-1}$ between 90°E and 95°E from 2000 LT to 0400 LT and dropping to approximately 12 $m{\cdot}s^{-1}$ at 110°E.

Correlation analysis reveals that, during the mei-yu season, the relationship between wind shear and precipitation is weak in the A region between 96°E and 104°E but stronger around 95°E and 105°E (Figure 11). In midsummer, a better correlation is observed at 104°E and 107°E, though it remains poor in other areas. This indicates two facts: (1) the precipitation difference between the northeastern TP and the eastward area is related to the wind shear at 200 hPa and 500 hPa on a large scale. (2) Wind shear is not the primary factor influencing precipitation differences between Xining and Lanzhou, as westerlies predominantly prevail during the mei-yu season. Instead, the variability in moisture conditions is the main factor affecting precipitation differences, with thermal and dynamic conditions being relatively similar in both regions. Xining and the adjacent Qinghai Lake are situated in a basin where diurnal variations in sensible heat flux are small over the lake but pronounced around Xining. This results in significant diurnal fluctuations in the lower-level vertical wind field, with valley and lake–land breezes enhancing convection. Precipitation is closely linked to CAPE. Water vapor is effectively intercepted near Xining, whereas, in Lanzhou, located at the base of a slope, descending air currents and high pressure result in dry, warm winds and reduced precipitation.

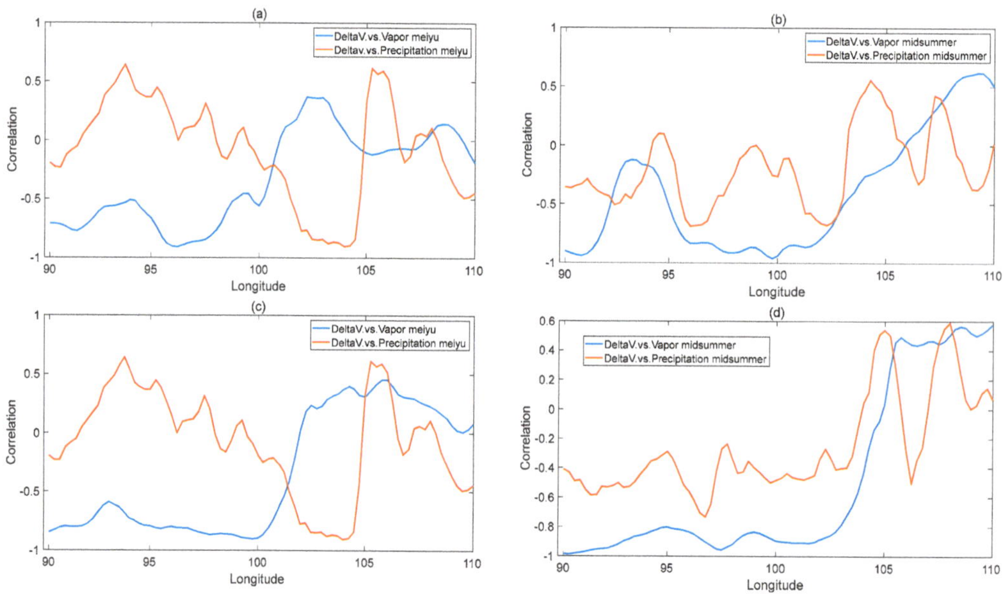

Figure 11. Correlation of ΔV and water vapor and precipitation during the mei-yu ((**a**) for Section A, (**c**) for Section B) and midsummer periods ((**b**) for Section A, (**d**) for Section B).

5. Conclusions

China's precipitation regime is characterized by three primary rainy seasons in the south and two distinct seasons over the Tibetan Plateau (TP), specifically the mei-yu and midsummer periods. Despite increasing warmth and moisture in Northwest China, the mechanisms governing land–atmosphere cloud–precipitation coupling remain inadequately understood, particularly regarding the disparities in precipitation benefits between the Loess Plateau (LP) and the Huangshui Valley (HV). This study employs the Fengyun-4 Satellite Quantitative Precipitation Estimation Product (FY4A-QPE) and ERA5 datasets to analyze cloud and precipitation characteristics, focusing on Lanzhou, a key city in the LP, and Xining, representative of the HV. Notably, these two cities are situated less than 200 km apart.

Our findings indicate that diurnal variations in precipitation are significantly less pronounced in the eastern TP than in its northeastern counterpart, attributed to fluctuations in convective available potential energy (CAPE) and wind shear between 200 hPa and 500 hPa. Notably, while both cities share similar wind shear patterns and moisture transport directions, Xining benefits from enhanced moisture availability due to snowmelt and effective water retention in surrounding mountains, resulting in higher precipitation levels. In contrast, despite high CAPE values, Lanzhou's moisture deficits limit cloud formation, leading to weaker precipitation dynamics.

As warming and moistening trends continue, enhanced precipitation is anticipated, particularly in the HV, with potential ecological and human habitat improvements. However, concentrated rainfall during summer afternoons and midnights raises concerns about extreme precipitation events, particularly given the region's susceptibility to geological hazards due to loess soils.

This study acknowledges limitations in latitude-averaged results and uncertainties related to the FY4A-QPE and ERA5 datasets. Future research should adopt advanced methodologies and higher-quality data to further elucidate the complex interactions driving precipitation dynamics in these regions.

Supplementary Materials: The following supporting information can be downloaded at https://www.mdpi.com/article/10.3390/rs16214059/s1: Figure S1: Diurnal variation in precipitation during the mei-yu (a: for Section A, c: for Section B) and midsummer periods (b: for Section A, d: for Section B) from the ERA5. Figure S2: Precipitation from FY4A-QPE (blue line) and ground observations (red line) over the (a) Lanzhou and (b) Xining. Blue shaded area is the standard error of precipitation from FY4A-QPE satellite estimates, and red shaded area is the standard error of precipitation from ground observations. Figure S3: Horizontal wind fields at 500 hPa at (a) 0800, (b) 1100, (c) 1400, (d) 1700, (e) 2000, (f) 2300, (g) 0200, and (h) 0500 BT from ERA5 data. The red dot represents the location of Yushu, Xining, and Lanzhou. Figure S4: Horizontal wind fields at 200 hPa at (a) 0800, (b) 1100, (c) 1400, (d) 1700, (e) 2000, (f) 2300, (g) 0200, and (h) 0500 BT from ERA5 data. The red dot represents the location of Yushu, Xining, and Lanzhou.

Author Contributions: Conceptualization, Y.L.; methods and software, B.C.; validation, X.Y.; writing—original draft preparation, B.C.; writing—review and editing, S.W.; visualization, S.W.; supervision, J.W.; project administration, X.Y.; funding acquisition, S.W. All authors have read and agreed to the published version of the manuscript.

Funding: This research was funded by the National Science Foundation of China (NSFC) (42465005, 42175174), "Kunlun Talents" talent introduction scientific research project of Qinghai Institute of Technology (2023-QLGKLYCZX-003), the Research start-up funding project of Qinghai Institute of Technology (2023021wys001), and the Second Tibetan Plateau Scientific Expedition and Research (STEP) program (2019QZKK0102).

Data Availability Statement: The ERA5 data utilized in this research can be accessed via the European Centre for Medium-Range Weather Forecasts (ECMWF) website at https://www.ecmwf.int/en/forecasts/datasets/reanalysis-datasets/era5, with the specific data retrieval date being 1 January 2024.

Conflicts of Interest: The authors declare no conflicts of interest.

References

1. Xu, X.; Lu, C.; Shi, X.; Gao, S. World water tower: An atmospheric perspective. *Geophys. Res. Lett.* **2008**, *35*, L20815. [CrossRef]
2. Fu, Y.; Liu, G.; Wu, G.; Yu, R.; Xu, Y.; Wang, Y.; Li, R.; Liu, Q. Tower mast of precipitation over the central Tibetan Plateau summer. *Geophys. Res. Lett.* **2006**, *33*, L05802. [CrossRef]
3. Boos, W.R.; Kuang, Z. Dominant control of the South Asian monsoon by orographic insulation versus plateau heating. *Nature* **2010**, *463*, 218–222. [CrossRef] [PubMed]
4. Fu, Y.; Ma, Y.; Zhong, L.; Yang, Y.; Guo, X.; Wang, C.; Xu, X.; Yang, K.; Xu, X.; Liu, L.; et al. Land surface processes and summer cloud-precipitation characteristics in the Tibetan Plateau and their effects on downstream weather: A review and perspective. *Natl. Sci. Rev.* **2020**, *7*, 500–515. [CrossRef] [PubMed]
5. Kukulies, J.; Chen, D.; Curio, J. The role of mesoscale convective systems in precipitation in the Tibetan Plateau region. *J. Geophys. Res. Atmos.* **2021**, *126*, e2021JD035279. [CrossRef]
6. Barton, E.; Taylor, C.; Klein, C.; Harris, P.; Meng, X. Observed soil moisture impact on strong convection over mountainous Tibetan Plateau. *J. Hydrometeorol.* **2021**, *22*, 561–572. [CrossRef]

7. Houze, R.A.; Wilton, D.C.; Smull, B.F. Monsoon convection in the Himalayan region as seen by the TRMM precipitation radar. *Q. J. R. Meteorol. Soc.* **2007**, *133*, 1389–1411. [CrossRef]
8. Chen, F.; Xie, T.; Yang, Y.; Chen, S.; Chen, F.; Huang, W.; Chen, J. Discussion of the "warming and wetting" trend and its future variation in the drylands of Northwest China under global warming. *Sci. China Earth Sci.* **2023**, *66*, 1241–1257. [CrossRef]
9. Zhao, P.; Xu, X.; Chen, F.; Guo, X.; Zheng, X.; Liu, L.; Hong, Y.; Li, Y.; La, Z.; Peng, H. The Third Atmospheric Scientific Experiment for understanding the earth-atmosphere coupled system over the Tibetan Plateau and its effects. Bull. *Am. Meteorol. Soc.* **2017**, *98*, 757–776. [CrossRef]
10. Tao, S.; Chen, L. A review of recent research on the East Asian summer monsoon in China. In *Monsoon Meteorology*; Chang, C.P., Krishnamurti, T.N., Eds.; Oxford University Press: Oxford, UK, 1987; pp. 60–92.
11. Ding, Y. Summer monsoon rainfalls in China. *J. Meteor. Soc. Jpn.* **1992**, *70*, 373–396. [CrossRef]
12. Chen, G. Large-scale circulations associated with the East Asian summer monsoon and the mei-yu over South China and Taiwan. *J. Meteor. Soc. Jpn.* **1994**, *72*, 959–983. [CrossRef]
13. Ding, Y.; Chan, J.C.-L. The East Asian summer monsoon: An overview. *Meteor. Atmos. Phys.* **2005**, *89*, 117–142.
14. Chen, G. Research on the phenomena of mei-yu during the past quarter century: An overview. In *The East Asian Monsoon*; Chang, C.-P., Ed.; World Scientific: Singapore, 2004; pp. 357–403.
15. Bai, A.J.; Liu, C.; Liu, X. Diurnal variation of summer rainfall over the Tibetan Plateau and its neighboring regions revealed by TRMM multi-satellite precipitation analysis. *Chin. J. Atmos. Sci.* **2008**, *51*, 518–529. [CrossRef]
16. Feng, J.; Liu, L.; Wang, Z. The statistic characteristics of radar echo and precipitation and some thermodynamic variables at Naqu in Qinghai-Xizang Plateau. *Plateau Meteor* **2002**, *21*, 368–374. (In Chinese)
17. Zhao, P.; Xiao, H.; Liu, J.; Zhou, Y. Precipitation efficiency of cloud and its influencing factors over the Tibetan plateau. *Int. J. Climatol.* **2022**, *42*, 416–434. [CrossRef]
18. Liu, L.; Zheng, J.; Zheng, R.; Cui, Z.; Hu, Z.; Wu, S.; Dai, G.; Wu, Y. Comprehensive radar observations of clouds and precipitation over the Tibetan Plateau and preliminary analysis of cloud properties. *J. Meteorol. Res.* **2015**, *29*, 546–561. [CrossRef]
19. Mei, Y.; Hu, Z.; Huang, X. A study of convective clouds in the Tibetan Plateau based on dual polarimeteic radar observations. *Acta Meteor. Sin.* **2018**, *76*, 1014–1028. [CrossRef]
20. Liu, L.; Feng, J.; Chu, R.; Zhou, Y.; Ueno, K. The diurnal variation of precipitation in monsoon season in the Tibetan Plateau. *Adv. Atmos. Sci.* **2002**, *19*, 365–378. [CrossRef]
21. Fujinami, H.; Yasunari, T. The seasonal and intraseasonal variability of diurnal cloud activity over the Tibetan Plateau. *J. Meteor. Soc. Jpn.* **2001**, *79*, 1207–1227. [CrossRef]
22. Ji, X.L.; Wu, H.M.; Huang, A.N.; Zhao, W.; Yang, W. Characteristics of the precipitation diurnal variation over Qinghai-Tibetan Plateau in summer. *Plateau Meteor* **2017**, *36*, 1188–1200. (In Chinese)
23. Zhao, C.; Liu, L.; Wang, Q.; Qiu, Y.; Wang, Y.; Wu, X. MMCR-based characteristic properties of non-precipitating cloud liquid droplets at Naqu site over Tibetan Plateau in July 2014. *Atmos. Res.* **2017**, *190*, 68–76. [CrossRef]
24. Xu, W.; Zipser, E.J. Diurnal variations of precipitation, deep convection, and lightning over and east of the eastern Tibetan Plateau. *J. Clim.* **2011**, *24*, 448–465. [CrossRef]
25. Wu, G.X.; Duan, A.; Liu, Y.; Mao, J.; Ren, R.; Bao, Q.; He, B.; Liu, B.; Hu, W. Tibetan Plateau climate dynamics: Recent research progress and outlook. *Natl. Sci. Rev.* **2015**, *2*, 110–116. [CrossRef]
26. Wang, Y.; Zeng, X.; Xu, X.; Welty, J.; Lenschow, D.H.; Zhou, M.; Zhao, Y. Why are there more summer afternoon low clouds over the Tibetan Plateau compared to eastern China? *Geophys. Res. Lett.* **2020**, *47*, e2020GL089665. [CrossRef]
27. Fu, Y.; Pan, X.; Xian, T.; Liu, G.; Zhong, L.; Liu, Q.; Ma, M. Precipitation characteristics over the steep slope of the Himalayas in rainy season observed by TRMM PR and VIRS. *Clim. Dyn.* **2018**, *51*, 1971–1989. [CrossRef]
28. Zhao, Y.; Li, J.; Zhang, L.; Deng, C.; Li, Y.; Jian, B.; Huang, J. Diurnal cycles of cloud cover and its vertical distribution over the Tibetan Plateau revealed by satellite observations, reanalysis datasets, and CMIP6 outputs. *Atmos. Chem. Phys.* **2023**, *23*, 743–769. [CrossRef]
29. Chen, G.; Sha, W.; Iwasaki, T. Diurnal variation of precipitation over southeastern China: 2. Impact of the diurnal monsoon variability. *J. Geophys. Res.* **2009**, *114*, D21105. [CrossRef]
30. Yu, R.; Li, J.; Chen, H.; Yuan, W. Progress in studies of the precipitation diurnal variation over contiguous China. *Acta Meteorol. Sin.* **2014**, *72*, 948–968. [CrossRef]
31. Cao, B.; Yang, X.; Li, B.; Lu, Y.; Wen, J. Diurnal variation in cloud and precipitation characteristics in summer over the Tibetan Plateau and Sichuan Basin. *Remote Sens.* **2022**, *14*, 2711. [CrossRef]
32. Sorooshian, S.; Hsu, K.L.; Gao, X.; Gupta, H.V.; Imam, B.; Braithwaite, D. Evaluation of PERSIANN system satellite-based estimates of tropical rainfall. *Bull. Am. Meteorol. Soc.* **2000**, *81*, 2035–2046. [CrossRef]
33. Huffman, G.J.; Bolvin, D.T.; Nelkin, E.J. Integrated Multi-Satellite Retrievals for GPM (IMERG) Technical Documentation. NASA/GSFC Code 612. 2015. Available online: http://pmm.nasa.gov/sites/default/files/document_files/IMERG_doc.pdf (accessed on 16 September 2019).
34. Huffman, G.J.; Bolvin, D.T.; Nelkin, E.J.; Wolff, D.B.; Adler, R.F.; Gu, G.; Hong, Y.; Bowman, K.P.; Stocker, E.F. The TRMM multisatellite precipitation analysis (TMPA): Quasi-global, multiyear, combined-sensor precipitation estimates at fine scales. *J. Hydrometeorol.* **2007**, *8*, 38–55. [CrossRef]

35. Joyce, R.J.; Janowiak, J.E.; Arkin, P.A.; Xie, P. CMORPH: A method that produces global precipitation estimates from passive microwave and infrared data at high spatial and temporal resolution. *J. Hydrometeorol.* **2004**, *5*, 487–503. [CrossRef]
36. Scofield, R.A.; Kuligowski, R.J. Status and outlook of operational satellite precipitation algorithms for extreme-precipitation events. *Weather Forecast.* **2003**, *18*, 1037–1051. [CrossRef]
37. Abhishek; Kinouchi, T.; Sayama, T. A comprehensive assessment of water storage dynamics and hydroclimatic extremes in the Chao Phraya River Basin during 2002–2020. *J. Hydrol.* **2021**, *603*, 126868. [CrossRef]
38. Guan, J.; Gao, Y.; Huang, C.; Li, J.; Zhang, L. Estimating FY-4A satellite precipitation based on a deep learning model Attention-Unet. *Acta Meteorol. Sin.* **2024**, *82*, 398–410. (In Chinese) [CrossRef]
39. Liu, J. Analysis on cloud microphysical property over Qinghai-Xizang Plateau using satellite data. *Plateau Meteor.* **2013**, *32*, 38–45. (In Chinese)
40. Guo, J.; Zhang, J.; Yang, K.; Liao, H.; Zhang, S.; Huang, K.; Lv, Y.; Shao, J.; Yu, T.; Tong, B.; et al. Investigation of near-global daytime boundary layer height using high-resolution radiosondes: First results and comparison with ERA-5, MERRA-2, JRA-55, and NCEP-2 reanalysis. *Atmos. Chem. Phys.* **2021**, *21*, 17079–17097. [CrossRef]
41. Cao, B.J.; Lyu, S.H.; Zhang, Y.; Yang, X.; Li, B.; Li, M. Factors influencing diurnal variations of cloud and precipitation in the Yushu area of the Tibetan Plateau. *J. Meteor. Res.* **2022**, *36*, 311–325. [CrossRef]
42. Craven, J.P.; Jewell, R.E.; Brooks, H.E. Comparison between observed convective cloud-base heights and lifting condensation level for two different lifted parcels. *Wea. Forecast.* **2002**, *17*, 885–890. [CrossRef]

MDPI AG
Grosspeteranlage 5
4052 Basel
Switzerland
Tel.: +41 61 683 77 34

Remote Sensing Editorial Office
E-mail: remotesensing@mdpi.com
www.mdpi.com/journal/remotesensing

www.ingramcontent.com/pod-product-compliance
Lightning Source LLC
Chambersburg PA
CBHW041409210926
51850CB00013B/3221

* 9 7 8 3 7 2 5 8 3 4 7 1 6 *